The Æ Model

Unification of Matter, Consciousness, and the Vibrational Field

Jesús Eduardo Jiménez de la Garza
(EduardoQuera4)

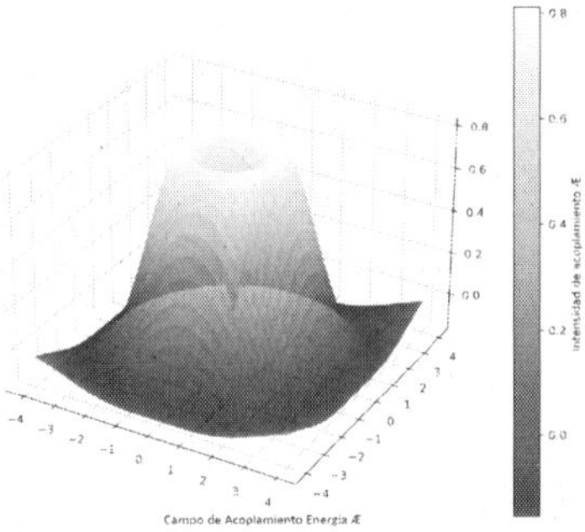

March 2026

The Æ Model

Unification of Matter, Consciousness, and the Vibrational Field

Typesetting: LaTeX

ISBN (Paperback Edition):
978-607-29-8542-1

First paperback edition, 2026
Printed in Mexico

Open Science Record
A preprint version of this work is archived in the Zenodo open-science repository operated by CERN, ensuring long-term preservation and transparent version tracking of the Æ Model.

Version 8 DOI (this release):

https://doi.org/10.5281/zenodo.19341565

Concept DOI (all versions):

https://doi.org/10.5281/zenodo.15313261

The Zenodo record provides a permanent, citable archive of the evolving Æ Model framework.

Scan for latest version:

Recommended Citation
JIMENEZ DE LA GARZA, J. E. (2026).
The Æ Model: Unification of Matter, Consciousness, and the Vibrational Field.
Zenodo.
https://doi.org/10.5281/zenodo.19341565

Everything that exists is a vibration,
and each vibration encodes information.

Para mis papá y mamá,

quienes entregaron su vida para darnos
amor y estabilidad a mi familia y a mí.

Y a Mau, Ceci y Rod.

Con todo mi amor.

Acknowledgements

This work was shaped by the ideas, discoveries, and efforts of countless thinkers, researchers, educators, and creators across many fields and generations.

From scientists and mathematicians whose work laid the foundations of modern understanding, to authors, teachers, musicians, and communicators who helped transmit knowledge in many different forms, their influence is deeply present throughout these pages.

Many contemporary researchers, educators, and science communicators have also influenced this work through their lectures, papers, books, and open discussions that helped illuminate different aspects of the questions explored in this work.

Some of these influences are cited explicitly throughout the text, while others appear more subtly through the intellectual tradition they represent.

I am also grateful for the tools, modern platforms and technologies that made the development of this work possible, including modern computational tools and artificial intelligence systems that helped organize, explore, and articulate many of the ideas presented here.

Finally, I extend my appreciation to all readers, thinkers, and future investigators who may engage with this framework, challenge it, refine it, or carry its questions further.

Abstract

This work introduces the Æ framework, a theoretical proposal that interprets complex physical, biological, and informational systems through the dynamics of phase coherence and topological persistence.

Within this framework, structures across multiple scales are modeled as coherent configurations embedded in an underlying dynamical field. Stable configurations of this field are described as *reality knots*, representing persistent topological structures maintained by coherence constraints and network interactions.

The formal development combines tools from dynamical systems theory, field theory, topology, and network science. In particular, phase dynamics and coherence–entropy relations are used to characterize the formation, stability, and interaction of coherent structures.

The framework is applied conceptually to several domains, including collective synchronization, biological organization, computational learning systems, and large-scale cosmological dynamics. Across these contexts, the model emphasizes coherence and phase structure as potential organizing principles for complex systems.

While the Æ framework remains exploratory in several aspects, it provides a unified language for describing how stable structures may emerge from interacting dynamical components. The work therefore proposes a research program aimed at investigating coherence-based mechanisms across physical, biological, and informational systems.

Future developments will require mathematical refinement, numerical modeling, and empirical investigation to determine the scope and applicability of the proposed framework.

How to Read This Book

This work presents the Æ Framework, a theoretical model that attempts to unify matter, information, and consciousness within a vibrational field description.

The document is structured in layers of increasing technical depth.

Readers interested primarily in conceptual understanding may focus on the introductory and interpretative sections, where the main ideas are described without requiring advanced mathematics.

Readers with a background in physics or mathematics may proceed directly to the formal sections, where the structure of the model is expressed through operators, dynamical equations, and statistical formulations.

The mathematical sections are written to remain compatible with standard theoretical physics notation whenever possible, while introducing new constructs only when required by the framework.

The reader is not required to follow the book linearly. Different sections may be explored independently depending on the reader's background and interests.

Research Status

The Æ Framework is presented as an exploratory theoretical program.

Several aspects of the model remain under development, and some mathematical connections proposed in this work require further investigation.

The purpose of this document is therefore to present a coherent conceptual and mathematical structure that may guide future research rather than a finalized theory.

Philosophical Position of the Framework

The Æ Framework is presented as a theoretical and conceptual investigation rather than a completed physical theory.

Its goal is to explore whether a unified description of matter, information, and consciousness can be formulated in terms of coherence structures emerging from a dynamical field.

In this sense, the framework occupies an intermediate position between formal physics and the philosophy of science. It proposes mathematical structures and dynamical relations, while also addressing broader questions about the ontological status of information, coherence, and organized complexity.

The work does not claim that the proposed structures represent a final description of physical reality. Instead, it should be understood as a research program that attempts to investigate whether coherence-based descriptions can provide useful explanatory tools across different domains.

Within this perspective, the Æ Framework is intended to remain compatible with established physical theories while exploring whether a deeper informational structure may underlie them.

Scope of the Framework

The Æ Framework proposes a structural description of reality in terms of phase relations, coherence measures, and informational structures embedded in a global vibrational field.

The aim of the model is not to replace established physical theories but to explore a broader formal structure in which known phenomena may appear as particular regimes of a more general informational dynamics.

Several aspects of the framework remain open to further investigation. Whenever a mathematical derivation is incomplete or a connection to known theory is still speculative, this is explicitly indicated in the text.

The work is therefore presented as an evolving research program rather than a finished theory.

Structure of the Book

The book is organized into several thematic parts.

The first part introduces the conceptual foundations of the framework and the intuition behind the main ideas.

The second part develops the mathematical structure of the model, including the operators, entropy measures, and dynamical equations governing phase coherence.

Within the mathematical sections, the framework is progressively constructed through a layered structure that introduces:

- the underlying geometric setting,
- the fields and informational variables of the model,
- the operators and dynamical relations governing phase behavior,
- the action principles and governing equations,
- and the observable consequences of the framework.

Later sections explore applications of the framework in different domains and discuss open problems, limitations, and directions for future investigation.

Notation and Conventions

Throughout this work the following conventions are used.

- Scalar fields are written in lowercase Greek or Latin letters.
- Phase variables are typically denoted by ϕ.
- Amplitude variables are denoted by A.
- The symbol Æ denotes the fundamental vibrational field introduced in this framework.
- Vectors are written in bold when needed.
- Operators are defined within the mathematical sections and follow standard conventions from theoretical physics whenever possible.

Unless otherwise specified, units are assumed to be dimensionless or expressed in natural units appropriate to the context.

Mathematical Preliminaries

The formal development of the Æ Framework relies on tools commonly used in theoretical physics and dynamical systems.

In particular, the mathematical language of the model draws from concepts in field theory, phase dynamics, topology, and information theory.

Complex scalar fields are used to describe the amplitude and phase structure of the underlying dynamical field, while coherence relations are used to characterize stable configurations within the system.

Only the minimum mathematical structure required to formulate the framework is introduced. Whenever possible, notation and conventions follow standard usage in theoretical physics.

Contents

Part I
Conceptual Foundations of the Æ Framework

1. Architectural Foundations

Before introducing the formal mathematical structure of the Æ framework, it is useful to clarify the conceptual architecture on which the theory is built. Scientific theories are not only collections of equations; they are also systems of assumptions about how reality is organized.

In physics, axioms serve as the structural backbone of a model. They define what kinds of entities exist, what relationships are considered fundamental, and how physical structures may evolve.

The Æ framework begins with a small set of foundational principles concerning fields, phase structure, coherence, and information. These principles are not arbitrary assumptions. They arise from observations across many domains of science, including field theory, nonlinear dynamics, information theory, biological synchronization, and complex systems.

The aim of this section is therefore to present the intuitive meaning of these principles before they are expressed in precise mathematical form. Each axiom introduced below captures a structural property that later appears in the equations governing the Æ field.

1.1 Flow Constraint (Structural Diagram)

Many physical systems can be understood as flows of structure through a medium. Water flowing through a river, electrical current moving through a conductor, and waves propagating through an ocean are all examples of patterns moving within an underlying substrate.

In the Æ framework, the evolution of physical systems is

interpreted in a similar way. Instead of describing isolated objects moving through empty space, the theory considers patterns of phase and coherence evolving within a continuous field.

The structural diagram introduced in the mathematical section represents this idea schematically. It shows how different quantities—phase, coherence, entropy, and informational structure—interact as a dynamical system.

An intuitive analogy is a system of currents within an ocean. Local vortices, waves, and turbulence may appear distinct, but they are ultimately manifestations of the same fluid. Similarly, the entities described in the Æ framework are interpreted as organized patterns emerging within a deeper informational field.

1.2 Axiom I — Existence of the Æ-Field

The first principle of the framework proposes the existence of a fundamental field that carries informational and phase structure. This field is referred to as the Æ-field.

The idea of a fundamental field is not new to physics. Modern theories already describe nature in terms of gravitational fields, electromagnetic fields, and quantum fields permeating space.

The Æ-field extends this perspective by emphasizing the role of phase relationships and informational organization within field dynamics.

An intuitive way to imagine this field is to think of the surface of a vast ocean. Waves, ripples, and currents can form complex patterns on the surface, yet they are all expressions of the same underlying medium.

In a similar way, the structures observed in physical systems may be interpreted as organized patterns within the Æ-field.

1.3 Axiom II — Phase Decomposition

Many natural processes exhibit oscillatory behavior. Examples include electromagnetic waves, mechanical vibrations, biological rhythms, and quantum phase evolution.

When oscillatory systems interact, their behavior can often be understood in terms of phase relationships. Even when the frequency of oscillation is the same, two systems may be aligned or misaligned depending on their relative phase.

The principle of phase decomposition states that complex dynamical processes may be analyzed by separating their phase structure into fundamental components.

An analogy can be found in music. A musical chord may contain several notes played simultaneously. By separating those notes, one can understand the harmonic structure of the sound.

Similarly, phase decomposition allows complex field dynamics to be analyzed as combinations of simpler phase relationships.

1.4 Axiom III — Relational Primacy of Phase

In many physical theories, quantities such as position, momentum, or energy are treated as primary descriptors of a system.

The Æ framework adopts a different viewpoint: it proposes that phase relationships play a more fundamental role in the organization of structure.

This means that what matters most is not the absolute value of a quantity, but how it relates to the phases of other interacting components.

A useful analogy is found in orchestral music. Each musician may produce a sound of similar intensity, but the musical structure emerges only when their timing is properly aligned.

If the timing between players is disrupted, the same collection of sounds becomes noise rather than music.

In the same way, the relational alignment of phase can determine whether a system produces stable structure or disordered dynamics.

1.5 Axiom IV — Coherence as Structural Measure

When oscillatory systems align their phases, coherence emerges. Coherence represents a degree of internal order in the dynamics of a system.

This phenomenon appears in many areas of science. Laser light, for example, is produced when electromagnetic waves become highly phase-aligned. Biological systems also exhibit coherent oscillations across networks of interacting cells and neurons.

In the context of the Æ framework, coherence provides a

measure of structural organization within the field.

High coherence corresponds to stable, organized patterns, while low coherence corresponds to disordered dynamics.

An intuitive picture is that of a large ensemble of pendulums. If each pendulum swings independently, the system appears chaotic. However, if their oscillations begin to synchronize, a coherent collective motion emerges.

1.6 Axiom V — Phase Entropy

While coherence describes the alignment of phases, entropy describes their dispersion.

When the phases of a system become increasingly uncorrelated, the system loses the capacity to maintain stable structure.

This concept parallels the role of entropy in thermodynamics. Just as thermal entropy measures the dispersal of energy, phase entropy measures the dispersal of phase relationships within a dynamical system.

An everyday analogy can be found in a crowded public space. If individuals move randomly with no coordination, the collective motion becomes unpredictable and chaotic.

Similarly, high phase entropy corresponds to disordered phase configurations within the field.

1.7 Axiom VI — Coherence–Entropy Duality

Coherence and entropy are not independent properties. They form a complementary pair that governs the organization of dynamical systems.

When coherence increases, entropy tends to decrease, and vice versa.

This duality reflects a deeper balance between order and disorder that appears throughout physical and biological systems.

For example, living organisms maintain local coherence by exporting entropy to their surroundings. Similarly, coherent field structures may persist only within certain ranges of phase organization.

The Æ framework treats this relationship as a central principle governing the stability of structures within the field.

1.8 Axiom VII — Stability Threshold

Not all configurations of phase and coherence are stable.

In many dynamical systems, there exists a threshold beyond which stable patterns can form. Below this threshold, structures dissolve rapidly due to fluctuations.

Examples appear in fluid turbulence, laser formation, and synchronization phenomena.

An intuitive analogy is the formation of a whirlpool in flowing water. If the flow is too weak, no vortex forms. Once the flow exceeds a critical intensity, a stable rotational structure suddenly appears.

The stability threshold described in the Æ framework plays a similar role in determining when coherent structures can persist within the field.

1.9 Axiom VIII — Informational Evolution

Physical systems are not static. Their configurations evolve over time as information propagates through interacting fields.

In this framework, evolution is interpreted as the continuous transformation of phase relationships and coherence structures.

Instead of viewing time merely as an external parameter, the evolution of a system can be understood as the reconfiguration of informational patterns within the field.

An analogy can be found in the evolution of weather systems in the atmosphere. Small perturbations may propagate, interact, and eventually generate large-scale structures such as storms or organized circulation patterns.

Similarly, informational evolution describes how local phase relationships can give rise to larger coherent structures over time.

1.10 Admissible Configuration Space

Not every possible arrangement of phases corresponds to a physically realizable configuration.

The framework therefore restricts attention to a subset of configurations that satisfy structural constraints such as coherence, bounded phase entropy, and consistent relational phase organization.

This admissible configuration space can be understood as the domain of physically meaningful states, where interactions between fields remain stable and well-defined.

Rather than considering all mathematically possible phase assignments, the theory focuses only on those that can sustain coherent dynamics over time.

In this sense, the configuration space acts as a selection principle: it defines which patterns of phase organization are capable of persisting, propagating, and interacting within the field.

All dynamical evolution of the system takes place within this constrained landscape.

1.11 Æ-Algebraic Primitives

To describe these structures mathematically, the framework introduces a set of algebraic primitives.

These primitives constitute the minimal building blocks of the theory. They encode the fundamental elements required to represent phase relations, coherence measures, and dynamical couplings between fields.

Each primitive captures a specific aspect of the system: local phase, relational alignment, or transformation rules under evolution.

More complex structures are constructed by combining these primitives, much like geometric objects emerge from simpler elements such as points and vectors.

Importantly, these primitives are not arbitrary. They are chosen to reflect the operational features observed throughout the framework: phase interaction, coherence propagation, and structural stability.

Subsequent sections provide their formal realization using geometric and algebraic tools.

1.12 Collective Criticality and Phase Synchronization

Many complex systems exhibit collective behavior when interactions between components reach a critical level.

Examples include synchronized oscillations in neural networks, collective motion in biological populations, and phase transitions in physical systems.

Near such critical regimes, the system becomes highly sensitive: small perturbations can propagate across large regions, allowing coordinated structures to emerge spontaneously.

Within the Æ framework, this corresponds to regimes in which phase synchronization becomes possible across extended domains of the field.

In these conditions, local phase relations are no longer isolated. Instead, they become dynamically linked, enabling the formation of coherent structures that span multiple scales.

This mechanism provides a natural pathway from local interactions to global organization, explaining how large-scale coherence can emerge from simple underlying dynamics.

Collective criticality therefore acts as a bridge between microscopic phase behavior and macroscopic structure.

2. Primitive Objects and Algebraic Structure

Once the foundational principles of the Æ framework have been established, the next step is to identify the primitive objects from which the mathematical structure of the theory is built.

In any scientific framework, primitive objects play the role of irreducible elements: they define what exists at the most basic level of description and determine how more complex structures arise.

In the Æ framework, these objects are designed to capture the essential features of phase, coherence, and relational dynamics that govern the field.

Rather than beginning with particles or localized objects, the framework starts from functional configurations of the field itself. These configurations form the fundamental objects upon which algebraic and geometric operations are defined.

The goal of this section is therefore to describe the conceptual meaning of these primitive structures before presenting the formal algebraic definitions that follow.

2.1 Æ Functional Space and Dual Pairing

In many areas of mathematics and physics, complex systems are described using functional spaces. Instead of representing a system as a single point or number, one considers entire functions or distributions that encode how quantities vary across space and time.

Within the Æ framework, configurations of the field are interpreted as elements of a functional space that captures the possible patterns of phase and coherence present in the system.

An intuitive analogy can be found in musical acoustics. A musical tone is not defined by a single value but by a waveform that varies continuously over time. Different waveforms correspond to different timbres and harmonic structures.

Similarly, different configurations of the Æ field correspond to different informational patterns.

Dual pairing arises naturally when one considers how such configurations interact or are measured. In simple terms, dual structures allow one to compare or evaluate patterns within the space of possible field configurations.

2.2 Algebraic Closure and Hopf-Like Structure

When mathematical objects are introduced, it becomes important to determine how they combine with one another.

An algebraic system is said to be closed if performing its allowed operations on valid objects always produces another object of the same type.

This property ensures that the system remains internally consistent under its own operations.

In the context of the Æ framework, field configurations may combine, transform, and interact in ways that resemble the structures encountered in Hopf algebras.

Hopf-like structures appear naturally in many areas of modern mathematics and theoretical physics, particularly in contexts where symmetry, composition, and decomposition of structures are important.

Conceptually, one may imagine complex patterns forming by combining simpler patterns, while also allowing those structures to split or transform under certain operations.

The algebraic framework introduced in the following sections captures this interplay between composition and decomposition.

2.3 Braiding and Non-Commutative Structure

In everyday arithmetic, the order of operations often does not matter. For example, adding two numbers produces the same result regardless of the order in which they are combined.

However, many physical and mathematical systems do not behave in this way.

When the order of operations changes the outcome, the system is said to exhibit non-commutative structure.

Braiding provides a useful way to visualize this phenomenon. Imagine several strands of rope being interwoven. The final pattern depends strongly on the order in which the strands cross each other.

Similarly, interactions between field configurations may depend on the sequence in which transformations occur.

Such braided interactions naturally lead to non-commutative algebraic structures, which play an important role in many areas of modern theoretical physics.

Within the Æ framework, these braided relationships capture how informational patterns intertwine and evolve within the field.

2.4 Transmission Operators

In order to describe how patterns propagate or transform within the field, the framework introduces transmission operators.

Operators can be understood as rules that map one configuration into another. They describe how a given state of the system changes under specific processes.

An intuitive analogy can be found in communication systems. When a signal passes through a transmission channel, it may be amplified, filtered, or transformed before reaching its destination.

Similarly, transmission operators describe how informational structures within the Æ field propagate and transform as the system evolves.

These operators form a central component of the mathematical formalism presented later in the work.

2.5 Algebraic Structure of the Æ-Space

Taken together, the primitive objects and operations introduced above form an algebraic structure associated with the space of possible field configurations.

This structure provides the mathematical environment in which phase relationships, coherence measures, and informational transformations can be expressed.

Rather than focusing on isolated elements, the emphasis is placed on the relationships between configurations and the rules governing their interactions.

In this way, the algebraic structure of the Æ-space provides a formal language for describing the dynamics of organized patterns within the field.

2.6 Ternary Structure (Nambu Seed)

While many physical theories rely on binary operations that combine two elements at a time, certain systems exhibit interactions that naturally involve three components.

Such ternary structures appear in various areas of mathematics and physics, including extensions of Hamiltonian mechanics and generalized bracket formulations.

Within the Æ framework, ternary relationships emerge as a natural extension of the relational perspective adopted by the theory.

Conceptually, one may think of three interacting processes whose balance determines the evolution of the system.

This triadic structure forms the conceptual seed of the Nambu-like formulations introduced later in the mathematical development.

2.7 Algebraic Closure

As the algebraic framework becomes more complex, it is important to ensure that the set of allowed operations remains internally consistent.

Algebraic closure guarantees that the system does not produce objects that lie outside the mathematical space initially defined.

From a conceptual standpoint, closure ensures that the rules of the framework are self-contained. All transformations and interactions described by the theory remain within the same structural environment.

This property is essential for maintaining mathematical coherence across the various layers of the formalism.

2.8 Coalgebra Structure

While algebraic operations describe how structures combine, coalgebraic operations describe how structures can be decomposed.

In many complex systems, both processes occur simultaneously. Patterns may merge to form larger structures, while also splitting into smaller components under certain conditions.

Coalgebraic ideas provide a natural language for describing these decompositions.

An analogy can be found in branching river systems. A river may divide into multiple channels, which in turn may merge again downstream.

Such processes illustrate the dual nature of combination and decomposition that appears throughout many dynamical systems.

2.9 Categorical Structure

As mathematical frameworks grow in complexity, it becomes useful to organize their components using the language of category theory.

Categories provide a way to describe relationships between objects and the transformations that connect them.

Rather than focusing solely on individual structures, categorical thinking emphasizes the network of mappings between them.

Within the Æ framework, categorical organization helps clarify how different algebraic components relate to one another and how transformations propagate through the overall system.

This perspective provides a unifying viewpoint for the various algebraic and geometric structures introduced throughout the theory.

2.10 Derived Differential Structure

Finally, the algebraic structures introduced in this section naturally give rise to differential structures that describe how field configurations vary continuously.

Differential relationships capture how small changes in the system propagate through the underlying space of configurations.

In physical terms, these structures allow the theory to describe dynamical evolution, stability, and response to perturbations.

The mathematical formalism presented in the following sections develops these differential relationships in precise terms, providing the analytical tools needed to study the evolution of structures within the Æ field.

3. Field Structure

Having introduced the primitive algebraic structures of the Æ framework, we now turn to the structure of the field itself.

Fields in modern physics are not static objects. They are dynamical entities capable of oscillation, rotation, propagation, and interaction. Electromagnetic waves, quantum wavefunctions, and many collective phenomena in physics can all be understood as manifestations of underlying field dynamics.

The Æ framework approaches field structure from the perspective of phase organization. Instead of beginning with particles or localized objects, the theory interprets physical structures as patterns of phase rotation within a continuous informational field.

In this view, the dynamics of the field are fundamentally

tied to rotational processes in phase space. The mathematical expressions introduced in the following sections formalize this idea by connecting phase rotation, temporal evolution, and energy relations within a unified framework.

3.1 Euler Identity (Rotational Seed)

One of the most remarkable relationships in mathematics is Euler's identity, which links exponential functions, imaginary numbers, and circular rotation.

At an intuitive level, this identity expresses a deep connection between oscillation and rotation. A point moving around a circle can be described equivalently as a sinusoidal oscillation in two orthogonal directions.

This observation provides a natural starting point for describing field dynamics in terms of phase rotations. Instead of treating oscillation and rotation as separate phenomena, they can be understood as different representations of the same underlying cyclic process.

Within the Æ framework, Euler's identity acts as a conceptual seed for describing how phase structures evolve within the field.

3.2 Field as Phase Rotation

Many physical phenomena that appear unrelated can be interpreted as manifestations of rotational dynamics in phase space.

For example, the oscillations of a quantum wavefunction, the propagation of electromagnetic waves, and the periodic motion of classical systems all involve cyclic evolution of phase.

In this perspective, a field configuration may be viewed as a collection of local phase rotations distributed throughout space and time.

An intuitive analogy can be found in a field of spinning gyroscopes. Each gyroscope rotates independently, yet interactions between them can produce collective patterns of motion.

Similarly, the Æ field can be imagined as a medium in which phase rotations occur continuously, interacting and organizing into structured dynamical patterns.

3.3 Pure Temporal Rotation

Temporal evolution itself can be interpreted as a form of phase rotation.

In many physical theories, time appears as an external parameter that measures the progression of events. However, in systems governed by oscillatory dynamics, the passage of time is closely tied to the evolution of phase.

For instance, the ticking of a clock corresponds to the periodic motion of a mechanical or electronic oscillator.

From this viewpoint, temporal evolution can be understood as a continuous rotation within an abstract phase space.

The formalism developed later in this section expresses this idea through operators that generate temporal phase rotation within the field.

3.4 Energy–Frequency Relation

One of the central insights of quantum theory is the deep connection between energy and frequency.

Systems with higher energy tend to oscillate at higher frequencies, while lower-energy configurations correspond to slower oscillatory behavior.

This relationship suggests that energy may be interpreted as a measure of dynamical activity within the field.

Within the phase-based interpretation adopted here, energy can therefore be viewed as a property associated with the rate of phase rotation.

This perspective naturally links energetic properties of a system with the underlying structure of its oscillatory dynamics.

3.5 Temporal Schrödinger Sector

The Schrödinger equation plays a fundamental role in quantum mechanics by describing how quantum states evolve in time.

Conceptually, it expresses how the state of a system rotates within an abstract space of possible configurations.

In the language of the Æ framework, this evolution can be interpreted as a specific realization of phase rotation within the field.

The temporal sector of the theory therefore recovers the familiar structure of quantum dynamics, while embedding it within a broader phase-based description in which evolution is understood as organized phase rotation rather than purely abstract state change.

3.6 Full Schrödinger Equation

While the temporal form of the Schrödinger equation captures the time evolution of quantum states, its full formulation includes spatial variation as well.

This extension allows wave-like structures to propagate through space, forming interference patterns and complex spatial configurations.

Within the Æ framework, such wave structures are interpreted as coherent phase patterns extending across the field, whose evolution reflects both local phase rotation and spatial coupling.

Conceptually, this behavior may be compared to the propagation of a structured wave on a surface, where local oscillations and spatial spreading combine to produce interference patterns.

The mathematical development that follows demonstrates how the familiar structure of the Schrödinger equation emerges as a particular regime of the broader phase-rotational dynamics described by the theory.

3.7 Temporal Operator Dictionary

To describe dynamical evolution rigorously, physical theories introduce operators that act on the states of a system.

These operators provide a dictionary connecting physical quantities such as energy, momentum, and temporal evolution to specific mathematical transformations.

Within the Æ framework, temporal operators are interpreted as generators of phase rotation associated with dynamical processes, linking observable quantities to underlying phase dynamics.

The formal dictionary introduced later establishes the precise correspondence between these operators and the transformations they induce on field configurations, ensuring consistency between the algebraic structure and the physical interpretation of the theory.

3.8 Hamiltonian Commutators (Ordering vs Dynamics)

In quantum mechanics, the order in which operators act on a state can affect the final outcome.

This non-commutative behavior reflects deep structural properties of dynamical systems.

Commutator relations capture how different operations interfere with one another and therefore constrain the allowed dynamics of the system.

Conceptually, this phenomenon can be compared to performing a sequence of rotations in three-dimensional space: rotating an object around one axis and then another generally produces a different orientation than performing the rotations in the opposite order.

Within the Æ framework, this non-commutativity can be understood as arising from the underlying phase geometry of the field, where different transformation sequences encode distinct pathways of phase evolution.

3.9 Quaternionic Proxy for Phase Entropy (Model Choice)

Phase entropy measures the dispersion or disorder of phase relationships within the field.

To describe this quantity mathematically, the framework adopts a representation based on quaternionic structures.

Quaternions can be understood as an extension of complex numbers that encode rotations not just in a plane, but in higher-dimensional spaces. While a complex number represents a rotation in two dimensions, a quaternion describes orientation and rotation in three dimensions in a unified algebraic form.

An intuitive picture is to think of each quaternion as specifying an orientation on a sphere. In this sense, phase configurations are not simple angles on a line, but points distributed over a curved space of possible orientations.

Another useful analogy is that of nested or layered structures, similar to Russian dolls or spheres within spheres: local phase relations are embedded within larger coherent structures, and quaternionic variables allow these nested orientations to be tracked consistently.

Within this representation, phase entropy corresponds to how spread out these orientations are across the configuration space. Highly aligned configurations correspond to tightly clustered orientations, while disordered states correspond to widely distributed orientations.

By representing phase organization through quaternionic objects, the model gains a compact and geometrically consistent way to describe both local alignment and global dispersion of phase structures within the field.

This choice acts as a mathematical proxy for measuring phase entropy in the dynamical system, linking algebraic representation to geometric intuition.

3.10 Temporal Action Ansatz (OPEN)

In many areas of physics, the evolution of a system can be derived from an action principle.

Such principles describe dynamics in terms of a quantity whose stationary behavior determines the physical trajectory of the system.

Conceptually, this can be understood as a selection rule: among all possible evolutions, the system follows those that are structurally consistent according to the action.

The temporal action introduced in the mathematical section represents a proposed extension of this idea to the phase-based structure of the Æ field.

In this context, the action governs not only trajectories in space, but the evolution of phase organization itself, including how coherence is maintained or lost over time.

The ansatz presented there serves as a candidate formulation for describing how phase configurations evolve, interact, and reorganize under dynamical constraints.

Note for further formalization. The precise structure, scope, and physical implications of this temporal action remain an open direction of investigation within the framework.

3.11 Partition Function on S^3

Statistical and quantum systems are often described using partition functions, which summarize the contributions of many possible configurations of a system.

Within the Æ framework, the configuration space associated with phase structures naturally leads to geometries that can be represented on higher-dimensional spheres.

In particular, the three–sphere S^3 provides a natural setting for describing phase configurations with rotational symmetry. While the ordinary sphere S^2 represents the set of directions in three-dimensional space, S^3 can be thought of as the space of normalized four-dimensional vectors, or equivalently as the manifold underlying unit quaternions.

Geometrically, this means that each point on S^3 encodes a possible orientation or phase configuration of the field. In this sense, the space of configurations is not a flat Euclidean space but a curved manifold where phase relations are naturally organized.

The formulation introduced in the mathematical section explores the possibility of defining a partition function over such a space.

This construction provides a bridge between the phase-based description of the field and statistical descriptions of ensemble behavior, allowing macroscopic properties to emerge from the collective organization of phase configurations.

In this sense, the partition function aggregates information over many possible phase arrangements, weighting them according to their structural coherence.

3.12 Æ-Product as a Binary Map

To describe interactions between field configurations, the framework introduces a binary operation known as the Æ-product.

This operation defines how two configurations combine to produce a new structure within the field.

Rather than a simple superposition, the Æ-product encodes interaction rules that depend on phase alignment and coherence structure.

Conceptually, one may imagine two interacting wave patterns whose relative phase determines whether the resulting configuration is amplified, suppressed, or reorganized.

An intuitive picture is that of two rotating vectors: depending on their relative orientation, they may reinforce each other, partially cancel, or produce a new direction altogether.

The Æ-product provides the formal rule governing such interactions within the algebraic structure of the theory.

3.13 Involution, Norm, and Inner Product

Mathematical structures describing physical systems often include notions of symmetry and measurement.

In the present framework, involutions represent transformations that reverse or conjugate phase-related properties of a configuration, capturing fundamental symmetry operations of the field.

Norms provide a measure of the effective magnitude or coherence of a configuration, while inner products quantify the degree of alignment between different phase structures.

Together, these concepts establish a consistent way to measure, compare, and transform configurations within the field, linking algebraic structure to physical interpretability.

An intuitive way to view this is that the norm measures how "strong" or coherent a configuration is, while the inner product measures how well two configurations are aligned in phase space.

3.14 Projector Invariants

Certain operators act as projectors, isolating specific components of a configuration.

Projectors can be used to extract particular modes, symmetries, or substructures within a larger system.

An intuitive analogy is that of a filter: given a complex signal, a projector isolates only the component aligned with a specific structure while discarding the rest.

The invariants associated with such projectors identify quantities that remain unchanged under specific transformations.

These invariants play an important role in characterizing stable and persistent structures within the field, particularly those that remain robust under dynamical evolution.

3.15 Æ-Functional Spaces and Weighted Metrics

The space of possible field configurations is not uniform. Different configurations may carry different degrees of coherence or informational organization.

Weighted metrics provide a way to measure distances between configurations while accounting for these structural differences.

Such metrics allow the theory to distinguish between highly organized configurations and those that are more disordered, effectively encoding coherence as a geometric property.

This distinction becomes essential when studying stability and dynamical evolution, as it determines which configurations can persist and which decay over time.

3.16 Nodal Realization of the Æ-Product

Complex field configurations often organize themselves around nodal structures.

Nodes may represent points, lines, or surfaces where certain properties of the field vanish or change qualitatively.

Within the Æ framework, the interaction rules encoded in the Æ-product can be realized through such nodal structures.

An intuitive picture is that of interference patterns: nodes correspond to regions where wave interactions either cancel or reorganize, creating stable geometric features that anchor the overall structure.

These nodes act as organizing centers for the formation and persistence of stable patterns within the field.

3.17 Algebraic Properties (Nodal Sector)

Finally, the nodal structures introduced above obey specific algebraic relations that determine how they interact and transform.

Understanding these relations is essential for describing how complex patterns form, merge, and evolve within the field.

In particular, these relations govern how nodal structures combine, split, or stabilize under interaction, providing a consistent framework for tracking structural evolution.

The mathematical section that follows develops these relations explicitly, establishing the formal properties of the nodal sector of the Æ framework.

These relations ultimately determine how coherent structures form, interact, and persist across different scales of the field.

4. Functionals

The previous sections established the structural and algebraic foundations of the Æ framework. We now turn to the dynamical quantities that govern how phase structures evolve within the field.

In mathematical physics, functionals provide a powerful tool for describing global properties of systems. While ordinary functions map numbers to numbers, functionals map entire configurations to scalar quantities that characterize their behavior.

Examples appear throughout physics. The action functional in classical mechanics summarizes the dynamics of a system, while entropy functionals measure the degree of disorder in thermodynamics and statistical mechanics.

Within the Æ framework, functionals are used to quantify the organization of phase structures. They provide measures of coherence, entropy, transmission, and memory across the field.

These quantities allow the theory to describe how local interactions between oscillatory components can generate large-scale organized structures.

4.1 Coherence Functional (Exponential Form)

Coherence represents the degree to which phases across a system become aligned.

In physical systems, coherence is often associated with the emergence of ordered behavior from many interacting oscillators.

For example, when a laser operates, billions of photons share a common phase structure. Similarly, biological systems often rely on coherent oscillations across large networks of interacting components.

The coherence functional introduced in the mathematical section provides a quantitative measure of this alignment.

Conceptually, one may imagine a field of rotating arrows. If the arrows point in random directions, the system is incoherent. As their orientations begin to align, a coherent structure emerges.

The functional captures the degree to which such alignment is present in the configuration of the field.

4.2 Phase Entropy Functional

While coherence measures alignment, phase entropy measures dispersion.

If the phases of a system become increasingly uncorrelated, the system loses its ability to sustain organized structures.

This concept parallels entropy in thermodynamics. Just as thermal entropy measures the spread of energy, phase entropy measures the spread of phase relationships.

High phase entropy corresponds to disordered oscillatory behavior, while low phase entropy corresponds to structured and coherent dynamics.

The entropy functional therefore provides a complementary measure to the coherence functional introduced above.

4.3 Kuramoto Phase Oscillator Dynamics

Many natural systems consist of interacting oscillators. Examples include power grids, neuronal networks, chemical reactions, and populations of biological cells.

The Kuramoto model provides one of the simplest mathematical descriptions of how such oscillators synchronize.

In this model, each oscillator evolves according to its own natural rhythm but is continuously influenced by the phases of neighboring oscillators.

As coupling increases, the system can transition from a disordered regime, where phases evolve independently, to a synchronized regime in which a large fraction of oscillators lock into a common phase evolution.

An intuitive picture is that of a group of metronomes: initially ticking at different tempos, but gradually adjusting until a shared rhythm emerges.

Within the Æ framework, Kuramoto-like dynamics provide a natural description of how phase alignment emerges within networks of interacting components, linking local interactions to global coherence.

4.4 Kuramoto Order Parameter

To quantify synchronization, the Kuramoto model introduces an order parameter that measures the collective alignment of phases.

When the oscillators are completely uncorrelated, this parameter approaches zero. As phase alignment increases, the parameter approaches unity.

This quantity therefore acts as a global indicator of coherence within the system.

Conceptually, it measures how much the individual oscillators behave as a single collective entity rather than as independent components.

In the context of the Æ field, the order parameter provides a natural measure of the degree to which phase structures have synchronized across the network, bridging local phase interactions with emergent large-scale coherence.

4.5 Illustrative Coherence System

To better understand these ideas, it is useful to consider simple illustrative systems.

Imagine a network of metronomes placed on a shared platform. Initially, each metronome swings independently, with no global coordination.

Through weak coupling transmitted across the platform, the metro- nomes gradually begin to influence one another. As their phases become more aligned, a collective rhythm emerges.

An intuitive way to characterize this behavior is to consider the average phase of the system. When the metronomes are completely uncorrelated, their contributions cancel out, and no global structure is observed. As synchronization develops, a coherent global oscillation becomes visible.

This collective alignment corresponds to a high value of the coherence functional, which measures how strongly the phases are clustered around a common orientation.

In addition, the stability of this synchronized state depends on how tightly the phases remain grouped. Systems with low phase dispersion tend to maintain coherence for longer periods, while systems with large fluctuations rapidly lose alignment.

This behavior motivates the introduction of a characteristic memory timescale, reflecting how long coherent phase structures can persist before being disrupted.

Such behavior is often discussed in the context of coherent media, where phase alignment and persistence are closely related to structural organization at the microscopic level.

4.6 Biological Stability Constraint

Biological systems provide striking examples of coherence maintained under complex and often noisy conditions.

From neuronal activity to cardiac rhythms and cellular signaling pathways, many biological processes rely on synchronized oscillatory behavior distributed across large networks of interacting components.

An intuitive picture is that of a living system constantly balancing order and fluctuation: too much disorder leads to loss of function, while excessive rigidity reduces adaptability.

To remain functional, biological systems must operate within a regime where coherent phase structures are preserved despite ongoing perturbations.

Within the Æ framework, this requirement is formalized as a stability constraint on phase configurations, ensuring that coherence remains above a critical threshold while phase entropy stays bounded.

These conditions determine which configurations of the field are physically sustainable, selecting those capable of maintaining organized behavior over time.

In this sense, biological systems can be viewed as naturally operating near optimal coherence regimes, where stability and adaptability coexist.

4.7 Biological Stability Constraint

Biological systems provide striking examples of coherence maintained under complex and often noisy conditions.

From neuronal activity to cardiac rhythms and cellular signaling pathways, many biological processes rely on synchronized oscillatory behavior distributed across large networks of interacting components.

An intuitive picture is that of a living system constantly balancing order and fluctuation: too much disorder leads to loss of function, while excessive rigidity reduces adaptability.

To remain functional, biological systems must operate within a regime where coherent phase structures are preserved despite ongoing perturbations.

Within the Æ framework, this requirement is formalized as a stability constraint on phase configurations, ensuring that coherence remains above a critical threshold while phase entropy stays bounded.

These conditions determine which configurations of the field are physically sustainable, selecting those capable of maintaining organized behavior over time.

In this sense, biological systems can be viewed as naturally operating near optimal coherence regimes, where stability and adaptability coexist.

4.8 Continuum Phase Transport

Phase relationships do not remain fixed in space. They propagate and evolve across the field.

This transport of phase information can be interpreted as a flow through the underlying medium, linking distant regions through dynamic phase coupling.

An analogy can be found in wave propagation on the surface of water: a disturbance created in one region generates waves that travel outward, carrying both energy and structural information.

Similarly, phase transport describes how oscillatory patterns propagate through the Æ field, transmitting coherence across the system.

When coherence is high, these patterns remain organized as they propagate. When coherence is low, the same transport leads to dispersion and loss of structure.

In this way, phase transport provides the mechanism by which local dynamics influence global organization.

4.9 Network Representation of the Æ-Field

In many situations, it is useful to represent the field as a network of interacting nodes.

Each node represents a local phase configuration, while links between nodes represent interactions or transmission pathways.

This network perspective allows complex systems to be analyzed using tools from graph theory and network dynamics.

Rather than focusing on individual components, this representation emphasizes how global structure emerges from local interactions.

Within this framework, coherence is not a property of any single node, but a collective feature arising from the alignment of phases across the network.

4.10 Graph Laplacian Dynamics

Graph Laplacians provide a mathematical tool for describing how quantities diffuse across networks.

They capture how local differences between neighboring nodes drive the evolution of the system.

An intuitive picture is that of a tension between connected nodes: when neighboring phases differ, the system evolves to reduce this mismatch.

In the context of the Æ framework, Laplacian dynamics describe how phase differences propagate and equilibrate through the network.

This process plays a central role in the formation and stabilization of coherent structures.

4.11 Propagation of Coherence

Once coherent structures emerge locally, they may spread through the network.

This propagation occurs when aligned regions influence neighboring regions, gradually extending the domain of coherence.

An analogy can be found in synchronized applause in a concert hall: initially individuals clap independently, but clusters of synchronized rhythm appear and expand as more participants align.

Similarly, in phase fields, coherence propagates as local alignment induces neighboring regions to synchronize.

This mechanism links local interactions to the emergence of large-scale coherent structures.

4.12 Coupled Node Evolution

Nodes within the network evolve under the influence of both internal dynamics and interactions with neighboring nodes.

These coupled dynamics determine whether coherence increases, decreases, or reorganizes across the system.

In particular, the balance between local tendencies and coupling strength governs whether the system approaches synchronization or remains disordered.

Understanding this evolution is essential for predicting the emergence, stability, and transformation of large-scale structures within the field.

4.12.1 Nonlocal Phase Interaction Kernel

In some systems, interactions are not restricted to nearest neighbors.

Nonlocal interaction kernels describe situations in which distant nodes can influence each other directly, with the interaction strength determined by a kernel function that depends on their separation.

Such long-range coupling allows phase information to propagate across large regions of the system rather than remaining confined to local neighborhoods.

As a consequence, the collective dynamics can change significantly, enabling phenomena such as large-scale coherence or global synchronization across extended domains.

4.12.2 Advective Phase Transport

In addition to diffusive processes, phase information may also be transported through advective flows.

In physical systems, advection occurs when a moving medium

carries quantities along with it—for example, when currents transport heat, particles, or momentum.

Within the Æ framework, advective transport represents situations where coherent phase structures propagate through the field as organized flows rather than spreading randomly.

In this regime, phase patterns behave similarly to vortices or wavefronts moving through a medium, preserving their internal organization while being transported across the field.

This type of transport contrasts with purely diffusive behavior, where phase information disperses and coherence is gradually lost.

4.13 Network Transmission Functional

Transmission functionals quantify how efficiently phase information propagates through the network.

Rather than measuring only connectivity, they capture the ability of the system to sustain and transmit coherent phase structures across its nodes.

This quantity plays a role similar to conductivity in electrical systems or permeability in fluid flow, but is modulated by coherence and phase alignment.

As a result, transmission is not uniform: certain pathways support stable propagation, while others lead to dispersion or decoherence.

4.14 Canonical Phase Transmission Operator

To describe phase transmission rigorously, the theory introduces a canonical transmission operator.

This operator acts on configurations of the field, mapping how phase structures propagate between different regions.

Its action encodes both the geometric connectivity of the system and the coherence constraints governing transport, linking structural organization with dynamical propagation.

4.14.1 Unitary Transport Representation

In certain regimes, phase transport preserves the norm of the configuration.

Such transport can be represented through unitary transformations, which preserve the internal structure and coherence of the system during propagation.

These regimes correspond to idealized conditions in which no information is lost and phase relationships remain fully consistent as structures move across the field.

4.14.2 Relation with Coherence

Transmission and coherence are closely related.

Efficient transmission pathways not only enable propagation but also sustain coherent structures over extended regions.

Conversely, when coherence is low, transmission becomes inefficient, leading to fragmentation or dispersion of phase patterns.

In this sense, coherence acts as a regulating factor that determines whether phase information is preserved or degraded during transport.

4.14.3 Network Transmission

In network representations, transmission processes can be understood as flows across the graph structure.

The mathematical formulation describes how these flows redistribute phase information throughout the system.

Rather than being purely topological, this transport is modulated by coherence: certain connections support stable propagation, while others lead to dispersion or loss of structure.

This distinction determines which pathways effectively carry coherent dynamics across the network.

4.15 Canonical Phase Entropy

The framework introduces a canonical definition of phase entropy that can be applied across several representations of the field.

These include continuous fields, discrete networks, and spectral decompositions.

The unified formulation allows entropy to be analyzed consistently across different scales, capturing the degree of phase disorder or desynchronization within the system.

4.16 Canonical Coherence Functional

Complementing the entropy measure, the canonical coherence functional provides a global measure of phase alignment.

While entropy quantifies disorder, coherence captures the extent to which phase structures are organized and mutually aligned.

Together, the entropy and coherence functionals define the fundamental thermodynamic-like quantities of the Æ framework, governing the balance between structure and dispersion.

4.17 Phase Memory Functional

Many dynamical systems retain traces of their past. Even when the instantaneous state changes, fragments of previous configurations remain encoded in the phase structure of the field.

In the Æ framework, memory is interpreted as the persistence of phase relationships through time. A system remembers its past when part of its phase configuration survives the evolution process.

This persistence can appear in many forms, from delayed responses in physical systems to long-term patterns in biological or cognitive dynamics. The phase memory functional provides a formal mechanism for describing how past configurations continue to influence the present state of the field.

4.18 Memory Kernels and Phase Retention

Memory kernels describe how past configurations influence present dynamics.

They determine the rate at which phase information is retained or dissipated over time, effectively weighting the contribution of earlier states to the current evolution of the system.

A useful analogy is that of ripples in water. Even after the original disturbance has passed, residual wave patterns can continue to influence the motion of the surface. In a similar way, memory kernels describe how earlier phase configurations leave traces that guide future dynamics.

4.19 Nonlocal Phase Memory Action

In some systems, memory effects extend over long temporal or spatial scales.

These systems cannot be fully described using purely local evolution laws, since their present state depends on an extended history of the field.

The nonlocal memory action introduced here provides a framework for describing such effects, allowing the dynamics to incorporate contributions from past configurations distributed across time or space.

Within this perspective, memory is not merely a stored variable but an active component of the dynamical structure of the field.

Within the Æ framework, correlations analogous to quantum entanglement emerge naturally from coherent phase alignment across the field.

Rather than requiring a separate postulate, such nonlocal correlations arise as a consequence of shared phase structure and coherent coupling between configurations.

4.20 Generic Memory Operator

To represent memory processes mathematically, the theory introduces operators that encode the influence of past configurations.

Rather than treating memory as an external mechanism, this formulation incorporates temporal persistence directly into the evolution of the field.

An intuitive way to understand this is to consider that the current state of the system is not determined only by its present configuration, but also by residual structures carried from previous states.

These operators act on the phase structure of the field, allowing past phase relations to influence ongoing dynamics.

In this way, memory effects arise naturally as part of the field evolution itself, linking local dynamics to accumulated phase history and enabling the persistence of coherent structures over time.

4.21 Long-Range Phase Memory Structures

Certain coherent structures can store phase information over extended durations.

When coherence is sufficiently strong, phase relationships can persist across long time scales, forming stable memory structures within the field.

Such long-range phase memory structures are relevant in a wide variety of systems, including biological regulation, information storage, and collective dynamics in complex networks.

In these cases the system does not simply react to immediate stimuli; instead it evolves in a way that reflects the accumulated history of its phase configuration.

4.22 Trauma as Persistent Phase Memory

In biological and cognitive systems certain experiences may generate persistent memory structures that continue to influence the dynamics of the system long after the original event has passed. In psychology these persistent imprints are commonly referred to as *trauma*.

Within the Æ framework trauma corresponds to a phase memory structure whose decay time greatly exceeds the normal relaxation timescale of the system.

Rather than being interpreted as a purely narrative construct, trauma is modeled here as a persistent phase configuration stored in the memory kernel of the system. Such structures act as dynamical constraints on future trajectories and may repeatedly bias the evolution of the local phase state.

Mathematically this corresponds to memory kernels whose decay is slow or incomplete, allowing the system to repeatedly revisit the phase structure associated with the original experience.

Resolution of trauma corresponds to the gradual reduction of this persistent phase constraint through repeated phase transport and entropy dissipation.

4.23 Geometric Memory Functional

Memory may also be encoded geometrically within the configuration space of the field.

In this perspective, memory is not stored as an isolated variable, but as accumulated structure along coherent trajectories of the system.

Geometric memory functionals describe how phase information is retained along paths in configuration space.

An intuitive picture is that of a trajectory traced on a curved manifold: as the system evolves, it accumulates information along its path, and this history influences future dynamics.

This idea is formalized through functionals defined over coherent paths, where memory corresponds to the integrated contribution of phase-retention structures along the trajectory.

In this way, memory is embedded in the geometry of the field, linking past evolution to present behavior through accumulated phase structure.

4.24 Structured Water Memory Extension

Some biological and chemical systems exhibit domains in which coherent phase patterns persist over time.

Within the Æ framework, such behavior can be described through phase fields whose coherence remains non-negligible across spatial regions.

In this context, the medium supports structured configurations in which phase alignment enables the retention and accumulation of information.

An intuitive interpretation is that coherence acts as a carrier of memory: when phase alignment is strong, contributions add constructively and persist over time; when alignment is weak, they cancel and memory is lost.

The corresponding formulation captures this behavior through coherence measures and time-integrated phase contributions, allowing memory to emerge as an accumulated effect of coherent phase dynamics.

In this way, biological memory can be understood as the superposition of multiple coherent subsystems, including neural, endocrine, and medium-based contributions.

4.25 Critical Coupling Threshold

Synchronization and coherence typically emerge only when coupling strength exceeds a critical threshold.

Below this threshold, oscillators behave largely independently and phase relations remain disordered. Above it, interactions become strong enough to induce collective alignment.

This transition is quantitatively characterized by a critical coupling value, which depends on the distribution of natural frequencies within the system.

Identifying this threshold is essential for understanding when a system transitions from incoherent behavior to synchronized dynamics, and therefore when large-scale coherence can emerge.

4.25.1 Phase Laplacian Representation

The Laplacian representation provides a compact mathematical description of how phase differences drive the evolution of the system.

In this formulation, local mismatches between neighboring phases generate dynamics that tend to reduce these differences.

The resulting operator incorporates both network structure and phase-dependent coupling, linking geometric connectivity with nonlinear interaction terms.

As a consequence, the phase Laplacian captures the tendency of the system to evolve toward alignment, playing a central role in the emergence of synchronized and coherent states.

4.26 Coherence Operators in the Æ Framework

The coherence of a configuration in the Æ-field is quantified by a set of operators collectively denoted $C_{Æ}$. These operators encode the degree of phase alignment relative to the entropy structure of the system.

An intuitive way to interpret coherence is as a measure of how aligned or organized phase configurations are. Highly coherent states correspond to tightly clustered phase relations, while incoherent states exhibit strong dispersion.

The canonical coherence measure is defined as

$$C_{Æ} = \frac{1}{1 + S_\phi} \tag{1}$$

where S_ϕ denotes the phase entropy. In this formulation, coherence decreases as phase disorder increases, providing a direct link between entropy and structural organization.

This quantity may be extended to describe local coherence across subregions of the field, as well as global and spectral coherence measures that capture large-scale organization.

Taken together, these coherence operators act as control parameters governing the stability and persistence of phase structures across the system.

They determine whether configurations remain organized, degrade into disorder, or transition between regimes, and therefore play a central role in the dynamical equations of the field.

5. Geometry

Up to this point, the Æ framework has been described in terms of algebraic structures, oscillatory dynamics, and functional descriptions of coherence and entropy. The next step is to understand the geometric structure underlying these objects.

Geometry plays a fundamental role in modern physics. Gravitational theory interprets spacetime as a curved geometric manifold, while gauge theories describe forces through geometric structures associated with internal symmetry groups.

In the Æ framework, geometry emerges from the organization of phase relationships within the field. Rather than treating geometry as a fixed background, the framework interprets geometric structure as a consequence of coherence and informational organization.

One may think of the field as a vast ensemble of oscillatory degrees of freedom. When these oscillations become partially synchronized, stable relationships begin to appear between different regions of the field. In this sense, geometry can be interpreted as the large-scale organization of phase coherence.

A helpful analogy is that of wave patterns on water. When many waves interact randomly, the surface appears disordered. However, when waves synchronize they can generate stable interference structures, such as standing patterns. In the Æ framework, coherent phase relations play a similar role: stable geometric relations emerge from the collective synchronization of oscillatory degrees of freedom.

The sections that follow develop the geometric language needed to describe these structures.

5.1 Coherence-Induced Metric

In many physical theories, a metric defines the notion of distance between configurations or points in a space.

Within the Æ framework, the concept of distance is not assumed a priori. Instead, it emerges from the coherence structure of the field itself.

Configurations that share aligned phase relationships can be interpreted as being closer in an informational sense, while configurations with disordered or uncorrelated phase relationships behave as if they were further apart.

Distance therefore acquires an informational meaning. Two configurations are close when their phase structures are strongly correlated, and distant when their phase relations are largely decorrelated.

An intuitive analogy can be drawn from neural systems. Neurons that fire in synchrony tend to form functional connections, while neurons with uncorrelated activity remain effectively distant within the network. The functional geometry of the brain therefore emerges from patterns of synchronization rather than from spatial proximity alone.

Similarly, in the Æ framework, coherence organizes the informational geometry of the field.

Conceptually, one may imagine a landscape in which regions of high coherence form valleys of stability where structures can persist, while regions of low coherence correspond to rugged terrain where configurations rapidly dissipate.

This leads naturally to a metric structure induced by coherence itself.

5.2 Quaternionic Structure

If the geometry of the field is determined by evolving phase relationships, the mathematical language used to describe phase rotations becomes essential.

Rotational structures in higher-dimensional phase spaces are most naturally described using quaternions.

Quaternions extend complex numbers by introducing three independent imaginary directions, allowing rotations in three dimensions to be represented compactly and without the singularities that arise in traditional coordinate descriptions.

Within the Æ framework, quaternionic structures provide a natural language for describing phase rotations that occur simultaneously across multiple degrees of freedom of the field.

A useful analogy is the problem of orientation in three-dimensional space. When an aircraft changes its orientation, the motion involves simultaneous rotations around multiple axes. Representing such motion using independent angles can lead to singularities and instabilities. Quaternion representations provide a stable and compact description of these multi-axis rotations.

In a similar way, quaternionic structures allow the framework to represent the coupled rotational dynamics of phase relationships across the field.

An intuitive way to view this is that phase configurations do not simply vary in magnitude, but also in orientation, and these orientations evolve through rotations in an internal space.

From this perspective, geometry is not imposed externally on the system. Instead, it emerges naturally from the organization and evolution of coherent phase relations within the field itself.

5.3 Exponential Map on S^3

When rotational structures are described using quaternions, their natural geometric setting becomes the three-sphere.

Rather than thinking of S^3 as an abstract object, it can be understood as the space of all possible orientations encoded by quaternionic phase configurations.

The exponential map provides a way to move from algebraic generators of rotation to actual rotational transformations within this space.

Conceptually, it converts an infinitesimal rotation (a local direction of change) into a finite motion along the curved geometry of the sphere.

Within the Æ framework, this map describes how local phase generators produce finite phase configurations across the field, linking local dynamics to global structure.

5.4 SU(2) Evolution

The group SU(2) plays a central role in many areas of theoretical physics, including quantum mechanics and spin dynamics.

Mathematically, SU(2) is equivalent to the set of unit quaternions and therefore shares the same geometric structure as the three-sphere.

An intuitive interpretation is that SU(2) describes all possible smooth rotations of phase configurations within this internal space.

In the context of the Æ framework, SU(2) evolution describes how phase structures rotate within the configuration space of the field.

This rotational evolution forms the geometric backbone of the dynamical structures introduced earlier, governing how coherence patterns transform over time.

5.5 Commutator Structure

In geometric formulations of physics, commutators often encode the curvature or nontrivial structure of the underlying space.

When transformations fail to commute, the resulting structure reflects the presence of geometric constraints or curvature.

An intuitive way to understand this is that performing two rotations in different orders leads to different results, revealing that the space itself has an internal structure.

Within the Æ framework, commutator relations describe how different phase transformations interact.

These relations reveal the underlying geometry of the configuration space in which the field evolves, and determine how sequences of transformations shape the resulting phase configurations.

5.6 Nambu Bracket (Internal Manifold)

While many physical theories rely on pairwise interactions, certain dynamical systems involve interactions among three quantities simultaneously.

The Nambu bracket provides a mathematical structure capable of describing such ternary dynamics.

An intuitive way to understand this is that the evolution of the system is not governed by a single interaction, but by the coupled influence of multiple quantities acting together.

Within the Æ framework, this structure appears naturally when considering interactions among phase, coherence, and entropy.

These three quantities jointly determine the dynamical organization of the field, forming a minimal set of interdependent variables that govern its evolution.

5.7 Information Geometry (Fisher Metric)

Information geometry studies the geometric structure of spaces formed by probability distributions or informational configurations.

One of the central objects in this field is the Fisher metric, which measures how distinguishable two configurations are.

Rather than thinking in terms of distance in physical space, the Fisher metric defines a notion of distance in the space of possible states.

In the Æ framework, phase configurations of the field can be treated as informational states.

The Fisher metric then provides a way to quantify how different two phase configurations are, and how easily one can be transformed into another within the space of possibilities.

5.8 Æ Differential Geometry

Combining the structures introduced above leads to a differential geometry associated with the Æ field.

This geometry integrates phase rotations, coherence structures, and informational relationships into a unified framework.

An important aspect of this construction is that the geometry is not imposed externally, but emerges from the internal organization of the field.

In this sense, curvature, structure, and connectivity are determined by how phase configurations interact and evolve, rather than by predefined spatial constraints.

5.9 Geometric Triple Product

Interactions among multiple components of the field often involve relationships beyond simple pairwise coupling.

Geometric triple products provide a compact way to describe interactions among three independent structures.

An intuitive picture is that of three directions or components combining to determine a new orientation or configuration that cannot be reduced to pairwise effects alone.

Within the Æ framework, such products play a role in describing higher-order interactions among phase structures, particularly in regimes where coherence, phase, and geometry interact simultaneously.

5.10 Layered Æ-Metric Ansatz

The geometry of the Æ field may be organized into multiple layers corresponding to different scales or degrees of coherence.

Rather than describing the field with a single metric, the layered ansatz introduces a hierarchy of geometric structures, each capturing a different level of organization.

An intuitive picture is that of successive levels of resolution: at coarse scales, only global structure is visible, while at finer scales, more detailed patterns emerge.

This can be understood as a form of geometric zoom, where each layer reveals additional structure without breaking consistency with the larger-scale geometry.

The layered metric ansatz introduced in the mathematical section provides a structured way to describe how these layers interact and remain coherently connected across scales.

5.11 Levi-Civita Connection

To describe how vectors and fields vary across a curved space, differential geometry introduces the concept of a connection.

The Levi-Civita connection provides the natural way to transport vectors while preserving the metric structure.

An intuitive interpretation is that it defines how a direction changes as it moves across a curved surface without introducing artificial distortion.

Within the Æ framework, this connection determines how phase structures evolve as they move across the geometric manifold associated with the field, ensuring that their evolution remains consistent with the underlying geometry.

5.12 Nodal Subspace Geometry

Certain configurations of the field concentrate around nodal structures.

These nodes correspond to regions where phase relationships undergo qualitative transitions, often acting as boundaries or organizing centers.

In such regions, the effective geometry may differ from that of the surrounding space, reflecting the local restructuring of phase relations.

Understanding this nodal geometry is essential for describing how stable structures form and persist, as these regions anchor the organization of the field.

5.13 Quaternionic Bundle Extension

When quaternionic structures vary across space, they form fiber bundles whose fibers carry internal rotational degrees of freedom.

An intuitive way to understand this is to think of each point in space as carrying its own internal orientation, described by a quaternion.

These orientations vary smoothly from point to point, forming a structured field of local rotations.

Such bundles allow local phase rotations to be attached consistently to different regions of the field, ensuring that internal phase dynamics remain compatible across space.

This geometric picture plays an important role in connecting local dynamics with global structure, linking pointwise phase behavior to the overall organization of the field.

5.14 Tetrahedral Structure on S^2

Certain stable configurations of rotational systems naturally organize themselves into tetrahedral arrangements.

These structures arise when multiple components distribute themselves as uniformly as possible on a spherical surface, minimizing mutual interference while maintaining symmetry.

An intuitive way to view this configuration is that each direction is maximally separated from the others, producing a balanced arrangement in which no single component dominates.

Within the Æ framework, tetrahedral structures provide a natural geometric representation of coherent phase configurations in equilibrium.

In such configurations, phase components are evenly distributed, leading to a state of balanced interaction and maximal structural stability.

This makes the tetrahedral arrangement a fundamental building block for describing symmetric and stable phase organizations within the field.

5.15 Hilbert Structure of the Æ Field

The space of possible field configurations can be treated as a Hilbert space, providing a natural framework for describing superposition, inner products, and orthogonality between states.

In this representation, each configuration of the Æ field corresponds to a vector in an abstract state space, while the inner product measures the degree of coherence between different configurations.

This interpretation connects the framework with the mathematical structures commonly used in quantum theory and functional analysis.

Conceptually, one may think of the Hilbert structure as the space of all possible phase configurations of the field. Superpositions correspond to the combination of multiple phase patterns, while orthogonality represents configurations whose phase structures are mutually uncorrelated.

This structure allows analytical tools from functional analysis, spectral theory, and quantum dynamics to be applied directly to the study of the Æ field.

5.16 Æ-Covariant Differential Structure

Covariant derivatives provide a way to differentiate fields while respecting the geometric structure of the space in which they live.

Rather than treating variations as purely local, the covariant formulation accounts for how the underlying geometry influences the evolution of field configurations.

In the Æ framework, this structure ensures that phase configurations evolve consistently with the geometry defined by the metric.

An intuitive interpretation is that phase structures are not simply changing in place, but are being transported across a curved space in a way that preserves their internal coherence.

The associated connection encodes how this transport occurs, while the resulting curvature captures how the geometry distorts phase relationships across the field.

5.17 Weighted Differential Operators

In systems where coherence varies across space, differential operators must account for local weighting factors.

These weights reflect the fact that not all regions of the field contribute equally to the dynamics.

Within the Æ framework, coherence acts as a modulating factor that influences how variations propagate.

As a result, differential operators acquire a weighted structure, linking geometric variation to the distribution of coherence across the field.

This allows the evolution of configurations to adapt dynamically to regions of high or low organization.

5.18 Metric Evolution and Interlayer Mediation

The geometry associated with the Æ field is not static. As coherence and entropy evolve, the metric structure itself may change.

This means that the space in which the field evolves is dynamically shaped by the state of the field.

An intuitive way to view this is that coherent configurations can deform the geometry, while geometric changes in turn influence the evolution of phase structures.

In this framework, different layers of the field interact through this evolving geometry, allowing structures at one scale to influence those at another.

This interlayer mediation provides a mechanism through which local coherence can affect global structure, and vice versa.

5.19 Differential Forms and Æ-Vorticity

The language of differential forms provides a natural way to describe rotational structures and circulation within the field.

Rather than tracking individual trajectories, this approach captures global patterns of flow and rotation in a coordinate-independent way.

Within the Æ framework, phase configurations can be associated with differential forms whose exterior derivatives describe rotational structure.

The resulting vorticity-like quantities measure how phase structures twist and circulate through the geometric manifold.

These objects play a central role in understanding the topological aspects of field dynamics, particularly in the formation of coherent vortical patterns and persistent structures.

6. Rotational Core

The structures introduced in the previous sections suggest a common underlying principle: many of the fundamental processes governing the Æ field can be interpreted as rotational dynamics.

Rotation appears repeatedly throughout modern physics. Spin systems, gauge symmetries, wave propagation, and phase evolution all involve cyclic transformations within internal spaces.

In classical mechanics, rotation describes the motion of rigid bodies. In quantum theory, however, rotation appears in a more abstract sense: states evolve as rotations within configuration spaces, and symmetries are encoded through group structures that generalize ordinary spatial rotations.

Within the Æ framework, this idea is extended further. Rotation is not just a feature of specific systems, but a central organizing principle of the field itself.

The fundamental degrees of freedom evolve through rotational structures operating within internal phase spaces, linking geometry, information, and dynamics into a unified description.

These rotations are not merely geometric motions in physical space. Instead, they represent transformations within the informational and phase structure of the field, through which coherent configurations evolve over time.

6.1 Quaternionic Field Constraint

To describe rotational dynamics consistently across multiple degrees of freedom, the framework adopts a quaternionic representation.

Quaternions provide a natural language for encoding three-dimensional rotations and their compositions. Unlike coordinate-based approaches, quaternionic rotations avoid singularities and offer a compact and globally consistent description of orientation.

Within the Æ framework, field configurations are constrained to lie on the unit quaternionic manifold, ensuring that the internal rotational degrees of freedom remain normalized and coherent.

These constraints guarantee that phase structures evolve through well-defined rotations, preserving the internal consistency of the field.

6.2 Primitive SU(2) Evolution

The group SU(2) is one of the most fundamental symmetry structures in modern physics, governing spin dynamics and appearing in a wide range of quantum and gauge theories.

Mathematically, SU(2) is equivalent to the group of unit quaternions, making it naturally compatible with the quaternionic structure introduced above.

In this framework, SU(2) represents the space of all possible internal rotations of phase configurations.

The fundamental dynamical law describes how these configurations evolve through continuous rotation within this space.

In the context of the Æ framework, SU(2) evolution provides the primitive mechanism by which the field changes its configuration, encoding time evolution as a rotation in the internal phase manifold.

6.3 Maurer–Cartan Structure

When studying continuous groups such as SU(2), differential geometry introduces the Maurer–Cartan forms as a way to describe infinitesimal transformations.

These forms capture how group elements change as one moves along the manifold of possible configurations, encoding the local structure of rotations at each point.

An intuitive way to interpret them is as a local "angular velocity field" defined over the space of configurations, describing how a small displacement translates into a rotation.

Within the Æ framework, the Maurer–Cartan structure provides the link between infinitesimal phase rotations and global dynamics.

It determines how local generators of rotation accumulate to produce the full evolution of the field.

6.4 Associator and Torsional Triple Product

In many algebraic systems, combining elements pairwise is sufficient to describe the dynamics.

However, certain structures exhibit higher-order interactions in which three elements participate simultaneously.

The associator measures how the grouping of operations affects the result of such combinations, revealing that the order in which interactions are applied can influence the final configuration.

An intuitive interpretation is that combining three rotational components may produce effects that cannot be reduced to a sequence of pairwise interactions.

Geometrically, these effects manifest as torsional structures within the underlying manifold, indicating that the space itself carries intrinsic twisting.

Within the Æ framework, the torsional triple product captures these higher-order interactions among phase structures, encoding how multi-component rotations generate additional geometric structure.

6.5 Torsional Geometry of the Æ-Manifold

When torsion is present in a geometric space, parallel transport around a closed loop may fail to return a vector to its original orientation.

This phenomenon reflects the presence of intrinsic twisting in the geometry, beyond simple curvature.

An intuitive picture is that of transporting an object along a loop and finding that it has rotated relative to its initial state, even though the path has closed.

Within the Æ framework, such torsional structures arise naturally from interactions between rotational degrees of freedom associated with phase dynamics.

These torsional effects influence how phase structures propagate, combine, and reorganize within the field.

As a result, torsion becomes a key ingredient in determining the global geometric organization of the Æ manifold and the persistence of coherent structures across it.

6.6 Dirac-Type Constraint

In relativistic quantum theory, the Dirac equation introduces constraints that connect spin structure, wave evolution, and relativistic symmetry.

Within the Æ framework, analogous constraints arise from the requirement that rotational phase structures remain compatible with the geometric organization of the field.

These constraints regulate the allowed configurations of phase structures and ensure consistency between the algebraic dynamics and the geometric structure of the manifold.

In this sense, Dirac-type constraints act as a bridge between the local rotational dynamics of the field and the global geometry of the Æ manifold.

6.7 Internal Nambu Dynamics

Standard Hamiltonian mechanics describes dynamical evolution through pairwise relationships between variables.

However, certain systems are more naturally described using ternary structures involving three interacting quantities.

Nambu dynamics provides a mathematical framework for such systems by introducing triple brackets that govern the evolution of the system.

In the Æ framework, interactions between phase, coherence, and entropy naturally form such triadic dynamical structures.

While Nambu dynamics was previously introduced in the algebraic formulation of the theory, it also admits a natural geometric interpretation.

From this perspective, the ternary relations between phase, coherence, and entropy describe internal rotational flows within the geometric structure of the Æ manifold.

The internal Nambu dynamics therefore capture the triadic organization of the field's evolution and provide a bridge between its algebraic dynamics and geometric structure.

6.8 Internal Polarities and Asymmetry

Rotational systems often exhibit internal polarities that distinguish different directions of evolution.

For example, clockwise and counterclockwise rotations represent distinct orientations within a rotational space.

Similarly, many physical processes involve asymmetries that determine preferred directions of evolution.

Within the Æ framework, internal polarities arise from the interplay between coherence, entropy, and rotational dynamics.

These asymmetries influence how structures form, persist, and transform within the field.

Understanding these polarities provides insight into the directional properties of dynamical processes within the Æ manifold.

7. Gauge Structure

Modern physics has revealed that many of the most fundamental interactions in nature arise from symmetry principles.

One of the most powerful of these principles is known as gauge symmetry.

Gauge symmetry expresses a simple but profound idea: certain transformations can be applied locally to the internal description of a system without changing the observable physics.

In other words, different mathematical descriptions can represent the same physical reality.

An intuitive way to understand this is that the internal "reference frame" of the system can be rotated independently at each point, provided that the laws describing the system adjust consistently.

Within the Æ framework, gauge structure emerges naturally from the rotational and phase-based nature of the field.

Because the fundamental degrees of freedom are phase-like quantities evolving within quaternionic spaces, local rotations of these phases must leave the physical content invariant.

The gauge sector formalizes how such local rotational freedom is consistently incorporated into the dynamics of the field.

7.1 Covariant Rotational Derivative

When a field evolves across space, one must compare its value at different locations.

However, if each point allows its own internal phase orientation, then ordinary derivatives are no longer sufficient.

A new type of derivative must be introduced that compensates for local rotations of the internal frame.

This construction is known as a covariant derivative.

An intuitive interpretation is that, before comparing two nearby configurations, one must first "rotate" them into a common frame.

Within the Æ framework, the covariant derivative tracks how phase structures change while accounting for the local rotational structure of the field.

The associated gauge field encodes how these local rotations vary across space, ensuring that the resulting dynamics remain consistent under local transformations.

7.2 Clifford Structure

To describe spin, rotations, and relativistic symmetries simultaneously, modern physics uses Clifford algebras.

These algebras provide a unified framework in which geometric transformations can be encoded algebraically.

Vectors, rotations, and reflections can all be represented within the same structure, allowing geometry and algebra to be tightly connected.

An important feature of this construction is that quaternionic structures arise naturally as part of the algebra, linking three-dimensional rotations with spinorial representations.

In the Æ model, the Clifford structure serves as the algebraic backbone that connects quaternionic rotations, spinorial dynamics, and gauge interactions into a single coherent framework.

7.3 Spinorial Sector

Spinors are mathematical objects that transform in special ways under rotations.

Unlike ordinary vectors, which return to their original configuration after a full rotation, spinors may require two complete rotations to return to their initial state.

This unusual property reflects the deeper structure of quantum rotations.

Within the Æ framework, spinorial structures naturally emerge from the quaternionic geometry of the field.

They provide a compact representation of internal rotational states.

7.4 Dirac–Æ Equation

One of the central equations of relativistic quantum theory is the Dirac equation, which describes particles such as electrons.

The Dirac equation combines spin, relativity, and wave dynamics into a single mathematical framework.

Within the Æ model, an analogous equation arises when the rotational structure of the field is combined with the gauge and spinorial framework.

This Dirac–Æ equation governs the evolution of spinorial states embedded within the Æ field.

7.5 Gauge-Coupled Derivatives

In many physical theories, taking a derivative is not simply measuring change, but measuring change relative to a local frame.

When gauge symmetries are present, these local frames can rotate or transform from point to point in space.

Derivatives must therefore incorporate additional structures that encode how these local frames relate to each other.

These structures are known as gauge connections.

Physically, they can be interpreted as the fields that mediate interactions, telling us how a field should adjust as it moves through space.

Within the Æ framework, gauge-coupled derivatives describe how phase structures evolve while continuously adapting to these connection fields.

7.6 Gauge Curvature

If a field is transported around a closed loop, one might expect it to return to its original state.

However, in the presence of a gauge connection, this is not always the case.

The field may come back rotated or shifted.

This mismatch encodes an intrinsic property of the space: its curvature.

Gauge curvature therefore measures how much the internal structure twists when transported through the field.

Physically, it corresponds to the field strength associated with an interaction.

In familiar physics, electromagnetic fields and other gauge fields can be understood precisely in terms of such curvature structures.

Within the Æ framework, this curvature is naturally embedded in the phase geometry, linking interaction strength directly to phase twisting.

7.7 Dirac–Æ Operator (Gauge-Coupled)

Once both spin structure and gauge interactions are present, they must be described within a single unified operator.

Combining spinorial structure with gauge-covariant derivatives leads to a generalized operator governing the evolution of the field.

This operator extends the standard Dirac operator by incorporating the internal phase geometry of the Æ framework.

An intuitive way to interpret this is that the operator not only describes how states move through space, but also how their internal phase orientation evolves under interaction.

In this sense, it encodes both motion and structure simultaneously.

It therefore becomes one of the central operators controlling the dynamics of spinorial states within the field, linking geometry, interaction, and phase evolution into a unified description.

7.8 Essential Spectrum and Resolvent

After defining an operator that governs the dynamics of a system, a natural question arises: what states are actually allowed?

Spectral theory addresses this by analyzing the eigenvalues and eigenstates of the operator.

These spectral properties determine the possible modes of excitation that the system can support.

An intuitive picture is that of the natural vibration modes of a structure: only certain patterns are stable, while others dissipate or fail to persist.

Within the Æ framework, the spectral structure of the Dirac–Æ operator reveals which phase configurations are stable, which are transient, and how coherent structures organize across different scales of the field.

7.9 Discrete Hodge and Internal Constraints

Geometric structures do not allow arbitrary field configurations.

Instead, they impose consistency conditions that restrict how fields can be arranged.

Hodge theory provides a systematic way of decomposing fields into components that respect these constraints.

An intuitive interpretation is that it separates configurations into those that are compatible with the geometry and those that are not.

In other words, it distinguishes between degrees of freedom that can evolve freely and those that are constrained by the structure of the space.

Within the Æ framework, discrete Hodge structures regulate how phase configurations organize across the manifold, ensuring consistency between local dynamics and global geometry.

7.10 Quaternionic Phase Normalization

Because the internal structure of the Æ field is described in terms of quaternionic phase configurations, it is necessary to impose normalization conditions that preserve their geometric consistency.

These conditions ensure that phase states remain on the unit quaternionic manifold, preventing unphysical distortions of the internal rotational structure.

In this way, normalization acts as a constraint that maintains coherence and stability of phase configurations as they evolve.

7.11 Phase Retention and Stability

For coherent structures to exist, it is not enough for phases to align momentarily; their relationships must persist over time.

Mechanisms of phase retention therefore play a critical role in the stability of the field.

They determine whether a configuration is transient or capable of sustaining itself.

These mechanisms are closely related to the coherence functionals introduced earlier, which quantify how well phase relationships are preserved.

7.12 Statistical Ensemble on S^3

When large collections of phase states are considered, tracking each one individually becomes impractical.

Statistical descriptions then become the natural language.

Within the quaternionic framework, these ensembles naturally live on the three-sphere S^3, which represents the space of normalized quaternionic states.

In this sense, S^3 is not just a geometric object, but the natural configuration space of internal phases.

Statistical distributions on this space allow the collective behavior of large numbers of oscillatory units to be analyzed in a unified way.

7.13 General Kuramoto Dynamics

Synchronization phenomena appear throughout nature.

Examples include flashing fireflies, neural networks, power grids, and chemical oscillators.

What these systems share is the ability of many independent units to spontaneously align their phases.

The Kuramoto model provides a mathematical description of how large populations of oscillators can achieve this synchronization through coupling.

Within the Æ framework, Kuramoto-like dynamics describe how local phase oscillators become synchronized through interactions, forming coherent structures across the field.

7.14 Order Parameter

To measure the degree of synchronization in a system of oscillators, the Kuramoto model introduces an order parameter.

This parameter compresses the collective state of the system into a single quantity representing phase alignment.

When phases are random, the order parameter is small, indicating low coherence.

When phases become aligned, the order parameter increases, signaling the emergence of collective structure.

Within the Æ framework, this parameter serves as a quantitative measure of coherence across the field.

7.15 Linearization Around Synchrony

Once a synchronized state is formed, a natural question arises: will it persist?

To answer this, one studies small perturbations around the synchronized configuration.

This procedure is known as linearization.

It reveals how the system responds to slight disturbances and whether it tends to return to synchrony or drift away from it.

In this way, linearization provides a local probe of stability.

7.16 Spectral Stability Condition

The stability of synchronization is not arbitrary; it is encoded in the spectral properties of the interaction structure.

Eigenvalues of the coupling operator determine whether perturbations decay or grow over time.

Negative or bounded spectra tend to stabilize the system, while unstable modes lead to desynchronization.

Within the Æ framework, these spectral conditions determine when coherent phase structures remain stable and when they fragment.

7.17 Synchronization Regimes

Depending on coupling strength and system parameters, different synchronization regimes may appear.

These include incoherent states, partial synchronization, and complete phase locking.

Rather than being abrupt transitions, these regimes often emerge gradually as system parameters vary.

Each regime corresponds to a distinct organizational structure within the oscillator network, reflecting different levels of coherence across the field.

7.18 Entropy Interpretation

Synchronization can also be understood from an informational perspective.

Highly disordered phase states correspond to high entropy.

As oscillators begin to align, the system loses randomness and entropy decreases.

Coherent synchronization therefore corresponds to lower entropy configurations.

Within the Æ framework, this relationship directly links phase dynamics to the entropy measures introduced earlier, connecting order, information, and stability within a single description.

7.19 Connection with Æ Coherence

The order parameter introduced in synchronization theory is not an isolated construct.

Within the Æ framework, it is directly related to the coherence functionals defined in earlier sections.

As oscillators begin to align, their collective phase configuration becomes more structured.

This increase in alignment is reflected as an increase in coherence.

In this way, the order parameter provides a dynamical manifestation of the same coherence measured geometrically within the field.

It therefore acts as a bridge between the dynamical oscillator picture and the underlying geometric structure of the Æ field.

7.20 Relation to the Phase Transmission Operator Δ

The phase transmission operator introduced earlier provides the mechanism through which phase information propagates across the field.

It determines how local phase variations influence neighboring regions and how alignment spreads.

Within networks of oscillators, this operator governs how phase synchronization propagates from local interactions to global order.

Rather than being imposed externally, synchronization emerges from the iterative action of this transmission process.

The operator Δ therefore plays a central role in connecting local phase dynamics with large-scale coherent organization.

7.21 Conclusion

The gauge structure of the Æ framework integrates rotational geometry, spinorial dynamics, and synchronization phenomena into a unified description.

What may initially appear as separate mechanisms—geometry, interaction, and collective alignment—are revealed to be different aspects of the same underlying process.

Through this structure, local phase interactions become capable of generating large-scale coherent organization.

Coherence, transmission, and synchronization are thus not independent features, but interconnected expressions of the same phase-based dynamics.

This mechanism forms one of the central pillars of the theory.

8. Dirac–Æ Operator

Up to this point, the Æ framework has introduced the fundamental ingredients required to describe its internal structure: phase dynamics, coherence functionals, geometric constraints, rotational structures, and gauge symmetries.

A natural question now arises:

What mathematical object governs the evolution of these structures?

In many areas of physics, such evolution is encoded through operators acting on spaces of states.

For example, in quantum mechanics the Hamiltonian operator determines how wavefunctions evolve in time.

In relativistic quantum theory, the Dirac operator plays a central role in describing spinorial particles.

Within the Æ framework, a related but more general structure emerges.

This structure is known as the Dirac–Æ operator.

It combines rotational geometry, gauge interactions, and phase dynamics into a single mathematical object that governs the evolution of states within the Æ field.

8.1 Weighted Nodal Hilbert Space

To define an operator, one must first specify the space on which it acts.

In quantum theory, states live in a Hilbert space, a mathematical structure that allows the definition of inner products, norms, and orthogonality.

Within the Æ framework, the relevant state space is a Hilbert space constructed from nodal configurations of the field.

These nodes represent localized phase structures that carry information about coherence and entropy within the system.

Because different nodes may possess different levels of coherence, the Hilbert space is naturally weighted.

The weighting reflects the influence of coherence and phase entropy on the stability of the states.

8.2 Definition of the Dirac–Æ Operator

Once the appropriate Hilbert space has been identified, the next step is to define the operator governing the evolution of the states.

The Dirac–Æ operator generalizes the familiar Dirac operator from relativistic quantum theory.

While the standard Dirac operator captures spin and relativistic wave propagation, the Dirac–Æ operator incorporates additional structures arising from the phase geometry of the Æ field.

In particular, it includes contributions related to coherence, entropy, and the internal rotational structure of the field.

Through this operator, the dynamics of the system become encoded in a compact algebraic form.

8.3 Klein–Gordon Emergence

One of the remarkable features of the Dirac equation in standard physics is that its square leads to the Klein–Gordon equation, which describes relativistic scalar particles.

A similar phenomenon occurs within the Æ framework.

When the Dirac–Æ operator is applied twice to the state space, an effective second-order dynamical equation emerges.

This resulting equation resembles a Klein–Gordon type structure.

This relationship provides an important consistency check for the framework, showing that familiar relativistic structures can emerge from the deeper phase-based dynamics of the field.

8.4 Effective Mass

In conventional particle physics, mass appears as a parameter in the Dirac and Klein–Gordon equations.

Within the Æ framework, mass is not introduced as an independent quantity.

Instead, it emerges from the internal structure of the field.

Interactions between phase coherence and entropy can generate effective mass terms within the dynamical equations.

From this perspective, mass becomes a manifestation of the

internal organization of the field rather than a fundamental primitive.

8.5 Dirac–Æ Operator Structure

The full structure of the Dirac–Æ operator reflects the combination of several ingredients introduced in earlier sections.

These include:

- the quaternionic rotational geometry,
- the gauge-covariant derivatives,
- the coherence and entropy functionals,
- and the internal phase dynamics of the field.

Together, these elements form a unified operator that governs how states evolve within the Æ manifold.

8.6 Projection onto the Physical Sector

Not every mathematical state in the Hilbert space necessarily corresponds to a physically realizable configuration.

The full space of solutions is typically much larger than what can manifest as stable or observable structures.

For this reason, projection mechanisms are introduced in physical theories to restrict attention to the relevant subspace of states.

These projections act as filters, selecting only those configurations that satisfy the required physical constraints.

Within the Æ framework, projections identify the subset of configurations that correspond to physically meaningful phase structures.

They enforce compatibility between the algebraic description and the coherence conditions of the field.

In this sense, projection is not merely a mathematical operation, but a selection principle that determines which phase configurations can persist as part of physical reality.

8.7 Resolvent and Spectral Structure

Once the operator governing the dynamics is defined, the question shifts from "how does the system evolve?" to "what behaviors are actually allowed?"

This information is encoded in the spectral properties of the operator.

The spectrum determines the possible excitation modes that the system can support, defining its natural dynamical responses.

The resolvent, in turn, provides a way to probe how the system reacts to external influences and perturbations.

It captures how disturbances propagate through the structure defined by the operator.

Within the Æ framework, the spectral structure of the Dirac–Æ operator plays a key role in understanding the organization and stability of coherent structures within the field.

It reveals which configurations are robust, which are transient, and how the field reorganizes under change.

9. Weighted Action

One of the most powerful principles in theoretical physics is the principle of least action.

Rather than describing dynamics directly through forces or differential equations, this principle states that the evolution of a physical system is determined by an extremum of a quantity known as the action.

In this perspective, the system does not simply follow local rules, but selects a trajectory that is globally consistent.

The action summarizes the behavior of the system over an entire trajectory, encoding both its structure and its evolution.

By applying a variational procedure to this action, the fundamental equations of motion of the theory can be derived.

Many of the most successful physical theories are formulated in this way, including classical mechanics, quantum field theory, and general relativity.

Within the Æ framework, the notion of action is extended to incorporate the phase, coherence, and entropy structures introduced in previous sections.

Because coherence plays a central role in the organization of the field, the action must reflect how coherence influences both the stability and the evolution of physical configurations.

In other words, not all trajectories are treated equally: those that preserve or enhance coherence may be dynamically favored.

For this reason, the framework introduces a weighted action, in which coherence and phase structure modify the effective dynamics of the system.

9.1 Extended Master Action (Dynamical Coherence Layer)

The weighted action extends the traditional action functional by incorporating a dynamical layer associated with coherence.

In contrast with standard formulations, coherence and phase entropy are not treated as fixed parameters, but as dynamical fields that evolve together with the rest of the system.

This layer captures how phase alignment, synchronization, and internal structure actively influence the evolution of the field.

An important consequence is that the geometry, interactions, and effective couplings become dependent on the coherence state of the system.

In physical terms, this means that the stability of field configurations depends not only on their energy, but also on their coherence.

Configurations that are highly coherent may persist longer or dominate the dynamics, even when energetically comparable alternatives exist.

Coherence thus acts as an additional organizing principle within the action, modifying both dynamics and effective geometry.

9.2 Operator Extraction (From the Variational Principle)

Once an action has been defined, the equations governing the system can be obtained through a variational procedure.

Small variations of the action with respect to the relevant fields produce the dynamical operators and equations that control the evolution of the system.

This procedure ensures that the resulting dynamics are consistent with the global structure encoded in the action.

Within the Æ framework, this variational process acquires additional significance, since coherence and phase entropy contribute directly to the structure of the operators.

As a result, operators such as the Dirac–Æ operator emerge naturally from the action, including effective mass terms and interaction contributions that depend on the coherence state.

This contrasts with standard approaches in which such structures are typically introduced by hand.

9.3 Variational Template (Reference Form)

To facilitate systematic analysis, the variational structure of the theory is presented in a general template form.

This template makes explicit the relationship between the action functional and the operators derived from it.

In particular, it clarifies how different contributions—energetic, geometric, gauge, and coherence-related—enter the action and interact within a unified structure.

An important feature of this formulation is that coherence-dependent terms modify both couplings and effective masses, introducing a dynamical feedback between phase structure and field evolution.

In this way, the template serves as a reference structure for constructing consistent dynamical models within the framework.

9.4 Variational Folding Principle (Topological Subsector)

Not all contributions to the dynamics arise from local terms.

Some emerge from global or topological properties of the field.

The variational folding principle captures how these non-local contributions can be consistently incorporated into the action.

It describes how global geometric structure can "fold" into local dynamics, influencing the evolution of the system even in regions where no explicit local interaction is present.

This provides a bridge between topology and dynamics within the framework.

9.5 Interaction-Specific Effective Actions

Different physical interactions may be described by effective actions derived from the master action.

These effective actions isolate the contributions associated with particular dynamical sectors.

For example, interactions related to phase transport, coherence propagation, or geometric curvature may each admit their own effective descriptions.

This decomposition allows complex dynamics to be studied in a modular way, while remaining consistent with the full theory.

9.6 Master Coherence-Dependent Action

At the center of the framework lies a master action that explicitly incorporates the dependence of the dynamics on coherence.

This action provides the unified starting point from which the various dynamical equations of the theory can be derived.

It encodes how energy, geometry, and coherence jointly determine the evolution of the field.

In this sense, the master action does not only govern motion, but also the selection of physically realized configurations.

It therefore plays a role analogous to the fundamental action in other field theories, while extending it to include the organizing influence of coherence.

9.7 Effective Mass Renormalization

Within the Æ framework, mass is not treated as a fixed primitive constant, but as a quantity that can emerge from internal interactions within the field.

As coherence and phase interactions modify the effective dynamics, they also influence how excitations propagate.

This change in propagation can be interpreted as a modification of the apparent mass.

In this way, mass becomes a dynamical property, dependent on the internal organization of the field.

This naturally leads to mechanisms of effective mass renormalization driven by coherence and phase structure.

9.8 Classical Consistency Limit

Any extended framework must reproduce known physical results in the appropriate limits.

In regimes where coherence effects become negligible or uniform, the additional structure introduced by the weighted action must reduce to standard forms.

Within this classical consistency limit, the weighted action recovers expressions compatible with established physical theories.

This ensures that the Æ framework does not contradict known physics, but extends it in a controlled and consistent way.

9.9 Variational Hierarchy

The variational structure of the theory is naturally organized into a hierarchy.

At the highest level sits the master action governing the full field.

From this unified structure, more specialized actions and equations can be derived for particular sectors or approximations.

This hierarchical organization allows the framework to remain flexible, while preserving a single coherent foundation across all levels of description.

9.10 Nonlocal Coherence Memory Sector

Not all contributions to the dynamics depend only on local field values.

Some effects arise from the history of the system, reflecting how past configurations influence present behavior.

These effects can be described as coherence memory.

Within the weighted action, such contributions appear as nonlocal terms that encode temporal correlations.

They capture how previously established phase relationships continue to shape the evolution of the field.

9.11 Modified Einstein Equations

Because the geometry of the manifold interacts with the coherence structure of the field, gravitational dynamics acquire additional contributions.

When the weighted action is varied with respect to the geometric degrees of freedom, these contributions naturally emerge.

The resulting equations extend the familiar Einstein equations by incorporating terms that depend on coherence and phase structure.

In this way, geometry is influenced not only by energy and momentum, but also by the organization of the field.

9.12 Coherence Field Equation

The variational procedure does not only produce geometric equations.

It also yields dynamical equations governing the evolution of the coherence field itself.

These equations describe how coherence is generated, propagated, and dissipated within the system.

They provide a direct dynamical description of how order emerges and evolves across the field.

9.13 Extended Coherence Layer in the Weighted Action

The coherence layer introduced in the action therefore plays a dual role.

On one hand, it influences the geometry of the manifold.

On the other, it governs the internal phase structure of the field.

Through this coupling, coherence becomes an active dynamical ingredient, linking geometry and phase dynamics into a single unified process.

9.14 Variational Origin of Weighted Operators

Finally, the variational framework provides the origin of the weighted operators used throughout the theory.

These operators are not introduced ad hoc, but arise naturally as functional derivatives of the action.

In this way, the entire dynamical structure of the Æ framework can be traced back to a single coherent variational principle, from which both operators and equations of motion emerge.

10. Equations of Motion

Once an action has been defined, the fundamental dynamical laws of the theory follow from the variational principle.

By considering small variations of the fields and requiring that the action remain stationary, one obtains the equations governing the evolution of the system.

These equations are known as the equations of motion.

In this sense, the action encodes all possible behaviors, while the equations of motion select those that are dynamically realized.

Within the Æ framework, the equations of motion describe how phase structures, coherence levels, and geometric properties of the field evolve together.

Because the theory integrates rotational geometry, gauge structure, and coherence dynamics, the resulting equations possess a layered structure.

Some equations describe the evolution of spinorial states, others regulate coherence dynamics, and still others constrain the geometric structure of the manifold.

Together, they define the full dynamical behavior of the Æ field.

10.1 Weighted Dirac Equation

The first dynamical equation arises from the variation of the weighted action with respect to the spinorial states.

This procedure produces a generalized Dirac-type equation that incorporates the coherence structure of the field.

In this equation, the evolution of spinorial states is influenced not only by geometry and gauge interactions, but also by the coherence weighting introduced earlier.

This means that phase propagation is modulated by how coherent the surrounding field is.

The resulting weighted Dirac equation therefore governs the propagation of phase structures within the Æ manifold, linking local dynamics to global coherence.

10.2 Nodal (Constant-Weight) Approximation

In certain regimes, the coherence weights may vary slowly or remain approximately constant within localized regions.

Under such conditions, the equations simplify significantly.

This leads to the nodal approximation, in which the dynamics can be studied around individual coherent nodes.

Within this approximation, the system behaves as if each node were an effective local center of organization.

This provides a useful bridge between the full theory and simpler models of localized excitations.

10.3 Gauge Current Closure

Gauge symmetries impose strong constraints on the dynamics of a system.

One important consequence is the conservation of associated currents.

Within the Æ framework, the closure of gauge currents ensures the internal consistency of the gauge structure.

It guarantees that the flow of phase and information within the field respects the symmetry principles established earlier.

In this sense, current closure acts as a consistency condition linking local dynamics to global symmetry.

10.4 Æ-Dynamics as a Hamiltonian Flow

The equations of motion may also be interpreted from the perspective of Hamiltonian dynamics.

In this formulation, the evolution of the system is generated by a Hamiltonian structure defined on the phase space of the field.

The resulting flow describes how the configuration of the system moves through its space of possible states over time.

Within the Æ framework, this Hamiltonian flow governs the coupled evolution of coherence, phase, and geometry, providing a continuous trajectory through the space of configurations.

10.5 Linearization and Spectral Modes

To understand the behavior of solutions, it is often useful to analyze small perturbations around equilibrium or coherent states.

This process is known as linearization.

By studying these perturbations, one can identify the spectral modes supported by the system.

These modes represent the natural vibrational patterns of the field and determine how the system responds to disturbances.

They provide insight into stability, resonance, and the propagation of fluctuations.

10.6 Nambu–Æ Equations of Motion

In addition to the Hamiltonian formulation, the dynamics of the Æ framework can also be expressed in terms of Nambu mechanics.

Nambu dynamics generalizes traditional Hamiltonian mechanics by allowing the evolution of the system to depend on multiple conserved quantities simultaneously.

Within the Æ framework, the Nambu equations describe ternary interactions between key dynamical quantities, such as phase, coherence, and geometric structure.

This formulation highlights the intrinsically multi-structured nature of the field, where different conservation principles interact to shape the evolution of the system.

It also prepares the ground for the emergence of higher-order invariants and conservation laws associated with the geometry of the phase space.

10.7 Casimir / Volume-Form Conservation

Certain quantities remain conserved throughout the evolution of the system.

These conserved quantities are known as Casimir invariants.

Unlike conserved quantities arising from explicit symmetries, Casimir invariants are tied to the intrinsic geometric structure of the phase space.

Their conservation reflects deep properties of the underlying manifold, independent of the specific dynamics.

Within the Æ framework, these invariants correspond to conserved structures associated with the volume form of the phase space.

They ensure that the global organization of the field remains consistent, even as local configurations evolve.

10.8 Local Equation for S_ϕ with Memory Kernel

One of the distinctive features of the framework is the central role played by phase entropy.

The evolution of the phase entropy field S_ϕ is not purely local.

It may depend not only on current field values, but also on the past evolution of the system.

This leads to the appearance of memory kernels in the dynamical equations.

Such kernels encode how past configurations continue to influence present behavior.

Through this mechanism, local dynamics become coupled to

temporal correlations, providing a bridge between immediate interactions and long-range coherence effects.

10.9 Renudation Condition

Under certain conditions, accumulated phase structures may lose stability.

When coherence is no longer sufficient to sustain a configuration, the system may undergo a topological reset.

This process, referred to as renudation, represents a reorganization of the underlying phase network.

The renudation condition defines the criterion under which such transitions occur.

It determines when a configuration can no longer persist and must reorganize into a new coherent structure.

This mechanism plays a key role in the adaptive and evolving nature of the field.

10.10 Æ Dynamical Evolution

Taken together, the equations of motion describe the full dynamical evolution of the Æ field.

They govern how phase structures form, interact, propagate, and reorganize across the manifold.

Through these equations, the combined effects of geometry, coherence, and gauge structure become dynamically realized.

The field is therefore not static, but continuously reshaped by the interplay of these fundamental components.

10.11 Nambu Equations of Motion

The ternary structure of Nambu dynamics provides an alternative representation of the evolution equations.

In contrast with standard Hamiltonian formulations, where dynamics are generated by a single function, Nambu dynamics are driven by multiple interacting quantities.

Within the Æ framework, these quantities can be naturally

associated with phase, coherence, and geometric structure.

This reflects the fact that the evolution of the field is not governed by a single constraint, but by the simultaneous interaction of several conserved or slowly varying components.

An intuitive interpretation is that the system evolves along directions defined by the interplay of these quantities, rather than following a single gradient flow.

This perspective highlights the inherently multi-layered nature of the dynamics, where different structural aspects of the field interact simultaneously to determine its evolution.

10.12 Volume Preservation in Nambu Dynamics

One of the defining properties of Nambu dynamics is the preservation of volume within phase space.

This property reflects deep conservation laws governing the system, ensuring that the global structure of the phase space remains invariant under evolution.

An intuitive way to understand this is that the dynamics rearrange configurations within the space without compressing or expanding the total set of accessible states.

Within the Æ framework, this preservation of phase-space volume guarantees that the overall organization of the field remains consistent, even as local configurations evolve and reorganize.

This provides a geometric foundation for the stability and continuity of the dynamics across scales, allowing coherent structures to transform without loss of global integrity.

11. Classical Limit

Any theoretical framework intended to describe physical reality must ultimately connect with the classical world that we observe in everyday experience.

While many physical theories are formulated in terms of abstract mathematical structures, their predictions must reproduce classical behavior under appropriate conditions.

This transition from microscopic dynamics to macroscopic stability is known as the classical limit.

Within the Æ framework, the classical limit arises when coherent phase structures stabilize into persistent configurations.

In this regime, fluctuations become small, phase relationships remain stable over long timescales, and the geometry of the field acquires an effectively smooth structure.

An intuitive interpretation is that rapidly varying internal dynamics average out, leaving behind stable, observable behavior.

The result is the emergence of classical objects, trajectories, and geometric relationships as effective descriptions of underlying coherent phase structures.

11.1 Nodal Stability

One of the key ingredients in the emergence of classical behavior is the stability of nodal structures.

Nodes represent localized regions in which phase relationships become strongly coherent.

When such nodes remain stable over time, they behave as persistent structures within the field.

An intuitive way to interpret this is that coherence acts as a binding mechanism, preventing the dispersion of the configuration.

These stable nodes can therefore serve as the building blocks of classical objects, providing localized structures that persist and interact in a predictable manner.

11.2 Quaternionic Classical Carrier

Because the internal geometry of the Æ framework is quaternionic, coherent nodes carry an intrinsic rotational structure.

Even in the classical limit, these internal rotations do not disappear. Instead, they manifest as effective physical quantities such as orientation, angular momentum, and spin-like properties.

An intuitive picture is that classical objects retain an internal rotational state, even if it is not directly observable.

The quaternionic structure therefore continues to play a role, acting as an underlying carrier of rotational information within classical behavior.

11.3 Metric Scaling and Emergent Classical Geometry

At microscopic scales, the geometry of the field may fluctuate due to variations in coherence and phase entropy.

However, as coherence increases and fluctuations become suppressed, these variations average out.

The resulting effective geometry becomes smooth and well-defined, allowing distances, trajectories, and curvature to be described in classical terms.

In this way, classical geometry emerges as a coarse-grained description of an underlying phase-coherent structure.

This transition provides the bridge between the microscopic dynamics of the Æ field and the macroscopic geometry observed in classical physics.

11.4 Pre-Metric Regime

Before the emergence of a stable metric geometry, the system may exist in a pre-metric regime.

In this regime, the relationships between points are determined primarily by phase interactions rather than by fixed geometric distances.

An important consequence is that the notion of distance itself is not yet well-defined. Instead, connectivity and interaction are governed by coherence structure.

This corresponds to a limit in which the effective metric becomes degenerate, and the system evolves purely through phase dynamics.

The geometry of the system therefore remains flexible and dynamically evolving.

The transition from this regime to a stable metric structure marks an important stage in the emergence of classical space.

11.5 Density vs. Dimension

Within the Æ framework, the distinction between density and dimension becomes important in understanding the classical limit.

Dimension refers to the ambient space in which configurations are embedded, determining the number of independent directions available.

Density, on the other hand, is associated with the degree of coherence of the field and controls how tightly phase structures are organized within that space.

An important point is that density does not change the underlying dimension, but it can modify the effective degrees of freedom available to the system.

In regions of high coherence density, phase configurations become strongly constrained, effectively reducing the accessible configuration space.

Conversely, in regions of low coherence, additional degrees of freedom become accessible, allowing more complex or less constrained behavior.

This can be interpreted geometrically as a density-dependent restriction of admissible states within the ambient space.

Within the quaternionic formulation, this is reflected in the fact that the radius of the coherence sphere depends on the local coherence, while the embedding dimension remains fixed.

Taken together, this interplay between density and effective degrees of freedom plays a central role in shaping the macroscopic structure of the field and the emergence of classical behavior.

12. Spectral and Scale Structure

Physical systems often reveal different behaviors when observed at different scales.

At microscopic scales, fluctuations and local interactions may dominate. At larger scales, collective structures and stable patterns emerge.

Understanding how a system changes as the observation scale

varies is therefore a central problem in modern theoretical physics.

One of the most powerful tools used to study such behavior is spectral analysis.

In this context, the term "spectrum" refers to the set of characteristic modes that a system can support.

These modes can be interpreted as the fundamental vibrations, excitations, or dynamical patterns of the system.

Within the Æ framework, spectral analysis plays a central role in understanding how coherence structures organize themselves across scales.

The spectral properties of the Dirac–Æ operator determine which phase structures can exist and how stable they are.

At the same time, scale-dependent flows describe how these spectral structures change as one moves between microscopic and macroscopic descriptions.

12.1 Beta Functions

When parameters of a theory depend on the scale at which the system is observed, their evolution is described by beta functions.

These functions measure how coupling constants and field parameters change as the observation scale varies.

An intuitive interpretation is that the system "looks different" depending on the resolution at which it is probed.

Within the Æ framework, beta functions capture how coherence-related quantities, such as phase entropy and alignment, evolve across scales.

This provides a quantitative description of how coherence structures reorganize between microscopic and macroscopic regimes.

12.2 Flow of the Dirac–Æ Operator

Because the Dirac–Æ operator governs the dynamics of the field, its structure may also change when the effective description of the system is examined at different scales.

The operator flow describes how the effective form of this operator evolves under scale transformations.

An important consequence is that both the kinetic and interaction components of the operator may acquire scale-dependent corrections.

Within the Æ framework, this reflects the fact that phase coherence and entropy modify the effective dynamics as one moves between scales.

12.3 Spectral Invariance Condition

Certain structural properties of the spectrum may remain invariant under changes of scale.

These invariances impose constraints on the allowed dynamics of the system.

An intuitive interpretation is that, while the detailed structure of the spectrum may shift, certain global features remain preserved.

Within the Æ framework, such spectral invariance ensures that the underlying organization of phase structures remains consistent across different scales.

This provides a stability condition linking microscopic dynamics with macroscopic behavior.

12.4 Effective Action Flow

The action introduced earlier can also depend on the scale at which the system is described.

As smaller-scale degrees of freedom are integrated out, the effective action governing the system changes.

This evolution is captured by the effective action flow.

An intuitive way to interpret this is that the theory is continuously re-expressed in terms of the most relevant variables at each scale.

Within the Æ framework, coherence and phase entropy contribute directly to this flow, modifying couplings, interactions, and effective geometry as the scale changes.

12.5 Coherence–Entropy Coupling Under Flow

Within the Æ framework, coherence and phase entropy form a coupled pair of quantities that influence the stability of the system.

Rather than evolving independently, these quantities are dynamically linked: changes in phase entropy affect coherence, and vice versa.

Under scale transformations, this coupling becomes particularly relevant, as it determines how structure is preserved or degraded across scales.

An intuitive interpretation is that coherence acts as an ordering tendency, while entropy introduces dispersion, and their balance governs the effective dynamics of the field.

12.6 Spectral Dimension

In certain systems, the effective dimension experienced by dynamical processes may vary with scale.

This concept is known as spectral dimension.

Unlike the topological dimension, which remains fixed, the spectral dimension reflects how diffusion, propagation, or phase transport occur within the system.

Within the Æ framework, the spectral dimension characterizes how phase structures spread across the manifold, revealing how accessible the space is to dynamical processes.

This provides a scale-dependent view of geometry, linking microscopic structure with macroscopic behavior.

12.7 Fixed Points and Stability

As parameters evolve under scale transformations, they may approach special configurations known as fixed points.

At these points, the system becomes invariant under further changes of scale.

An intuitive interpretation is that the system has reached a form of self-consistency across scales, where its structure appears the same regardless of the level of resolution.

Such fixed points often correspond to particularly stable or universal physical regimes, acting as anchors for the behavior of the system.

12.8 Topological Charges

Some properties of the system are protected by topology.

These quantities, known as topological charges, remain unchanged under continuous transformations of the field.

An intuitive way to understand this is that these structures depend on global features of the configuration, rather than on local details.

Within the Æ framework, topological charges act as robust markers of coherent structures, ensuring that certain features persist even under strong dynamical evolution.

12.9 Spectral Stability and Correlation Length

The stability of coherent structures is closely related to the correlation length of the system.

This length measures how far information or phase alignment can propagate across the field.

An intuitive interpretation is that it defines the scale over which different regions of the system remain effectively connected.

Spectral properties of the operator determine how this correlation length behaves, linking stability to the structure of the spectrum.

In the Æ framework, this provides a bridge between local coherence and global organization of the field.

12.10 Kernel Formalism and Spectral Decomposition

Many dynamical processes can be described using kernel operators that encode interactions between different points of the system.

These kernels act as generalized interaction maps, determining how information or phase structure propagates across the field.

Spectral decomposition allows these kernels to be expressed in terms of their fundamental modes.

An intuitive interpretation is that the dynamics can be decomposed into a set of basic patterns, each evolving independently.

This representation provides a powerful way to analyze the system, as complex interactions can be understood as combinations of simpler spectral components.

12.11 Prime-Indexed Oscillatory Sector

Certain oscillatory structures may be naturally organized through discrete indexing schemes.

One possible structure involves sequences related to prime numbers, which provide a non-redundant and irreducible indexing of modes.

An intuitive interpretation is that prime indices act as fundamental building blocks within the spectral decomposition, analogous to how prime numbers generate integer structures.

Within the Æ framework, such indexing schemes may encode hierarchies of resonant modes, organizing spectral components in a way that reflects deep structural properties of the field.

12.12 Mellin Transform Representation

Integral transforms provide useful tools for analyzing spectral structures.

The Mellin transform is particularly well suited for studying scale-invariant phenomena, as it relates multiplicative structure to additive representations.

An intuitive way to interpret this is that scaling behavior can be translated into a spectral domain where patterns become more transparent.

Within the Æ framework, Mellin representations provide a natural tool for analyzing how structures behave across scales, potentially revealing hidden symmetries of the spectral kernel.

12.13 Dual Space Structure

Spectral analysis naturally leads to the consideration of dual spaces in which dynamical operators become diagonal or simplified.

In such representations, complex interactions in the original space are transformed into simpler, often decoupled structures.

An intuitive interpretation is that the system is being viewed from a different perspective in which its underlying organization becomes more apparent.

Within the Æ framework, dual space representations allow phase dynamics, coherence, and spectral structure to be analyzed in a more transparent way, providing insight into the organization of the field across scales.

12.14 Running of the Coherence Field

Just as coupling constants may evolve with scale, the coherence field itself may exhibit scale dependence.

This running reflects how the organization of phase structures changes as the effective description of the system is modified.

An intuitive interpretation is that coherence reorganizes across scales, with different structures becoming dominant depending on the level of resolution.

In this way, coherence is not a fixed property, but a scale-dependent feature of the field.

12.15 Flow of the Memory Kernel

Memory effects introduced earlier may also evolve under scale transformations.

As the system is described at different scales, the memory kernel acquires its own effective structure, reflecting how past configurations influence present dynamics.

An important consequence is that memory becomes scale-dependent, with different temporal and spatial correlations becoming relevant at different levels of description.

This provides a mechanism through which long-range coherence can be preserved or reorganized across scales.

12.16 Scale-Dependent Spectral Dimension

At different scales, the system may effectively behave as if it possesses different dimensional properties.

This scale-dependent spectral dimension reflects how propagation, diffusion, and phase transport occur within the system.

An intuitive interpretation is that the field may appear more constrained or more accessible depending on the scale at which it is probed.

This provides a dynamic notion of geometry, linking spectral behavior with the structure of the field.

12.17 Transcendental Fixed Point

In some cases, the flow of parameters may approach particularly special fixed points characterized by nontrivial mathematical structure.

These transcendental fixed points represent regimes in which the system exhibits scale invariance together with deeper structural constraints.

An intuitive interpretation is that the system reaches a configuration that is stable across all scales, acting as an organizing center for the dynamics.

Within the Æ framework, such fixed points may encode highly ordered states of coherence and phase structure.

12.18 Spectral Gap and Coherence Stability

A spectral gap refers to a separation between the lowest eigenvalues of an operator and the rest of the spectrum.

Such gaps often indicate the presence of stable coherent structures within the system.

An intuitive interpretation is that the gap protects certain configurations from small perturbations, allowing them to persist over time.

Within the Æ framework, spectral gaps act as indicators of coherence stability, linking spectral structure to the robustness of phase configurations.

12.19 Hierarchical Spectral Stability

Finally, spectral stability may appear at multiple hierarchical levels.

Local coherence structures may stabilize first, followed by larger-scale organizations.

An intuitive interpretation is that structure emerges progressively, with small-scale coherence providing the foundation for larger, more complex patterns.

This hierarchical emergence provides a pathway through which complex and persistent structures develop within the Æ field.

Taken together, these mechanisms describe how coherence, scale, and spectral structure interact to produce stable organization across multiple levels of the system.

13. On Knots

Many physical systems contain structures that persist not because of energetic barriers, but because of their topology.

Topology studies properties that remain unchanged under continuous deformations. Objects may stretch, bend, or twist, but certain global structures remain preserved.

Although the term may sound unfamiliar, topology can be understood intuitively as the study of how things are connected.

Knots provide one of the simplest and most intuitive examples of such topological structures.

If a loop of string forms a knot, the knot cannot disappear unless the string is cut or passed through itself.

This type of stability is known as topological protection.

Importantly, knots are not only a visual or intuitive concept. They are studied within a rigorous mathematical framework known as knot theory.

Within this field, knots can be classified, compared, and even combined through well-defined operations.

For example, there exist notions analogous to "prime" knots, which cannot be decomposed into simpler structures, and "composite" knots, which arise from combinations of simpler ones.

There are also algebraic structures associated with knots, allowing them to be studied using tools similar to those of group theory and invariants.

These developments show that knots are not merely visual curiosities, but deeply structured mathematical objects.

Within the Æ framework, coherent configurations of the field may form structures that behave similarly to topological knots.

These configurations arise from organized phase structures and may persist over long periods of time.

13.1 Nodes as Quasi-Stable Configurations

Nodes introduced earlier represent localized regions of coherence, where the field organizes itself in a more structured manner.

Under certain conditions, these nodes do not simply fluctuate and dissolve, but instead become quasi-stable structures within the field.

Their stability does not come from being static, but from a dynamic balance between coherence generation and entropy production.

In this sense, a node remains stable as long as this balance is maintained.

13.2 Knots as Topological Invariants

When coherent phase structures form closed loops, they may acquire topological invariants.

Intuitively, this means that the configuration cannot be continuously deformed into a different one without breaking its structure.

These invariants classify the possible knot configurations that the field can support.

Because these quantities cannot change continuously, the associated structures become robust and resistant to deformation.

13.3 Topological Protection

Topological invariants protect certain configurations from dissolving under small perturbations.

Small fluctuations may distort the shape of a structure, but they do not change its topological class.

As long as the topological class remains unchanged, the structure persists.

This mechanism provides a natural explanation for the stability of many coherent structures within the field.

13.4 Memory as Topological Persistence

One of the most interesting consequences of topological structures is their ability to store information.

A knot retains information about how the field was configured in the past, not through a stored signal, but through its very form.

Within the Æ framework, this leads to an interpretation of memory as topological persistence.

The configuration itself becomes a record of the system's history.

13.5 Memory as Path Dependence and Holonomy

Holonomy describes how an object changes when transported along a closed path in a curved space.

In simpler terms, the final state depends on the path taken, not only on the starting point.

Within the Æ framework, the phase structure of the field may retain information about the paths taken through its configuration space.

This path dependence provides another mechanism through which memory can arise, complementing its topological interpretation.

13.6 Casimir Invariants of the Nambu–Æ Bracket

The Nambu dynamics introduced earlier possess conserved quantities known as Casimir invariants.

These invariants remain constant throughout the evolution of the system, regardless of the specific dynamics taking place.

In more intuitive terms, they represent features of the system that cannot change under the allowed transformations.

Within the knot interpretation, such invariants may correspond to conserved topological quantities, such as linking or global structure.

This establishes a direct bridge between dynamical conservation laws and topological stability.

13.7 Topological Helicity Interpretation

Helicity measures the linking structure of field lines.

In fluid dynamics and plasma physics, helicity describes how vortex lines twist, coil, and link with each other over time.

Within the Æ framework, helicity provides a natural measure of the topological complexity of phase structures.

In this sense, it captures not only how much structure is present, but how deeply intertwined that structure has become.

This makes helicity a natural candidate for encoding global topological information within the field.

13.8 Node–Knot Transition Map

Nodes can evolve into knots when phase structures begin to loop, fold, and reconnect.

This transition does not require external intervention, but emerges from the internal dynamics of the field.

The node–knot transition describes this transformation between localized coherent structures and extended topological configurations.

Conversely, when coherence is lost beyond a critical threshold, these structures may collapse, leading to the disappearance of their associated topology.

13.9 Structural Triangle

Many of the structures described in the framework can be organized around three interacting elements:

coherence, geometry, and topology.

Coherence determines whether a structure can persist, geometry defines how it is shaped, and topology determines what cannot change.

This triadic relationship forms a structural triangle governing the formation and persistence of stable configurations.

Rather than acting independently, these three aspects continuously influence one another, defining the possible states of the system.

13.10 Loop-Dependent Holonomy

Closed loops within the field may accumulate phase shifts when transported through the manifold.

These accumulated shifts represent holonomy effects.

In intuitive terms, the final state of a loop depends not only on where it starts, but on the path it follows.

Such effects play an important role in the formation and stability of knots, as different loops may encode different phase histories.

13.11 Fundamental Group Structure

Topology provides powerful tools for classifying loops.

One of these tools is the fundamental group, which describes how loops can be continuously deformed into each other without breaking.

Loops that can be deformed into one another belong to the same class, while those that cannot represent distinct topological structures.

Within the Æ framework, the fundamental group helps classify the possible knot structures supported by the field.

13.12 Internal Helicity Field

Helicity can also be viewed as a field describing the twisting and linking of phase structures.

Rather than being a single quantity, it may be distributed throughout the system, capturing local and global patterns of entanglement.

This helicity field interacts with the coherence structure of the system, influencing how knots form, evolve, and persist.

13.13 Holonomy Criterion

Certain holonomy conditions determine whether a knot-like structure can remain stable.

Not all loops lead to persistent configurations; only those that satisfy specific phase relations can reinforce themselves.

When these conditions are met, phase loops effectively sustain their own configuration over time.

13.14 Identity–Return Mechanism

Some configurations allow a phase structure to return to its original state only after multiple rotations.

In other words, a single loop is not enough to recover the initial configuration.

This behavior resembles the double-cover structure found in spinorial systems, where a full return requires two complete rotations.

Within the framework, this mechanism contributes to the stability and richness of certain knot structures.

13.15 Localized Coherence Nodes and Reality Knot Encoding

Reality Knots (NR), or "Nudos de Realidad," are not purely abstract structures.

Under certain conditions, they may localize within the system as persistent distortions of coherence.

These localized configurations correspond to regions where phase organization is incomplete, and the system fails to return to a trivial configuration after cyclic evolution.

In more precise terms, they can be associated with nontrivial holonomy, where closed loops in the field do not map back to the identity.

Within the Æ framework, such regions act as coherence nodes capable of storing and re-expressing unresolved phase structures over time.

Rather than storing information in a symbolic or static way, these nodes retain information through their topological and dynamical configuration.

This provides a natural mechanism through which repeated patterns may emerge.

Recurrent behavioral, emotional, or perceptual patterns can then be interpreted as the reactivation of underlying phase configurations associated with persistent holonomy.

From this perspective, repetition is not redundancy, but the continued unfolding of a structure that has not yet reached closure.

Reality Knots are therefore stable configurations maintained by both elevated phase entropy and nontrivial holonomy.

At larger scales, these structures may form interconnected networks, where individual knot trajectories interact and resonate with one another.

In such cases, configurations with similar holonomy may reinforce each other, giving rise to historical resonance across the field.

Within this context, renudation corresponds to the progressive reduction of nontrivial holonomy.

As the system approaches closure, phase relations become consistent, and the structure moves toward a trivial configuration.

This process does not erase the past configuration, but integrates it into a coherent global state.

In this sense, Reality Knots provide a bridge between topological structure, dynamical evolution, memory, and large-scale organization of the field.

13.16 Quaternionic Periodicity

Because the internal geometry of the field is quaternionic, periodicity properties arise naturally.

These periodicities are not limited to simple cycles, but may involve more complex return conditions associated with higher-dimensional rotations.

As a result, phase structures may require multiple rotations to return to their original state.

These periodicities influence how phase structures wrap around the manifold and contribute to the formation of stable configurations.

13.17 Collective and Embedded Knots

Knots need not exist in isolation.

Multiple knots may interact, overlap, or become embedded within larger structures.

In such cases, individual knots do not lose their identity, but become part of a more complex configuration.

These collective arrangements can produce networks of coherence that extend across different scales of the system.

13.18 Reality Knot Network Structure

At larger scales, knots may form interconnected networks.

These networks organize coherent regions of the field into structured patterns, linking local configurations into global architectures.

Rather than isolated objects, knots become nodes within a larger network that defines the structure of the field itself.

13.19 Temporal Knot Projection

Knots are not purely spatial objects.

They may also evolve in time, forming trajectories through the configuration space of the field.

In this sense, a knot is not just a structure, but a process that unfolds dynamically.

Temporal projections of these structures describe how they form, evolve, and potentially dissolve over time.

13.20 Composite Knots and Protein Structure

In biological systems, knot-like structures appear in molecules such as proteins and DNA.

These structures provide clear evidence that topology plays a role in real physical systems.

Some proteins, for example, contain knotted configurations that influence their stability and function.

Within the Æ framework, similar principles may apply to coherent structures in the field, linking biological organization with topological constraints.

13.21 Holonomy as Interaction Carrier

Holonomy effects can mediate interactions between different structures.

Phase shifts accumulated along closed paths influence how coherent regions affect each other.

In this way, holonomy acts as a carrier of relational information between different parts of the field.

13.22 Residual Holonomy

Even when large-scale configurations change, residual holonomy effects may remain.

These residual structures encode traces of previous configurations, persisting beyond the original state.

As a result, past dynamics can continue to influence future evolution, providing a mechanism for long-term memory within the system.

13.23 Local Knot Locking

Some knot configurations become locally locked.

In such cases, the structure cannot be simplified through local adjustments alone.

Any attempt to modify the configuration requires a global

reorganization of the field.

This reflects the fact that the stability of the knot is distributed across its entire structure, rather than concentrated in a single region.

13.24 Topological Stability

Topological stability arises because the structure cannot be removed through small perturbations.

Local fluctuations may alter the geometry of the system, but they do not change its topological class.

As a result, the configuration persists even in the presence of external disturbances.

This stability provides a natural mechanism for the persistence of complex and highly organized structures.

13.25 Holonomy Return and Double-Cover Structure

In certain systems, returning to an initial configuration may require more than a single cycle.

In particular, some structures require two complete cycles before the original state is recovered.

This behavior reflects the double-cover structure associated with spinorial geometry.

Within the Æ framework, such properties influence how phase structures evolve and stabilize over time.

13.26 Casimir Invariants and Linking Relations

The conserved quantities associated with Nambu dynamics can be interpreted as measures of linking and twisting between phase structures.

Rather than being abstract constants, these invariants encode global relationships within the field.

Through these relations, it becomes possible to classify different knot configurations based on their internal linking structure.

13.27 Temporal Reprojection of Knot States

As knot structures evolve, their configurations may be reprojected into different regions of the field.

This reprojection does not destroy the structure, but translates it across the configuration space.

In this sense, a knot may reappear in different contexts while preserving its underlying topological identity.

13.28 Topological Memory Surface Flux

Surface fluxes associated with topological invariants can encode persistent information about the field.

These fluxes capture how topological structures extend across regions of the system.

As a result, they act as carriers of topological memory, allowing information to be preserved beyond local dynamics.

13.29 Holonomy Criterion for Knot Persistence

The persistence of a knot requires that holonomy conditions remain satisfied.

In particular, nontrivial holonomy must be maintained along the relevant loops.

When these conditions fail, the structure loses its topological support and may reorganize or decay into simpler configurations.

Thus, holonomy acts as a consistency condition for the long-term stability of knot structures.

13.30 Topological Classification via $\pi_3(S^3)$

Higher homotopy groups provide tools for classifying complex topological structures that go beyond simple loops.

While the fundamental group describes how one-dimensional loops behave, higher homotopy groups capture how higher-dimensional structures wrap and organize within a space.

Within the quaternionic geometry of the Æ framework, the

space S^3 naturally arises as the domain of normalized quaternionic states.

In this context, the group $\pi_3(S^3)$ plays a central role.

Intuitively, it characterizes how three-dimensional structures can wrap around the three-sphere in nontrivial ways.

This provides a natural language for classifying coherent configurations that cannot be reduced to simple loops.

Within the framework, knot configurations may therefore be understood not only as one-dimensional objects, but as projections of higher-dimensional topological structures.

As a result, $\pi_3(S^3)$ offers a deeper classification scheme for the possible states of the field, providing a complementary perspective to traditional knot theory.

14. Merkaba Projection

The geometric structures introduced in the previous sections suggest that the internal phase dynamics of the Æ field possess a rich higher-dimensional organization.

In particular, the knot configurations described earlier may be understood as projections of more fundamental structures defined in higher-dimensional spaces.

While these structures naturally live in quaternionic spaces such as the three-sphere S^3, physical processes are often observed through lower-dimensional projections.

Understanding how information encoded in higher-dimensional phase structures becomes visible in lower-dimensional geometry is therefore an important problem.

Within the Æ framework, this process is described through the Merkaba projection.

Here, the term "Merkaba" is used in a geometric sense, referring to a structured mapping that transforms phase information from S^3 into a distributed representation on a lower-dimensional manifold.

More precisely, this mapping may be viewed as a composition of projections that first relate quaternionic states to points on S^2, and then distribute this information across a simplex structure.

This projection does not destroy the original information; rather, it reorganizes it into a set of weights that encode how phase structure is distributed.

In this sense, observable configurations may be viewed as lower-dimensional expressions of higher-dimensional phase organization.

14.1 Hopf Projection $S^3 \to S^2$

One of the most remarkable constructions in geometry is the Hopf fibration.

This mapping describes how the three-sphere S^3 can be decomposed into circles that project onto points of a two-sphere S^2.

Each point on S^2 corresponds to an entire fiber in S^3, meaning that different internal configurations may share the same observable projection.

Within the Æ framework, this mapping provides a natural mechanism for relating internal phase states to observable configurations.

14.2 Tetrahedral Decomposition

To analyze the structure of the projection more concretely, it is useful to decompose the sphere into simpler geometric regions.

One natural decomposition involves a tetrahedral partition of S^2.

In this representation, each point on the sphere is associated with a set of weights that describe its relation to four reference directions.

These weights define a simplex structure that encodes how phase information is distributed across the system.

Within the Æ framework, this tetrahedral geometry provides a minimal yet structured way of organizing phase information.

It also allows phase transport and holonomy to be represented as distributions over interacting regions rather than as a single localized quantity.

14.3 Merkaba as Phase Transduction Operator

The Merkaba can be interpreted as an operator that converts internal phase relationships into geometric configurations.

In more precise terms, it acts as a phase transduction mechanism, mapping phase information defined on S^3 into a distributed representation on a lower- dimensional space.

Through this transformation, the internal rotational structure of the field becomes encoded in a set of geometric weights.

In this sense, the Merkaba does not merely project the field, but reorganizes its phase structure into an interpretable geometric form.

14.4 Nodal Stability via Merkaba

Nodes introduced earlier may become stabilized when their phase relationships align with the geometric structure defined by the Merkaba projection.

In such cases, the projection reinforces coherence across the node by maintaining consistent phase distributions.

This alignment effectively locks the node into a stable configuration, reducing internal phase mismatch.

As a result, the Merkaba structure acts as a stabilizing framework for coherent regions of the field.

14.5 Explicit Tetrahedral Geometry on S^2

When the Hopf projection is expressed in terms of tetrahedral decomposition, the resulting geometry on S^2 reveals a highly symmetric structure.

Each point on the sphere can be associated with a set of weights relative to four reference directions defined by the tetrahedron.

This symmetry organizes the possible orientations of phase states and provides a natural coordinate system for analyzing their interactions.

Within this representation, phase information is no longer localized, but distributed across the vertices of the tetrahedron.

14.6 Merkaba Projection Operator

Mathematically, the Merkaba projection can be represented as an operator acting on phase states defined on S^3.

This operator may be understood as a composition of mappings that first project quaternionic states onto S^2, and then encode them as weight distributions over a simplex.

As a result, the higher-dimensional phase structure is transformed into a structured geometric representation that preserves global relationships.

14.7 Phase Distribution and Entropy on Δ_3

The tetrahedral representation naturally introduces a simplex structure denoted Δ_3.

Within this simplex, phase states may be interpreted as normalized distributions across the four vertices of the tetrahedron.

In this context, entropy measures acquire a direct geometric interpretation.

High entropy corresponds to a more uniform distribution across the simplex, while low entropy reflects concentration near specific vertices.

Thus, coherence and entropy can be visualized as geometric properties of the phase distribution.

14.8 Nonlocal Phase Transport

Because the Hopf projection preserves global relationships among phase states, transport processes may exhibit nonlocal features.

Phase information transported along one fiber does not remain confined locally, but contributes to the global structure of the projection.

As a result, changes in one region may influence configurations elsewhere in the projected geometry, reflecting the underlying connectivity of the field.

14.9 Algebraic Structure of the Æ Product

Interactions between projected phase structures can be expressed through an algebraic product defined within the framework.

This product encodes how multiple phase contributions combine within the projected geometry, preserving both their relative phase relations and their geometric distribution.

Rather than acting as a simple multiplication, the Æ product captures the interaction between phase-aligned structures as they are mapped onto the projected space.

In this sense, it reflects how coherent configurations interfere, reinforce, or redistribute across the simplex representation.

The product is therefore not only algebraic, but also geometric, as it operates on phase information that has already been transduced into spatial structure.

This provides a natural mechanism for describing interactions between nodes, knots, and larger coherent configurations within the field.

14.10 Merkaba Projection Operator

The projection operator associated with the Merkaba structure therefore acts as a bridge between internal phase space and observable geometry.

It translates quaternionic phase information into geometric relations within the projected manifold.

In combination with the Æ product, this operator defines how phase structures are both represented and combined in the observable domain.

Together, they form a minimal operatorial framework for describing the emergence of structured configurations from internal phase dynamics.

14.11 Viviani–Hopf Geometry of the Merkaba

The merkaba projection admits a geometric interpretation that connects its structure to classical results in triangular geometry.

Each face of the tetrahedral merkaba corresponds to an equilateral triangle.

Within such a triangle, the theorem of Viviani states that the sum of the perpendicular distances from any interior point to the three sides remains constant.

In barycentric coordinates, this condition is expressed as the normalization constraint

$$\lambda_1 + \lambda_2 + \lambda_3 = 1. \tag{2}$$

The merkaba projection naturally produces a tetrahedral set of weights

$$\{w_i\}_{i=1}^4, \qquad \sum_{i=1}^{4} w_i = 1, \tag{3}$$

which act as barycentric coordinates on the tetrahedron.

When restricted to any triangular face, these weights satisfy a Viviani-type relation.

This implies that while the distribution of coherence across the vertices may vary, the total projected coherence remains conserved.

From this perspective, the Viviani relation provides a geometric interpretation of coherence conservation within the merkaba representation.

Thus, a classical geometric constraint emerges naturally from the projection of quaternionic phase structure onto the tetrahedral simplex.

14.12 Geometric Encoding of Phase

Ultimately, the Merkaba projection illustrates a general principle of the framework.

Phase information encoded in higher-dimensional spaces can be translated into geometric patterns in lower-dimensional manifolds.

These patterns are not arbitrary, but reflect the underlying phase relationships and their algebraic interactions.

Through this mechanism, the internal dynamics of the Æ field become visible as structured geometric configurations, where geometry, topology, and algebra interact as a unified system.

15. Differential Geometry of the Æ-Space

The structures introduced throughout the previous sections suggest that the Æ framework is naturally described using the language of differential geometry.

In classical physics, geometry often provides the stage upon which physical processes occur.

For example, in general relativity, the geometry of spacetime determines the motion of matter.

Within the Æ framework, geometry plays a similar but expanded role.

Rather than describing only the structure of spacetime, the geometry of the Æ-space encodes the organization of phase, coherence, and entropy.

In this sense, geometry is not merely a background, but an active structure that governs how information is distributed and evolves.

This space may therefore be understood as an extended geometric manifold in which information and dynamics are intrinsically intertwined.

The differential geometry of this manifold describes how phase structures propagate, interact, and reorganize across the field.

15.1 Nodal Torsion and Æ-Connection

When transporting vectors or fields across a curved manifold, one must specify a connection that determines how the transport is performed.

Within the Æ framework, this connection is not purely metric-compatible in the classical sense, but includes torsional contributions associated with nodal structures.

In particular, coherence gradients induce torsion, modifying the way structures are transported across the manifold.

As a result, nodes act as localized sources of geometric deformation, influencing both curvature and torsion.

This provides a geometric interpretation of how phase, coherence, and entropy affect the evolution of the system.

15.2 Hopf-Algebraic Closure

The quaternionic and Hopf structures introduced earlier imply the presence of algebraic closure relations within the geometry of the field.

These relations ensure that phase transformations combine in a consistent and structured manner.

In particular, the composition of transformations follows algebraic rules that preserve the internal coherence of the system.

Rather than allowing arbitrary combinations, the evolution of the field is constrained by these closure relations.

In this sense, the Hopf-algebraic structure provides a natural framework for unifying geometric and algebraic aspects of the Æ-space.

15.3 Quasi-Hopf Structure

In more general settings, the algebraic structure of the system may deviate from strict associativity.

This leads to a quasi-Hopf structure, in which the combination of transformations depends on an additional associator.

Such behavior reflects the fact that the ordering and grouping of phase interactions can influence the resulting configuration.

Within the Æ framework, this provides a refined description of how complex phase structures interact and reorganize.

15.4 Inter-Layer Transmission and Admissibility

The Æ framework also allows for the interaction between different effective layers of description.

Information transfer between layers is not trivial and requires specific conditions to be satisfied.

In particular, coherence must remain above a critical threshold for transmission to occur without loss.

Below this threshold, information may degrade, leading to effective decoherence between layers.

As a result, admissibility conditions govern when and how structures can be transmitted across different levels of the system.

This provides a natural mechanism for understanding hierarchical organization and the emergence of effective sectors within the field.

15.5 Mental Sector as a Restricted Coherent Subspace

Within the Æ framework, the mental layer can be described as a restricted coherent subspace of the broader astral functional space.

Rather than representing a separate entity, it corresponds to a region of the field where coherence is sufficiently high and phase entropy remains constrained.

In this sense, membership in the mental sector is determined by coherence and entropy conditions that select a more ordered subset of configurations.

Thought may then be interpreted as a trajectory within this restricted subspace.

These trajectories evolve dynamically and, in stabilized regimes, may exhibit quasi-periodic behavior, reflecting persistent patterns of phase organization.

Importantly, configurations in the mental sector remain coupled to other layers of the system.

In particular, phase structures may circulate between astral and mental representations, and under appropriate conditions, may transduce into physical expression.

This establishes a continuous relation between internal phase dynamics and observable outcomes.

15.6 Coherent Solution Sectors and Effective Basins

Let the evolution of the system be described as trajectories within a solution space of the Æ framework.

Within this space, coherent basins can be defined as regions toward which trajectories converge under the dynamics of the system.

Each basin corresponds to a stable or metastable pattern of organization characterized by a specific coherence profile.

The relative dominance of these basins is determined by their associated phase entropy.

Configurations with lower entropy tend to be more stable and therefore more likely to govern the effective behavior of the system.

In this sense, the system naturally selects dominant configurations through an entropy-driven mechanism.

Distances between basins can be defined in terms of differences in their coherence profiles, providing a geometric notion of separation between possible states.

Furthermore, basins may exhibit a nested structure, reflecting hierarchical organization within the solution space.

Compatibility between different basins requires a minimal level of coherence overlap, ensuring that transitions between them remain dynamically admissible.

15.7 Spectral Dressing by Coherence and Phase-Entropy

In the spectral-localized sector, it is often convenient to redefine the field variables so that coherence and phase-entropy effects are absorbed into the representation of the state.

This transformation effectively redistributes the influence of the Æ-weight into the dynamical variables.

As a consequence, derivative operators acquire additional contributions that depend on spatial variations of coherence and phase-entropy.

These contributions can be interpreted as drift-like terms that modify how fields evolve across the manifold.

In particular, the effective mass becomes dependent on gradients of both coherence and phase-entropy.

This implies that mass is not a fixed quantity, but a functional that reflects the local structure of the field.

Regions with strong entropy gradients or coherence variations induce corrections to the effective mass, altering the local dynamics of the system.

In this sense, spectral properties are not intrinsic only to the field itself, but are dynamically dressed by the underlying phase organization.

15.8 Metric Response and Interlayer Geometric Mediation

Beyond the static conformal and torsional structure, the Æ-metric may be regarded as dynamically responsive to coherent field configurations.

Rather than remaining fixed, the geometry adapts to the distribution of phase and coherence within the field.

In particular, variations in effective curvature or coherence loading may induce changes in the metric structure.

This leads to a picture in which geometry and field dynamics are mutually coupled.

Furthermore, inter-layer transmission is not purely an algebraic or projection-based phenomenon.

Instead, the transfer of information between layers may also induce local geometric responses within the Æ-space.

In this interpretation, coherent gradients act as sources of geometric deformation, shaping the local structure of the manifold.

Conversely, incoherent or highly entropic configurations contribute weakly, producing diffuse or dissipative geometric effects.

Thus, the metric of the Æ-space becomes an active participant in the dynamics, mediating interactions between layers and reflecting the underlying coherence structure of the field.

16. Effective Sectors and Emergent Reductions

The complete configuration space of the Æ framework is generically extremely large.

It contains all possible configurations of phase, coherence, and nodal structure compatible with the underlying dynamical laws.

In principle, this space encodes every admissible state of the system.

However, physical systems rarely explore this full configuration space.

Instead, observable phenomena typically occur within restricted regions that emerge naturally through dynamical constraints, stability conditions, or coherence selection.

Intuitively, although many configurations are mathematically allowed, only a subset is dynamically accessible or stable.

These restricted regions define what we refer to as *effective sectors*.

Such sectors may therefore be interpreted as emergent reductions of the full theory.

This is analogous to effective field theories in conventional physics, where the full microscopic description is reduced to the subset relevant for a given regime.

In the Æ framework, these reductions are not imposed externally, but arise from the intrinsic structure of coherence and phase dynamics.

16.1 Full Configuration Space

The full configuration space of the Æ framework may be viewed as the set of all admissible field configurations consistent with the dynamical equations and geometric constraints.

Each point of this space corresponds to a complete specification of the metric, gauge fields, spinor content, coherence, and phase entropy.

In this sense, it represents the total state space of the theory.

Effective sectors correspond to subsets of this space in which additional constraints or simplifications apply.

16.2 Effective Action (Path Integral Definition)

Within a given sector, the effective dynamics may be obtained by integrating out degrees of freedom that are not relevant at the scale of interest.

This leads to an effective action that captures the dominant behavior within the selected sector.

In intuitive terms, the system is described using only the variables that remain active in that regime, while others are absorbed into effective corrections.

16.3 Scalar–Coherence Sector

One important sector arises when coherence and phase entropy are treated as dynamical fields.

In this regime, the geometry induced by coherence interacts directly with scalar degrees of freedom.

The resulting dynamics describe how coherence and entropy propagate and self-organize within the field.

This sector provides a natural framework for understanding large-scale organization and memory effects.

16.4 Spectral-Localized Sector

In the spectral-localized sector, the dynamics are dominated by eigenmodes of the Dirac–Æ operator.

Each mode represents a localized pattern of phase organization.

The effective mass associated with these modes depends on coherence and phase-entropy, leading to a dynamically dressed spectral structure.

This provides a bridge between local field behavior and global spectral properties.

16.5 Biological Resonant Regime

In certain regimes, the system exhibits collective synchronization phenomena.

When frequencies of different components become sufficiently close, coherent oscillations may emerge.

These dynamics are well captured by phase-coupled systems, where interactions between components lead to synchronization.

Such regimes provide a natural description of biological coherence and collective behavior.

16.6 Hierarchical Embedding

Effective sectors are not isolated, but are organized hierarchically.

Higher-level descriptions emerge from more fundamental ones

through successive reductions.

In this sense, the theory admits a layered structure, where each level captures relevant dynamics at a given scale.

16.7 Nested Quaternionic Layer Structure

The total state of the system may be decomposed into multiple interacting layers.

Each layer corresponds to a different level of organization and admits both phase and quaternionic representations.

These layers are not independent, but embedded within one another, forming a nested structure.

Transformations between layers define how information propagates across different levels of the system.

In this way, the Æ framework provides a unified description of hierarchical organization, where phase, coherence, and geometry are consistently linked across scales.

16.8 Admissible Layer Domains

Not all configurations of the field are equally accessible at every level of description.

Each effective layer admits only those states that satisfy specific coherence and entropy conditions.

In this sense, every layer defines its own domain of admissibility, selecting configurations that are stable and dynamically consistent within that regime.

The total state of the system must therefore belong to the intersection of all admissible domains across layers.

This ensures that the global configuration remains coherent when viewed simultaneously from multiple levels of description.

16.9 Physical Effective Sector

The physical sector corresponds to the subset of configurations that can be projected into observable spacetime.

In this regime, coherence must remain sufficiently stable, and phase variations must be slow enough to support persistent structures.

Only under these conditions do configurations become observable as matter, fields, or classical phenomena.

The geometry perceived in this sector is therefore not fundamental, but induced from the underlying Æ-structure.

Stability in this regime corresponds to energetic equilibrium, ensuring that observable configurations remain dynamically consistent.

16.10 Etheric Fractal Sector

Beyond the physical layer, the etheric sector may be understood as a fractal support structure underlying observable configurations.

In this regime, the geometry is no longer strictly integer-dimensional, but exhibits scaling behavior characteristic of fractal systems.

Structures in this layer are generated through iterative refinement processes, leading to self-similar patterns across scales.

The etheric sector acts as an intermediary, translating deeper phase organization into structures that can be projected into the physical domain.

16.11 Astral Geodesic Sector

The astral sector describes configurations that evolve along intrinsic trajectories of the configuration space.

These trajectories follow geodesic-like behavior, reflecting paths of minimal deformation within the underlying geometry.

In this regime, configurations may evolve without direct coupling to the physical layer, allowing for dynamics that are not immediately observable.

This sector therefore captures the evolution of phase structures at a level where geometry and dynamics are more directly aligned.

16.12 Explicit Layer Mediation Operators

The different layers of the system are not isolated, but connected through structured mediation processes.

Each lower layer emerges from a higher one through specific transformations that preserve coherence while adapting the representation of the state.

These mediation processes define how configurations are translated between layers, allowing information to propagate across different levels of description.

In this way, the hierarchical structure of the system is not static, but dynamically maintained through continuous inter-layer interaction.

16.13 Transduction and Layer Coupling

Coupling between layers may be understood as a transduction process.

Information encoded in one layer is transformed into a representation suitable for another layer, while preserving its essential structure.

These transductions determine how coherence, phase relationships, and structural patterns propagate across the system.

They provide the mechanism through which higher-order organization becomes accessible at lower levels.

16.14 Nested Effective Layers

The observable universe may therefore be described as a system of nested effective layers.

Each layer represents a different level of organization of the same underlying field.

The physical layer corresponds to observable matter and interactions, while deeper layers encode progressively richer structures of coherence and phase organization.

These layers are not independent, but embedded within one another, forming a hierarchical structure through which information flows and reorganizes.

In this sense, reality may be understood as a multi-layered manifestation of a single underlying field, observed through different effective reductions.

16.15 Density and Dimensional Representation

Effective layers may also be characterized by their dimensional structure and density properties.

In the Æ framework, different regimes of the field may be associated with distinct effective dimensions, reflecting how phase information is organized and distributed.

Observable configurations can then be understood as reductions of higher-dimensional structures into lower-dimensional representations.

In this sense, dimensionality is not fixed, but depends on the level of description and the coherence structure of the system.

A given configuration may therefore encode both its dimensional behavior and its effective density within the same underlying structure.

16.16 Ontological Layer Structure

The nested sectors introduced above may also be interpreted as ontological layers corresponding to different regimes of the underlying Æ-field.

Each layer represents a distinct mode of description of the same fundamental configuration.

These layers are organized hierarchically, with each level providing an effective description appropriate to a given scale or regime.

Rather than representing separate realities, these layers correspond to different perspectives on a single underlying structure.

Transitions between layers therefore reflect changes in the effective variables used to describe the system, rather than the creation of new independent entities.

16.17 Etheric Layer Representation

Among the nested effective layers, the etheric sector plays a distinct role in mediating coherence between physical and higher-order configurations.

In this regime, structure is not purely continuous, but may exhibit fractal or hierarchical organization.

Such behavior can be understood as a distribution of coherence across nested subsets of the configuration space.

This makes the etheric layer particularly suitable for describing intermediate-scale structures that are neither fully localized nor fully delocalized.

As a result, it acts as a bridge between the physical and astral sectors, allowing coherence to be transmitted and reorganized across layers.

16.18 Multi-Layer Projections and Observables

Observables in the Æ-framework arise through projections of the underlying field configuration onto specific effective layers.

Each projection selects a subset of variables and suppresses others, producing a description adapted to a given observational regime.

In this way, the same underlying configuration may give rise to different observable phenomena depending on the layer through which it is examined.

Rather than being absolute, observables are therefore layer-dependent expressions of a deeper structure.

This provides a unified perspective in which multiple descriptions of reality coexist as projections of a single underlying field.

16.19 Æ-Covariant Structure

Within the Æ framework, the notion of a covariant derivative extends naturally to the full configuration space.

This derivative defines how structures evolve consistently within the geometry of the field.

In particular, it preserves the internal algebraic structure associated with the Æ-product, ensuring that interactions between configurations remain compatible under transport.

From this perspective, curvature emerges as a measure of how phase structures fail to return to their original configuration after parallel transport.

This curvature is not merely geometric, but encodes the stability properties of the system.

Regions of positive curvature correspond to convergent dynamics and stable configurations, while negative curvature indicates divergence and instability.

16.20 Geodesics in the Conformal Æ-Metric

The geometry induced by coherence defines a natural notion of distance within the Æ-space.

In this setting, geodesics represent the trajectories that minimize deformation within the field.

These paths describe how configurations evolve when no external forcing is present.

Because the metric depends on coherence, geodesic motion is influenced directly by the distribution of phase and entropy.

As a result, the natural evolution of the system follows paths determined by its internal coherence structure.

16.21 Nambu–Æ Structure

The dynamical evolution of the system can also be expressed in terms of generalized bracket structures.

Within the Æ framework, these take the form of a Nambu-type structure, where evolution depends on multiple generating functions.

Intuitively, instead of being driven by a single quantity, the system evolves under the simultaneous influence of several guiding structures.

One may think of this as a flow determined not by a single direction, but by the interaction of multiple gradients.

In this sense, the evolution resembles the motion of a point constrained by several surfaces at once, where the resulting trajectory emerges from their intersection.

This formulation naturally preserves volume in the configuration space, reflecting the conservation of information within the system.

Rather than a single Hamiltonian flow, the dynamics arise from the interaction of multiple generators, capturing the inherently multi-dimensional nature of phase evolution.

16.22 Invariant Measure

The Æ-space admits a natural measure that combines coherence and phase-entropy into a single weighting structure.

This measure determines how different configurations contribute to the overall dynamics of the system.

Regions of high coherence and low entropy contribute more strongly, while highly disordered configurations are suppressed.

In this sense, the measure encodes a probabilistic structure that favors coherent organization.

It provides a unifying description linking geometry, dynamics, and statistical behavior within the field.

17. Fundamental Forces from Æ Field Dynamics

One of the central questions in modern theoretical physics concerns the origin and possible unification of the fundamental interactions.

Within the standard model and general relativity, the electromagnetic, strong, weak, and gravitational interactions are described using distinct mathematical frameworks.

Gauge theories account for the electromagnetic, weak, and strong forces, while gravity is described geometrically through the curvature of spacetime.

The Æ framework provides a broader dynamical setting in which these interactions may be interpreted as effective sectors emerging from a unified underlying field.

Rather than postulating independent fundamental forces, the

framework suggests that the familiar interactions correspond to distinct regimes of phase organization, coherence structure, and geometric coupling within the Æ field.

In this sense, the forces do not originate as separate entities, but appear as emergent responses of the field under different symmetry and stability conditions.

The purpose of this section is therefore not to replace the successful theories describing the fundamental forces, but to show how analogous interaction sectors arise naturally within the Æ framework.

17.1 Unified Covariant Structure

The formulation of the Æ dynamics employs a covariant structure that extends the standard notion of gauge connections.

In this unified description, phase transport, coherence gradients, and geometric connections are combined into a single operatorial framework.

This structure allows the dynamics of the field to remain invariant under transformations that preserve both its geometric and algebraic organization.

From this perspective, what appear as distinct interactions correspond to different components of a single underlying covariant structure.

17.2 Coherence Field Notation

To describe the interaction sectors emerging from the Æ field, it is convenient to introduce a coherence field representation.

This field encodes the local degree of synchronization between phase structures within the manifold.

Different interaction regimes can then be associated with distinct coherence configurations and symmetry properties.

In this sense, the coherence field plays a role analogous to an order parameter, organizing the emergence of effective interaction sectors.

17.3 Electromagnetic Sector

The electromagnetic interaction may be understood as one of the simplest coherent regimes of the Æ field.

In this case, the interaction arises from phase differences between coupled field components.

Intuitively, it corresponds to how oscillatory structures align or misalign with one another.

When phase differences remain small, coherence is preserved and interaction is efficient.

When phase differences increase, coherence is reduced and the interaction weakens.

In this sense, electromagnetism reflects the degree of phase alignment between field configurations.

17.4 Electromagnetic Resonance Coupling

At a more detailed level, electromagnetic interaction can be understood as a resonance phenomenon.

Systems with compatible internal frequencies exchange energy efficiently through coupling with field modes.

This is analogous to two oscillators tuning into the same frequency and amplifying each other's response.

Within the Æ framework, such resonance processes provide a bridge between field dynamics and structured systems, such as molecules, where transitions occur between discrete configurations.

17.5 Strong Sector

The strong interaction corresponds to a regime in which coherence is highly constrained and tightly bound.

In this case, field configurations do not simply align, but become locked into stable structures.

The interaction is therefore non-linear and self-coupled, reflecting the fact that the field interacts with itself.

Intuitively, this can be seen as a regime where phase structures are confined, forming persistent and tightly bound configurations.

17.6 Strong Interaction Sector

Within this regime, the dynamics are governed by internal symmetries that enforce conservation and confinement.

The interaction does not diminish with separation in the same way as simpler forces, but instead maintains strong binding within localized regions.

This reflects a highly structured coherence pattern in which phase configurations remain tightly coupled.

17.7 Weak Sector and Symmetry Breaking

The weak interaction emerges in regimes where symmetry is partially broken.

In this context, the field admits configurations that are not fully stable, but can transition between states.

This introduces asymmetry into the system, allowing processes such as decay or transformation to occur.

Intuitively, this corresponds to a situation in which phase structure becomes unbalanced, enabling changes that would otherwise be forbidden.

17.8 Weak Interaction Sector

The weak interaction therefore governs transitions between different configurations of the field.

These transitions are mediated by intermediate structures that exist only temporarily, reflecting the unstable nature of the underlying symmetry.

This sector captures how coherent structures can reorganize when stability conditions are relaxed.

17.9 Emergent Gravitational Sector

Gravity emerges as a large-scale geometric response of the Æ field.

Rather than being driven by local phase alignment, it reflects the cumulative effect of coherence distribution across the field.

In this sense, gravity corresponds to how the geometry of the system adapts to the presence of structured configurations.

17.10 Gravitational Sector

Within the Æ framework, gravitational interaction arises as curvature induced by coherence and memory effects.

Persistent configurations influence the geometry of the field, creating pathways that guide the motion of other structures.

Intuitively, one may think of this as the field remembering where coherence has accumulated, and shaping its geometry accordingly.

17.11 Total Æ-Lagrangian

The full Æ Lagrangian combines all interaction sectors into a single unified description.

Each term corresponds to a different aspect of field behavior, including gauge interactions, scalar dynamics, and entropy contributions.

Together, they describe a system in which geometry, phase, coherence, and interaction are not separate ingredients, but different expressions of the same underlying structure.

17.12 Memory Stress-Energy Contribution

Within the Æ framework, persistent coherence structures contribute directly to the energetic content of the field.

These contributions arise from spatial variations of coherence, which encode how structured configurations are distributed across the system.

In this sense, memory is not an abstract concept, but a physical component that carries energy and influences the dynamics of the field.

Regions where coherence changes rapidly contribute more strongly, reflecting the presence of structured or accumulated configurations.

Thus, memory becomes an active ingredient in the energetic and geometric behavior of the system.

17.13 Unified Lagrangian (Extended Template)

The Æ framework admits a unified Lagrangian description in which geometric, scalar, and entropy contributions are combined into a single structure.

This formulation extends the usual gravitational sector by incorporating higher-order curvature terms and coherence-dependent contributions.

In this sense, the dynamics of the field are governed by a balance between geometry, phase organization, and entropy.

Rather than treating these aspects independently, the Lagrangian encodes them as different expressions of the same underlying structure.

17.14 Gravitational Interaction

17.14.1 Coherence-Induced Geometry

Within the Æ framework, gravity emerges as a response of geometry to variations in coherence.

Instead of being driven solely by mass-energy, curvature is influenced by how coherence is distributed across the field.

Regions of lower coherence effectively rescale the geometry, altering distances and trajectories.

In this sense, geometry adapts dynamically to the organization of the field.

17.14.2 Effective Einstein Sector

This leads to a modified gravitational description in which coherence contributes directly to the effective Planck scale.

As a result, the strength of gravitational interaction is not fixed, but depends on the local coherence structure of the field.

Additional terms arise from coherence variations, modifying the standard Einstein equations.

These terms reflect how the field reorganizes its geometry in response to internal structure.

17.14.3 Newtonian Correspondence

In appropriate limits, the model reproduces classical gravitational behavior.

In particular, coherence variations give rise to an effective potential, providing a bridge to the Newtonian description.

This establishes consistency with known physical regimes while extending the interpretation of gravity to include coherence effects.

17.14.4 Note for Further Formalization

A complete derivation of the gravitational sector, including full coupling to matter and cosmological consistency, remains an open direction for future work.

17.15 Electromagnetic Interaction

17.15.1 Phase Connection and Gauge Structure

Within the Æ framework, electromagnetism emerges as a rotational interaction associated with phase structure.

The gauge connection represents how phase varies across the manifold, acting as a mediator of interaction between configurations.

17.15.2 Relation to Phase Structure

In highly coherent regions, the connection may be locally approximated by simple phase gradients.

However, nontrivial electromagnetic behavior arises when the phase structure develops defects or global inconsistencies.

These defects generate curvature in the connection, producing observable electromagnetic effects.

17.15.3 Electromagnetic Action

The resulting dynamics follow a structure analogous to classical electromagnetic theory, ensuring compatibility with established physical laws.

17.15.4 Left-Right Phase Differential

Electromagnetic polarity may be interpreted as a difference between complementary phase components.

When this imbalance becomes nontrivial, it generates circulatory structures that manifest as electromagnetic fields.

17.15.5 Note for Further Formalization

A complete embedding into a fully developed gauge framework, including quantization and renormalization, remains a subject for future work.

17.16 Strong Interaction

17.16.1 Topological Binding and Nodal Confinement

Within the Æ framework, the strong interaction may be understood as a mechanism of topological binding.

Rather than arising from simple force exchange, it corresponds to configurations of the field that become locked through their internal structure.

In particular, phase channels may twist, braid, and interlink in such a way that they cannot be separated without a global reconfiguration.

This produces a form of confinement, where coherent structures remain bound together.

Intuitively, one may think of this as knots in the field: once formed, they cannot be undone by small deformations.

The stability of these configurations is determined by topological invariants, which remain approximately constant over time.

As long as these invariants are preserved, the structure remains stable.

However, increasing phase entropy weakens this binding.

As coherence is lost, the effective confinement strength decreases, allowing the structure to destabilize.

17.16.2 Strong-Sector Action

The dynamics of this sector follow a non-linear self-interacting structure, reflecting the fact that the field interacts with itself.

This non-linearity is essential for maintaining topological confinement and generating stable bound configurations.

17.16.3 Note for Further Formalization

A complete mapping between these topological structures and experimentally observed particles, as well as a derivation of confinement from first principles within the Æ framework, remains an open direction for future work.

17.17 Weak Interaction

17.17.1 Irreversible Transitions and Symmetry Breaking

Within the Æ framework, the weak interaction may be understood as a mechanism of transformation between coherent configurations.

Unlike more stable sectors, this regime allows structures to change form through processes that are not fully reversible.

These transitions are associated with symmetry breaking, where previously equivalent configurations become distinguishable.

Intuitively, this corresponds to a situation in which a balanced phase structure becomes unstable, allowing the system to move into a new configuration.

Such processes are inherently directional, reflecting an increase in phase entropy.

17.17.2 Transition Dynamics

The evolution in this sector includes both coherent propagation and transition terms that modify the structure of the state.

These transition processes are linked to entropy production, indicating that the system is moving toward a different organizational regime.

17.17.3 Spontaneous Symmetry Breaking

A key feature of this sector is the emergence of preferred configurations.

The system selects a specific state from a set of equivalent possibilities, breaking the underlying symmetry.

This selection introduces effective mass scales and defines the structure of the interaction.

17.17.4 Weak-Sector Action

The dynamics combine gauge-like interactions with scalar field behavior, capturing both propagation and transformation within the system.

17.17.5 Note for Further Formalization

A full connection with observed decay processes, as well as a precise formulation of irreversibility within the Æ framework, remains an open problem.

17.18 Unified View of Æ-Forces

17.18.1 Common Structural Origin

Within the Æ framework, the four fundamental forces can be understood as different manifestations of a single underlying interaction structure.

Rather than arising independently, each force reflects a specific way in which the field organizes its phase, coherence, and geometry.

Gravity corresponds to gradients of coherence that reshape the geometry of the field.

Electromagnetism emerges from rotational structure in phase, producing circulating interactions.

The strong interaction arises from topological binding, where phase structures become locked into stable, confined configurations.

The weak interaction corresponds to transitions between configurations, driven by symmetry breaking and irreversibility.

In this sense, the forces are not separate entities, but different lenses through which the same field dynamics are expressed.

This unification is structural rather than reductionist: it does not eliminate the distinctions between forces, but explains them as different regimes of a common underlying system.

17.19 Master Action and Field Equations

The Æ framework admits a unified action in which geometry, gauge interactions, scalar fields, and entropy are combined into a single structure.

In this formulation, the standard ingredients of modern physics appear naturally, but are modulated by coherence and phase-entropy.

Gravity is governed by a coherence-dependent effective scale, linking geometry directly to the internal organization of the field.

Gauge interactions retain their fundamental structure, but their propagation is influenced by coherence and entropy through effective weighting factors.

Scalar fields mediate symmetry breaking and transitions, while fermionic fields encode localized excitations.

A key addition is the presence of memory effects, which introduce a causal structure linking past configurations to present dynamics.

As a result, the system is not only local in time, but retains a structured dependence on its own history.

The resulting field equations describe a system in which geometry, interaction, and information are fully coupled.

Rather than separate components, they appear as different aspects of a single dynamical framework.

18. Oscillatory Patterns and Resonance

Oscillatory behaviour plays a fundamental role in many physical systems.

From electromagnetic waves to quantum phase evolution and biological rhythms, oscillations provide a natural mechanism through which information and energy propagate.

Within the Æ framework, oscillatory patterns arise directly from the time evolution of phase.

As phase rotates, the field naturally generates periodic behaviour, producing interference patterns, resonances, and coherent structures.

In this sense, oscillations are not an additional feature of the theory, but a direct consequence of its internal dynamics.

They represent the observable expression of continuous phase evolution within the field.

18.1 Oscillatory Æ Field

The Æ field contains intrinsic phase components that evolve continuously in time.

When expressed in observable form, this phase evolution appears as oscillatory behaviour.

Intuitively, one may think of this as a rotating phase being projected into a measurable signal.

As the phase advances, it traces out periodic patterns, giving rise to waves and oscillations.

Thus, oscillatory phenomena emerge naturally from the underlying phase structure of the field.

18.2 Phase Synchronization

When multiple oscillatory components interact, their phases may influence one another.

If the coupling between them is sufficiently strong, their phase differences stabilize over time.

This leads to synchronization, where multiple components oscillate in a coordinated manner.

Within the Æ framework, this process is governed by coherence: higher coherence promotes alignment, while lower coherence leads to desynchronization.

18.3 Lissajous Resonance

When oscillatory modes with different frequencies interact, their combined motion may form stable geometric patterns.

These patterns, known as Lissajous figures, arise when the frequencies satisfy simple ratios.

They can be understood as visual representations of resonance between oscillatory components.

Within the Æ manifold, such patterns reflect stable interference structures generated by coupled phase dynamics.

18.4 Pattern Stability

Not all oscillatory configurations remain stable.

Stability requires that phase differences between interacting components remain bounded over time.

When coherence is sufficiently high, these phase differences remain small, allowing synchronized patterns to persist.

As coherence decreases, phase deviations grow, leading to desynchronization and eventual collapse of the pattern.

From a dynamical perspective, stability is also related to how perturbations evolve: stable systems naturally damp deviations, while unstable ones amplify them.

18.5 Phase Transmission

Oscillatory phase patterns may propagate across the field through structured transport mechanisms.

This propagation allows coherent information to move between different regions or layers of the system.

Intuitively, this can be seen as the transmission of phase alignment from one region to another.

When phase mismatch is small, transmission is efficient.

When mismatch or entropy increases, transmission degrades.

18.6 Operator Formulation

The evolution and transmission of oscillatory structures can also be described in operator form.

In this formulation, phase evolution corresponds to transformations generated by dynamical operators.

These operators encode how states evolve and how phase relationships are transported across the configuration space.

18.7 Network Coherence

Large collections of oscillatory elements may form coherent networks.

In such networks, synchronization extends across many interacting components, producing large-scale patterns of coordinated behavior.

Examples of similar phenomena appear in neural activity, laser systems, and collective oscillations in physical systems.

Within the Æ framework, these correspond to regions of the field where coherence, phase alignment, and nodal stability reinforce each other, allowing global synchronization to emerge.

19. Time as an Emergent Process

Time is often treated as a fundamental parameter in physical theories.

Within the Æ framework, however, time is not taken as a primitive quantity, but as an emergent manifestation of deeper dynamical processes.

In particular, time arises from the evolution of phase, coherence, and entropy within the field.

Rather than a single universal clock, the system admits multiple temporal descriptions, each corresponding to a different level of organization.

In this sense, time is not a fixed background, but a structured process that depends on how the field evolves.

19.1 Coexisting Temporal Variables

Within the Æ framework, different notions of time coexist in a hierarchical structure.

Some correspond to intrinsic evolution of the field, while others emerge as observable or subjective descriptions.

These temporal variables are not independent, but related through a process of emergence.

Intuitively, one may think of them as different "projections" of the same underlying temporal flow.

At deeper levels, time is tied directly to phase evolution.

At higher levels, it appears as physical time, and ultimately as subjective experience.

Thus, time is not a single quantity, but a layered structure reflecting the organization of the field.

19.2 Æ-Time as Phase–Topological Time

Within the Æ framework, the most fundamental notion of time is associated with phase evolution.

As phase rotates, it generates a natural ordering of events.

In this sense, time may be understood as a measure of accumulated phase change.

Intuitively, one may think of this as counting how many "turns" the system has undergone.

This evolution is not purely local, but may also accumulate along trajectories, giving rise to global phase structures.

These accumulated phase changes define a form of topological time, in which the history of the system is encoded in its phase configuration.

Thus, time is not external to the system, but emerges from its internal dynamics.

19.3 Entropy, Coherence, and Temporal Orientation

The direction of time is closely related to the evolution of entropy and coherence.

As phase entropy increases, coherence tends to decrease, and the system evolves toward more disordered configurations.

At the same time, memory accumulates, encoding the history of the system.

This creates a natural temporal orientation, in which the system moves from more coherent states to less coherent ones, while retaining information about its past.

In this sense, the arrow of time is not imposed, but emerges from the internal balance between entropy, coherence, and memory.

Time flows in the direction in which this balance evolves.

19.4 Deformed Temporal Algebra

The different notions of time in the Æ framework can be described through a set of operators that generate temporal evolution.

These operators are not independent, but related through transformations involving phase, coherence, and entropy.

In particular, physical time may be understood as a rescaled version of a more fundamental temporal parameter, modulated by the coherence and entropy of the system.

As a result, the algebra of temporal operators is deformed relative to the standard formulation.

Intuitively, this means that time does not flow uniformly, but depends on the internal state of the field.

Regions of high coherence and low entropy evolve differently from regions of low coherence and high entropy.

Thus, time becomes a dynamical quantity shaped by the structure of the system itself.

19.5 Unified Temporal Generator

Within the Æ framework, time is not generated by a single operator, but by a combination of processes.

These include intrinsic evolution, phase rotation, and entropy dynamics.

The unified temporal generator therefore combines these contributions into a single structure.

Intuitively, one may think of time as emerging from three simultaneous effects:

the internal progression of the system, the rotation of phase, and the accumulation of entropy.

Rather than a single flow, time is the result of their interaction.

Different regimes of the system may be dominated by different contributions, leading to variations in how time is experienced or measured.

In this sense, time is not universal, but depends on the internal state of the field.

19.6 Variational Formulation of Emergent Time

The emergence of time can also be described through a variational principle.

In this formulation, time arises from the balance between phase evolution, entropy dynamics, and their interaction through the coherence structure.

Intuitively, the system evolves along trajectories that optimize this balance.

Phase tends to evolve smoothly, entropy tends to increase, and coherence constrains the relationship between them.

The resulting dynamics define the effective flow of time.

From this perspective, time is not imposed externally, but emerges as the path that the system follows in its own configuration space.

19.7 Temporal Generators and Loop Constraints

The evolution of states in the Æ framework may also be described in terms of loops in configuration space.

As the system evolves, phase may accumulate along closed trajectories.

These loops encode memory and constrain the possible evolution of the system.

In particular, certain closed paths lead to quantized conditions, reflecting the discrete nature of accumulated phase.

Intuitively, one may think of this as the system "remembering" how many times it has rotated or cycled through a configuration.

These loop constraints provide a bridge between continuous evolution and discrete structure.

19.7.1 Hamiltonian Compatibility Conditions

For the dynamics to remain consistent, the generators of time must be compatible with the Hamiltonian structure of the system.

In particular, the intrinsic phase-based notion of time preserves the underlying dynamics.

However, when additional temporal structures are introduced, such as entropy-driven evolution, the effective Hamiltonian may itself evolve.

This reflects the fact that, within the Æ framework, the system is not governed by a fixed background, but by a dynamically changing structure.

As a result, time and dynamics become interdependent, rather than independent concepts.

19.8 Phase Differences and Interaction

Phase differences encode the relative state between oscillatory components of the field.

They determine how different regions evolve with respect to one another.

Intuitively, a phase difference represents a temporal offset between configurations.

As these differences evolve, they define how interaction strength emerges and how synchronization is maintained or lost.

In this sense, temporal structure is encoded directly in phase relationships.

19.9 Phase Coherence and Entropic Stability

Coherence measures the degree to which phase remains aligned across the field.

High coherence corresponds to stable temporal relations, while increasing phase entropy leads to temporal dispersion.

As entropy grows, synchronization is lost and the system evolves toward less ordered configurations.

This establishes a direct link between entropy growth and the degradation of shared temporal structure.

Thus, time is not only a measure of change, but also of how coherence is maintained or lost.

19.10 Nonlocal Memory Kernel

The evolution of the field depends not only on its present state, but also on its history.

This memory is encoded through nonlocal interactions that couple past configurations to present dynamics.

As a result, the system retains information about previous states, influencing its future evolution.

Time, in this sense, is not purely local, but carries a structured dependence on history.

19.11 Holonomy Memory Conservation

Memory may also be encoded geometrically through holonomy.

As phase evolves along closed paths, it accumulates a net shift that cannot be removed by local changes.

These accumulated shifts represent persistent temporal structures.

In this sense, time can be understood as the accumulation of phase cycles within the field.

19.12 Spectral Memory Operator

Memory can also be represented through spectral structures.

Stable modes encode persistent patterns that resist perturbation.

These modes act as carriers of temporal information, preserving structure across evolution.

A spectral gap ensures that these configurations remain stable over time.

19.13 Entropic Filtering and Forgetting

Not all information is preserved indefinitely.

Entropy acts as a filtering mechanism, limiting the amount of memory that can be retained.

Only configurations that remain sufficiently coherent survive this filtering process.

This introduces a natural form of forgetting, ensuring that the system does not retain all past information.

Thus, memory is both preserved and selectively erased.

19.14 Discrete Nodal Representation

The system may also be described as a network of interacting nodes.

Each node carries its own phase, frequency, coherence, and intrinsic temporal parameter.

The global behavior emerges from the interaction of these local temporal structures.

In this sense, time is not a single global quantity, but a distributed property of the system.

19.15 Graph Structure of the Network

The system of interacting nodes may be represented as a graph, where nodes correspond to localized configurations and edges represent their coupling.

This structure defines how information and phase relationships propagate across the field.

Intuitively, the network determines who influences whom, and how strongly.

The associated Laplacian encodes the connectivity of the system, governing how phase differences spread and relax over time.

In this sense, phase evolution across the network follows a diffusion-like process, modulated by internal frequencies.

19.16 Spectral Mode Decomposition

The dynamics of the network can be decomposed into independent modes.

Each mode represents a collective pattern of oscillation across the system.

These modes evolve with different rates, depending on the

structure of the network.

Low-frequency modes correspond to large-scale, slow variations, while high-frequency modes describe localized, rapidly changing patterns.

This decomposition provides a natural way to understand complex dynamics as a superposition of simpler oscillatory structures.

19.17 Coherence Evolution in the Network

Each node carries a local coherence value that evolves dynamically.

Coherence decreases with entropy and increases through interaction with neighboring nodes.

The global coherence of the system emerges from the combined contribution of all nodes.

Propagation of coherence follows the network structure, allowing regions to influence one another.

In this sense, coherence behaves as a distributed quantity that flows and reorganizes across the graph.

19.18 Quaternionic Phase Velocity

Each node carries an internal phase state with a quaternionic structure.

The evolution of this state defines a local phase velocity, representing how the internal configuration rotates in time.

This rotation is not merely scalar, but encodes multidimensional phase dynamics.

Intuitively, each node behaves like a rotating entity in an internal space, with its own direction and speed.

19.19 Causal Coupling Constraint

Interactions between nodes are constrained by causality.

Information can only propagate along paths that respect the underlying geometric structure.

This ensures that phase relationships do not propagate instantaneously, but follow well-defined causal connections.

In addition, coupling strength decreases with distance, reflecting the finite range of interaction.

Thus, the network is both structured and causally constrained.

19.20 Æ Coupling Strength

The effective interaction between nodes depends on several factors.

It increases with coherence, decreases with entropy, and is modulated by phase alignment.

Nodes that are highly coherent and well aligned in phase interact more strongly.

Conversely, large phase differences or high entropy reduce the coupling.

In this sense, interaction strength reflects the degree of compatibility between nodes.

19.21 Propagation Speed Bound

The transfer of information across the network defines an effective propagation speed.

This speed is bounded by the causal structure of the system.

As a result, information cannot propagate arbitrarily fast, but is limited by intrinsic constraints of the field.

This ensures consistency with the underlying geometry and preserves causal order in the evolution of the system.

19.22 The Self as a Trajectory in Quaternionic State Space

Within the Æ framework, the notion of self may be understood as a trajectory through the space of admissible field configurations.

Rather than being a fixed entity, the self corresponds to a continuous path traced by the evolving state of the system.

In this sense, identity is not static, but defined by the history of its evolution.

19.23 Coherent Sectors in Solution Space

The space of possible trajectories can be organized into coherent sectors.

Each sector represents a family of trajectories that converge toward similar asymptotic configurations.

These sectors act as attractors, organizing the long-term behavior of the system.

Different trajectories may evolve differently at early stages, but converge toward the same coherent structure.

19.24 Admissibility and Timeline Set

Not all trajectories are dynamically allowed.

Admissible trajectories are those that maintain sufficient coherence and remain within stability bounds.

The set of such trajectories defines the space of possible evolutions of the system.

In this sense, what may be interpreted as different "timelines" correspond to different admissible paths within the same underlying space.

19.25 Internal Parametrization and Arc-Length Time

Time may be defined intrinsically along a trajectory.

Rather than being imposed externally, it corresponds to the progression of the system along its path in state space.

Intuitively, this may be understood as the "distance" traveled by the configuration.

Thus, time reflects how much the system has changed, rather than how much an external clock has advanced.

19.26 Energetic Modulators of Trajectory Stability

The stability of a trajectory depends on how rapidly the system evolves.

Large variations in the state correspond to higher internal activity, which may destabilize coherence.

Conversely, smoother evolution promotes stability.

In this sense, internal energy acts as a modulator of trajectory coherence.

19.27 Optimal Trajectories (Variational Characterization)

Among all possible trajectories, some are dynamically preferred.

These optimal paths balance internal evolution and structural stability.

The system naturally tends to follow such trajectories, which minimize instability while preserving coherence.

19.28 Temporal Algebra of the Æ-Field

Time evolution may be described as the action of a generator acting on field configurations.

In this representation, time measures the ordered transformation of the system.

Observables evolve according to this structure, reflecting the internal dynamics of the field.

19.29 Recurrence and Temporal Structure

In highly coherent regimes, the system may exhibit recurrence.

Configurations return approximately to previous states after a characteristic interval.

These cycles generate periodic structures that contribute to the observed temporal behavior.

19.30 Æ-Time Relations

The rate at which time is experienced depends on the evolution of coherence.

Highly coherent configurations evolve more slowly, while rapidly changing configurations correspond to faster effective time.

In this sense, time is directly linked to the internal dynamics of the system.

19.31 What Time Means in the Æ Framework

Throughout this section, we have described time not as a fundamental background, but as something that emerges from the internal dynamics of the field.

This idea can sound abstract at first, but it can be understood through simple intuition.

When we say that time is related to phase, what we mean is that time measures how much a system has "rotated" or changed internally.

Just as a clock measures time through repeated cycles, the Æ framework describes time as the accumulation of internal cycles of the field.

When we say that time depends on coherence, we mean that systems that are more ordered and synchronized tend to evolve more smoothly and slowly.

In contrast, systems that are disordered or rapidly changing experience time as faster or more fragmented.

When we say that time is linked to entropy, we are describing the direction of change.

As disorder increases, the system moves forward in a well-defined direction, giving rise to what we perceive as the flow of time.

When we say that time depends on memory, we mean that the present is not independent of the past.

The system carries traces of previous configurations, and these influence how it evolves.

This gives time a sense of continuity and history.

Finally, when we describe the self as a trajectory, we are saying that time is not something external that happens to us.

Instead, it is the path we trace as our state evolves.

In this sense, time is not something that exists on its own, but something that emerges from change, structure, and experience.

From this perspective, the flow of time is not imposed from outside the system.

It is the natural result of how the field evolves, organizes, remembers, and transforms.

Thus, time in the Æ framework can be understood as:

the measure of change, the accumulation of cycles, the loss of coherence, the persistence of memory, and the trajectory of the system through its own state space.

All of these are different ways of describing the same underlying process.

Time is not a separate entity.

It is what the system does.

20. Chemistry and Atomic Structure

20.1 Atomic Orbitals as Phase-Coherent Modes

Within the Æ framework, atomic orbitals may be understood as stable patterns of phase organization.

Rather than being static objects, they correspond to coherent modes of the field that remain stable over time.

Intuitively, one may think of an orbital as a standing wave of phase.

When the phase evolution is balanced, the structure remains stable, forming what we recognize as an atomic bound state.

In this sense, atoms are not rigid particles, but coherent configurations of the field.

20.2 Energy Spectrum with Coherence Correction

The energy associated with these configurations depends not only on their spatial structure, but also on their coherence.

Highly coherent configurations are more stable and require less energy to maintain.

As phase entropy increases, stability decreases, and the effective energy of the state rises.

Thus, energy levels may be understood as reflecting both geometric structure and coherence properties.

20.3 Spectral Kernel Representation (Minimal Form)

The evolution of atomic configurations can also be described through a kernel structure.

This kernel encodes how the state of the system at one moment influences its future evolution.

Intuitively, it acts as a rule that determines how patterns persist and transform over time.

Stable configurations correspond to modes that are preserved under this evolution.

These modes form the spectral structure of the system, providing a natural explanation for discrete energy levels.

In this sense, atomic stability arises from the existence of persistent spectral patterns within the field.

21. DNA as Long-Range Phase Memory Structure

DNA is conventionally understood as the molecular carrier of genetic information.

Beyond its role as a sequence of nucleotides, the DNA molecule also possesses a highly structured helical geometry that supports complex dynamical interactions with its surrounding environment.

Within the Æ framework, DNA may be interpreted as a long-range phase memory structure.

In this view, the helical configuration of the molecule allows phase information to be distributed and stabilized along its length.

Rather than functioning solely as a static information storage system, DNA may therefore participate in dynamical phase organization processes within biological systems.

21.1 Helical Phase Manifold

Within the Æ framework, DNA is not only a chemical structure, but also a geometric and dynamical one.

It may be understood as a helical path along which phase is organized and propagated.

Intuitively, one may think of DNA as a spiral track, like a twisted wire, along which a signal travels.

However, instead of carrying electrical current, it carries phase information.

As the phase advances along the helix, it completes cycles, much like a rotating wave moving through a spiral structure.

This makes DNA a natural medium for storing and transmitting structured information.

21.2 Phase Memory Functional

A key idea in this framework is that DNA can retain phase information over long distances.

Intuitively, this can be compared to a long string of oscillators.

If the oscillators remain synchronized, the system preserves a coherent pattern across its entire length.

If they fall out of sync, the pattern becomes noisy and the information is lost.

The memory functional measures how well this coherence is preserved.

High values correspond to stable, organized patterns, while low values indicate randomization and loss of structure.

In this sense, DNA acts as a long-range memory system, storing information not only in its sequence, but in its phase organization.

21.3 Non-Local Æ Kernel on DNA

DNA is not only local in its behavior.

Different regions of the strand can influence each other, even if they are not adjacent.

This can be understood as a non-local coupling along the helix.

Intuitively, it is similar to a string where vibrations at one point can affect distant regions through the overall structure.

In this way, DNA behaves as a distributed system in which information is shared across its entire length.

21.4 DNA Phase Waveguide

DNA can also be interpreted as a waveguide.

Just as optical fibers guide light, DNA may guide phase information along its structure.

Signals propagate along the strand, but are affected by entropy and environmental noise.

If coherence is maintained, the signal travels efficiently.

If entropy increases, the signal degrades.

This provides a natural mechanism for information transmission and regulation.

21.5 Relation to Generic Long-Range Memory

The behavior of DNA can be seen as a specific example of a more general principle: long-range memory.

Systems that preserve coherence over time and space can store information beyond their immediate state.

DNA is one such system.

It accumulates information, preserves patterns, and allows them to influence future configurations.

In this sense, it acts as a bridge between past and future states of the organism.

21.6 DNA Resonant Modes and Expression Coupling

The helical structure of DNA supports vibrational and oscillatory modes associated with its geometry.

Such resonant modes may interact with the biochemical processes responsible for gene expression.

Within the Æ framework, these interactions may be interpreted as coupling between phase dynamics and molecular activity.

21.7 DNA Resonant Modes and Expression Coupling

Within the Æ framework, DNA can be described as a collection of resonant modes.

Each mode represents a pattern of phase organization along the strand.

These modes can be activated or suppressed depending on internal and external conditions.

Intuitively, this is similar to a musical instrument, where different notes can be played depending on how the system is excited.

Gene expression may then be understood as the activation of specific modes.

Only certain patterns become active, while others remain latent.

21.8 DNA as a Coherent Information Lattice

DNA may also be viewed as an information lattice.

Rather than a simple linear code, it forms a structured network in which information is distributed across multiple scales.

Each point along the strand carries both amplitude and phase, contributing to a larger coherent pattern.

In this sense, DNA is not just storing information, but organizing it dynamically.

21.9 DNA–Protein Coupling

DNA does not act in isolation.

It interacts with proteins and other molecular structures.

Within the Æ framework, this interaction can be understood as a coupling between different coherent systems.

Proteins can modify the phase structure along DNA, while DNA influences protein configurations.

This creates a feedback loop in which structure and function co-evolve.

21.10 Phase Retention Along the Helix

For DNA to function as a memory system, it must retain phase information over time.

This retention depends on the balance between coherence and entropy.

If coherence is preserved, patterns remain stable.

If entropy dominates, patterns decay.

This defines the limits of information storage and transmission within the system.

21.11 Fractal Antenna Hypothesis

At larger scales, DNA may exhibit fractal-like properties.

This means that similar patterns repeat across different scales of organization.

Intuitively, this is similar to an antenna that can receive and transmit signals across multiple frequencies.

When phase remains aligned across scales, the system supports coherent interaction.

When this alignment breaks, coherence is lost.

21.12 Fractal Antenna Hypothesis

The spatial organization of DNA exhibits a multiscale structure consistent with fractal-like behavior and electromagnetic responsiveness.

Rather than functioning solely as a biochemical information carrier, DNA may be interpreted as a geometrically structured medium capable of interacting with coherent electromagnetic fields.

Its helical and folded configurations enable the coupling between local molecular processes and extended field dynamics, suggesting that information propagation is not purely chemical but also phase-mediated.

Within the Æ framework, this behavior is naturally described in terms of phase coherence distributed across scales, where the geometry of DNA supports the stabilization and transmission of structured phase information.

This interpretation aligns with the view that biological systems maintain coherence not only through local interactions but also through their ability to resonate with extended fields via structured geometries.

Reference Note on DNA Electromagnetic Response

The interpretation of DNA as a geometrically structured and electromagnetically responsive system is inspired by prior studies suggesting that DNA may exhibit antenna-like behavior due to its spatial configuration and conductive properties.

In particular, it has been proposed that the helical and fractal-like organization of DNA enables interactions with electromagnetic fields across multiple scales, supporting the possibility of resonance phenomena beyond purely biochemical dynamics.

Within the Æ framework, this perspective is not adopted as a definitive claim, but rather as a motivating analogy that informs the modeling of DNA as a multiscale coherence-supporting structure. The role of DNA is thus interpreted in terms of phase organization and geometric mediation of coherence, consistent with the broader formalism of phase entropy and nodal coupling.

Disclaimer. The referenced interpretation of DNA as an electromagnetic or fractal antenna is based on exploratory and non-consensus research directions. Its inclusion here serves as conceptual inspiration and does not constitute an established or experimentally validated component of the Æ framework.

22. Proteins and Topological Stability

Proteins are complex macromolecules whose structure and function are intimately linked.

The folding of a protein chain produces three-dimensional configurations that determine its biochemical activity.

Within the Æ framework, proteins may be interpreted as dynamical structures whose stability is influenced by both energetic and topological constraints.

In this perspective, the geometry of the folded chain may be viewed as a topological configuration capable of supporting coherent phase and resonance structures.

22.1 Proteins as Composite Knots

Within the Æ framework, proteins can be understood as structured curves embedded in space.

As these curves fold and loop around themselves, they form configurations that can be classified topologically as knots.

Intuitively, one may think of a protein as a piece of string that folds into a very specific shape.

Once this shape is formed, it cannot be changed without cutting or unfolding the structure.

Complex proteins can be seen as combinations of simpler knots joined together.

This provides a natural way to describe the complexity of protein structures in terms of topological composition.

22.2 DNA–Protein Phase Coupling

Protein folding is not random.

Within the Æ framework, it is guided by phase information originating from DNA.

DNA provides a structured phase pattern, and proteins emerge as folded realizations of that pattern.

Intuitively, DNA acts like a musical score, while proteins are the performed sound.

The phase structure encoded in DNA guides how the protein folds into its final configuration.

22.3 Topological Casimirs and Æ-Helicity

Once a protein has folded into a specific shape, certain properties of that configuration become conserved.

These conserved quantities are related to how different parts of the structure are linked or twisted.

Intuitively, if two loops of a structure are linked together, that linking cannot be removed without breaking the structure.

These invariants provide a natural explanation for the robustness of protein configurations.

22.4 Topological Stability Functional

Not all folded shapes are equally stable.

Some configurations are energetically and topologically preferred.

Stability depends on factors such as how entangled the structure is, how curved it is, and how complex its topology becomes.

Intuitively, a protein settles into a shape that balances tension, structure, and coherence.

This balance defines its functional form.

22.5 Variational Folding Principle

Protein folding can be understood as an optimization process.

The system evolves toward configurations that minimize instability while preserving coherence.

Intuitively, this is similar to how a hanging chain naturally settles into a stable shape.

The final folded state represents a balance between chemical interactions and topological constraints.

22.6 Protein States, Perturbations, and Metastability

Proteins are not rigid.

They can exist in multiple configurations, some of which are more stable than others.

External influences, such as molecules binding to the protein, can modify its state.

These changes can shift the protein into different configurations.

Some of these states are metastable: they persist for a time, but can transition into more stable configurations.

This dynamic flexibility is essential for biological function.

22.7 Spectral Topological Stabilization

Protein stability can also be understood in terms of spectral properties.

Certain configurations are protected by structural gaps that prevent small perturbations from destabilizing them.

Intuitively, this is similar to how a musical note remains stable when it is well defined within a harmonic structure.

These spectral gaps contribute to the robustness of protein configurations.

22.8 Molecular Resonance, Water Coupling, and Electromagnetic Response

Proteins are not static structures.

They behave as dynamic systems capable of oscillating between different configurations.

Within the Æ framework, these oscillations correspond to transitions between phase-coherent states.

Intuitively, one may think of a protein as a musical instrument that can resonate in different ways.

Each configuration corresponds to a "note", and the system can transition between these notes depending on internal dynamics and external influences.

Water plays a key role in this process.

Rather than being a passive environment, it acts as a medium that couples to the protein's phase structure.

If the phase of the protein and the surrounding water become aligned, interaction is strengthened.

If they become misaligned, coherence is reduced.

External electromagnetic fields may also influence these dynamics.

They can shift the phase of the system, effectively tuning how the protein evolves.

In this sense, protein behavior emerges from a combination of internal resonance, environmental coupling, and external perturbations.

22.9 Quantum Phase Dynamics of Protein States

At a deeper level, protein configurations may be described as superpositions of different states.

Each state carries its own phase and amplitude, and the overall configuration evolves through their interaction.

Intuitively, one may think of this as multiple possible shapes coexisting, with the system continuously transitioning between them.

These transitions are influenced by both energy differences and phase relationships.

Random fluctuations may also play a role, introducing small variations that can shift the system from one configuration to another.

In this sense, protein dynamics combine structured evolution with controlled variability.

22.10 Water and Electromagnetic Coupling

Proteins interact continuously with their environment.

Water acts as a structured medium that can either support or disrupt coherence.

When the phase structure of water aligns with that of the protein, the system becomes more stable.

External electromagnetic fields can further modify this interaction.

They may enhance or suppress certain configurations by altering the phase relationships within the system.

Intuitively, this is similar to tuning a system by applying an external signal.

22.11 Protein Resonance Spectrum

The behavior of a protein can be analyzed in terms of its resonance spectrum.

Each stable configuration contributes to a set of characteristic frequencies.

These frequencies define how the system responds to internal and external perturbations.

A functional protein exhibits a clear and well-defined resonance profile.

This means that its dynamics are concentrated around specific modes rather than being spread randomly.

In this sense, biological function is associated with structured resonance.

22.12 Protein Phase-Coherent Structures

Proteins may be understood as phase-coherent structures distributed along their molecular backbone.

Each point along the structure carries both amplitude and phase information.

When this phase structure remains aligned, the protein maintains its functional form.

If coherence is lost, the structure becomes unstable and may transition to a different configuration.

Thus, protein stability is directly linked to the maintenance of phase coherence.

22.13 Topological Stability of Protein Structures

Stable protein configurations correspond to topologically robust structures.

Once a coherent configuration is formed, it tends to persist over time.

This persistence requires both structural stability and phase alignment.

Intuitively, the protein must maintain its "shape" and its internal "rhythm" simultaneously.

If either is disrupted, the system may transition to a different state.

22.14 Topology–Biology Correspondence

Topological structures in the Æ framework have direct biological counterparts.

Protein configurations may be viewed as physical realizations of underlying topological patterns.

In this sense, biological form reflects deeper geometric and dynamical structures of the field.

The mapping between topology and biology provides a bridge between abstract mathematical structures and observable biological systems.

22.15 Protein Resonance Dynamics

Protein dynamics may also be described in terms of forced oscillations.

External influences act as driving forces that interact with the natural resonance of the system.

If the external driving matches the natural frequency, resonance occurs and the response is amplified.

If not, the effect is weaker.

Intuitively, this is similar to pushing a swing: if timed correctly, the motion increases; if not, it dissipates.

Thus, protein behavior depends on how internal and external dynamics align.

23. Biological Coherence Sector

Biological systems exhibit remarkable levels of organization across many spatial and temporal scales.

From molecular processes inside the cell to the coordination of organs within an organism, living systems maintain a form of dynamical order that is robust against external perturbations.

Within the Æ framework, biological activity may be interpreted as occurring within a coherence-dominated sector of the dynamical configuration space.

In this regime, coherent phase relationships across molecular, cellular, and systemic levels contribute to the stability and function of living organisms.

23.1 Helical Phase Configuration

Within the Æ framework, biological systems may be understood as organized phase structures.

A simple but fundamental pattern is the helical configuration.

Intuitively, this is similar to a spiral wave that propagates along a structure, such as DNA or biological fibers.

This helical organization allows phase to evolve in a stable and repeatable way, supporting long-range coherence.

23.2 Canonical Coherence and Admissibility

Coherence measures how ordered and synchronized a system is.

When coherence is high, the system behaves in a stable and coordinated way.

When coherence is low, the system becomes disordered.

Biological systems operate within a narrow window: they must remain sufficiently coherent to function, but flexible enough to adapt.

Intuitively, this is like a group of musicians: if they are too disorganized, there is noise; if they are too rigid, there is no adaptability.

Life exists in the balance between these extremes.

23.3 Bio-Photon Emission (Effective Law)

Biological systems can emit weak light signals, often referred to as bio-photons.

Within the Æ framework, this emission can be understood as a byproduct of rapid changes in coherence or phase.

Intuitively, when a system reorganizes quickly, it releases energy.

This release may manifest as light.

Thus, bio-photon emission reflects internal dynamical activity.

23.4 Biological Effective Action

The behavior of biological systems can be understood as a balance between three tendencies:

to maintain structure, to preserve coherence, and to dissipate disorder.

Intuitively, the system tries to stay organized while constantly interacting with its environment.

This balance determines the evolution of the organism over time.

23.5 Reality Knots Embedded in the Body (Organ/Hormonal Loops)

Certain persistent patterns within the body can be understood as localized coherence structures.

These may correspond to stable loops of phase that remain over time.

Intuitively, they are like knots within the system that store information.

Some of these structures support healthy function, while others may correspond to unresolved or persistent patterns.

When coherence is restored, these loops can relax and reintegrate into the global system.

23.6 Cancer as Multiscale Coherence Disruption (Minimal Criterion)

From this perspective, disease may be understood as a disruption of coherence.

In particular, cancer can be interpreted as a loss of alignment between local and global structures.

Cells no longer follow the collective pattern of the organism.

Instead, they evolve independently.

Intuitively, it is like a musician in an orchestra who stops following the rhythm of the group.

This loss of coordination leads to instability at larger scales.

23.7 Endocrine Slow Coherence Field

The endocrine system acts as a slow, global modulator of coherence.

Hormones do not act instantly.

Instead, they gradually influence the state of the organism.

Intuitively, this is like a background signal that slowly adjusts the behavior of the system.

This slow modulation helps coordinate long-term processes such as growth, adaptation, and recovery.

23.8 Neural–Non-Neural Integration

The nervous system operates on much faster timescales.

It allows rapid communication between different parts of the organism.

Within the Æ framework, neural and endocrine systems interact to form a unified network.

Intuitively, the neural system is like a fast communication channel, while the endocrine system is a slow regulatory field.

Together, they create a multiscale coordination mechanism.

Dual Bioelectric Channels and Longitudinal Phase Transport

In addition to localized coherence nodes, the biological system requires a mechanism to distribute phase information across the entire organism. This is achieved through continuous longitudinal flows that connect distant regions into a unified dynamic structure.

Rather than operating as isolated components, different parts of the body are linked through directional channels that carry bioelectric activity, phase information, and coherence signals. These channels enable coordination between neural, endocrine, and cellular processes.

Within this framework, phase does not remain confined to local regions. Instead, it propagates along preferred directions, allowing patterns formed in one part of the system to influence others. This propagation provides the physical basis for the integration of distributed processes into a single coherent state.

When these longitudinal channels are balanced, phase differences remain stable, coherence is preserved, and the organism behaves as an integrated whole. Information flows smoothly, and local dynamics remain synchronized with global structure.

However, when this balance is disrupted, phase propagation becomes irregular. Misalignment between channels increases phase entropy, reduces coherence, and weakens the coupling between different biological subsystems.

This disruption may contribute to the persistence of localized distortions, as phase information is no longer efficiently redistributed across the system. In such cases, previously described coherence nodes may become more rigid, reinforcing recurrent patterns.

Restoring longitudinal balance allows phase information to flow again across the organism, supporting reintegration and enabling coherence to recover at both local and global scales.

Thus, these dual channels act as a structural backbone for biological coherence, linking distributed nodes into a continuous phase-transport system.

Axial Dual-Channel Structures (Interpretative Correspondence)

In various historical and cross-cultural frameworks, axial structures within the body have been described as dual channels supporting longitudinal flow and functional polarity.

These descriptions, while differing in terminology, often referred to in traditional systems as dual axial channels, associated with complementary dynamics such as activation and regulation, expansion and contraction, or outward and inward flow.

Within the present framework, such structures may be interpreted as emergent representations of dual bioelectric channels supporting phase transport along the organism.

Rather than invoking specific traditional models, this correspondence is understood here as a phenomenological alignment between observed bioelectric dynamics and historically described axial dual-channel systems.

Disclaimer. The correspondence drawn here is interpretative and does not assume direct equivalence with any specific traditional or cultural system. It is included as a conceptual bridge linking modern bioelectric models with historically reported descriptions of axial organization.

23.9 Cellular Coherence Dynamics

At the cellular level, biological systems can be understood as coherent subsystems composed of many interacting internal modes.

Each internal component contributes to the overall state of the cell through its phase and coherence.

Intuitively, a cell behaves like a small network of oscillators that must remain synchronized.

When internal coherence is maintained, the cell functions properly.

When coherence breaks down, cellular function is compromised.

Cellular Transport and Memory Flux

Cells are not isolated.

They continuously exchange information with their surroundings.

This exchange can be understood as a flow of coherence and phase information across the cell boundary.

Intuitively, this is like a constant dialogue between the cell and its environment.

This flow allows the cell to adapt, respond, and maintain internal organization.

However, it also introduces entropy, which must be managed to preserve stability.

Cytoplasmic Phase Forces

Inside the cell, internal forces guide the organization of its components.

These forces can be understood as arising from gradients in phase and coherence.

Intuitively, the cell reorganizes itself in response to internal imbalances, much like a fluid adjusting to pressure differences.

This internal dynamics allows the cell to maintain structure and respond to external stimuli.

Cell Cycle Constraints

Cellular processes follow cyclic patterns.

These cycles represent the repeated evolution of the cell through specific states.

For the cycle to remain stable, coherence must be preserved throughout the process.

If coherence remains high, the cell can divide and reproduce.

If coherence drops below a critical level, the cycle becomes unstable.

Intuitively, the cell must "complete its rhythm" to continue functioning.

23.10 Unified Multiscale Stability Condition

Biological systems exist across multiple scales: molecular, cellular, tissue, neural, and endocrine.

Each level contributes to the overall coherence of the organism.

The system remains stable only if coherence is maintained across all these levels.

Intuitively, this is like a multi-layered system where failure at one level affects the whole.

Thus, global stability depends on the collective balance of all subsystems.

23.11 Organism as a Coherent Entity

A living organism may be understood as a unified coherent structure.

All subsystems are coupled and contribute to a shared global state.

When coherence is maintained, the organism functions as an integrated whole.

When coherence decreases, subsystems begin to decouple.

Intuitively, the organism behaves like an orchestra.

When all instruments are synchronized, there is harmony.

When synchronization is lost, the system becomes disordered.

23.12 Cellular Life (Minimal Sector)

Within the Æ framework, a cell may be understood as a coherent system composed of many interacting subsystems.

Each internal component contributes to a shared state through its phase, frequency, and coherence.

Intuitively, a cell behaves like a small orchestra, where many elements must remain synchronized.

The membrane plays a special role.

It acts as an interface between internal and external states, regulating the exchange of information.

This exchange is not purely material, but also involves phase and coherence.

The membrane potential reflects the difference between internal and external states, and evolves dynamically over time.

Inside the cell, internal forces organize structure.

These forces arise from gradients in phase and coherence, guiding the system toward stable configurations.

The cell cycle can be understood as a repeated process in which the system evolves through specific states.

For the cycle to remain stable, coherence must be preserved.

If coherence exceeds a certain threshold, the system may divide, producing two coherent subsystems.

If coherence drops below a critical value, the system loses structure and becomes non-viable.

Thus, cellular life is defined by the balance between coherence, entropy, and exchange.

23.13 Quaternionic Coherence Measure (Protein-Level Proxy)

At the molecular level, coherence may be quantified through averaged phase structure.

Intuitively, this measures how aligned a system is within its internal configuration space.

Higher alignment corresponds to more stable and functional structures, while lower alignment indicates disorder.

23.14 Life as a Multiscale Process

Life is not confined to a single level of organization.

It emerges from the interaction of multiple scales: molecular, cellular, tissue, and organism.

Each level contributes to the global state of the system.

Intuitively, life is like a layered structure in which each level supports and influences the others.

The organism exists only when coherence is maintained across all these layers.

23.15 Cross-Scale Synchronization

For biological systems to function, coherence must be maintained not only locally, but across different scales.

This requires synchronization between subsystems.

Intuitively, this is like different sections of an orchestra staying in rhythm with one another.

When this synchronization is preserved, the system behaves as a unified whole.

When it breaks down, coordination is lost.

23.16 Biological Coherence Stability

A biological system is stable when its coherence remains approximately constant over time.

At the same time, internal disorder must remain bounded.

If coherence fluctuates too strongly, or entropy grows beyond a critical level, stability is lost.

This leads to degradation of function.

Thus, life corresponds to a dynamically stable regime rather than a static state.

23.17 Phase Entropy as Thermodynamic Variable

Within the Æ framework, phase entropy plays the role of a fundamental thermodynamic quantity.

It measures how distributed or disordered phase configurations become.

Intuitively, it captures how much structure has been lost.

23.18 Coherence–Entropy Duality

Coherence and entropy are complementary.

When coherence is high, entropy is low.

When entropy increases, coherence decreases.

This duality governs the evolution of the system.

Intuitively, it reflects the balance between order and disorder.

23.19 Effective Temperature

An effective temperature can be defined in terms of how entropy changes with energy.

This temperature reflects the internal activity of the system.

Highly active systems correspond to higher effective temperatures, while stable systems remain at lower values.

23.20 Vibrational Free Energy

The system evolves toward configurations that balance energy and entropy.

This balance defines a free energy.

Stable states correspond to configurations where this quantity is minimized.

Intuitively, the system settles into states that are both energetically efficient and structurally organized.

23.21 Coherence-Induced Geometry (Thermo–Gravitational Coupling)

At larger scales, variations in coherence may influence the geometry of the system.

Regions with different coherence levels can generate effective curvature.

Intuitively, this means that structure and geometry are not independent.

Instead, the organization of the field shapes the space in which it evolves.

In this way, thermodynamics, coherence, and geometry become interconnected aspects of a single dynamical framework.

24. Definition of the Self

The concept of the self has been interpreted in many different ways across philosophy, psychology, and neuroscience.

Within the Æ framework, the self is not treated as a static entity but as a persistent dynamical structure emerging from the evolution of coherent processes within the system.

From this perspective, identity corresponds to a trajectory in the configuration space of the underlying dynamical field.

The persistence of this trajectory arises from stability conditions that allow the system to maintain coherent organization over extended periods of time.

24.1 Historical Attractors

Over time, certain patterns tend to repeat or stabilize.

These patterns act as attractors in the space of possible configurations.

Intuitively, they are like habits or tendencies that the system naturally returns to.

They encode accumulated history and influence how the system evolves in the future.

In this sense, the self is shaped not only by its current state, but by the patterns it has reinforced over time.

24.2 Trajectory Persistence

For a self to exist, its trajectory must persist.

This means that the system must remain sufficiently coherent across time.

If coherence between past and present states is maintained, the trajectory continues smoothly.

If this coherence is lost, the sense of continuity breaks down.

Intuitively, identity depends on remembering and remaining connected to previous states.

Without this connection, the trajectory fragments.

24.3 Ontological Layer Interpretation

The trajectory of the self is not confined to a single level.

It extends across multiple layers of the system.

These include physical, intermediate, and more abstract levels of organization.

Intuitively, the self can be seen as a thread that runs through all these layers at once.

What we perceive as identity is the coherent projection of this thread across them.

Thus, the self is not located in one place, but emerges from the coordinated evolution of multiple interacting layers.

25. Dream Regime as Reduced Coherence Sector

During sleep, biological systems undergo significant changes in their dynamical organization.

In particular, sensory coupling with the external environment becomes strongly reduced while internal neural activity remains highly active.

Within the Æ framework, this situation may be interpreted as a reduced coherence sector in which the coupling between internal and external phase structures is partially decoupled.

In this regime, the system evolves primarily under the influence of internal dynamical processes.

25.1 Coherence Decoupling Condition

Within the Æ framework, the dream state is not a shutdown of the system, but a reorganization.

What changes is not whether the system is active, but where its coherence is directed.

During waking life, coherence is strongly coupled to the physical world.

During dreaming, this coupling is reduced.

Intuitively, it is like unplugging from external input while remaining fully active internally.

The system stops prioritizing external alignment and shifts toward internal coherence.

Different subsystems respond differently.

Neural coherence associated with external processing decreases, while internal and regulatory processes remain active or even increase.

In this sense, the dream state is not inactivity, but a redistribution of coherence.

Dream Memory Accumulation

Dreaming is also a period of internal memory processing.

As external input is reduced, internal patterns can reorganize and interact freely.

Intuitively, it is like replaying and mixing past experiences without interruption.

These internal dynamics contribute to memory formation and restructuring.

Rather than simply storing information, the system reorganizes it into coherent patterns.

25.2 Internal Rotational Dominance

In the dream state, internal phase dynamics dominate.

External sensory influence becomes weak, and the system evolves primarily according to its internal structure.

Intuitively, this is like a system spinning freely without external constraints.

This allows the emergence of complex and sometimes unusual configurations, which appear as dream imagery.

The reduction of sensory influence can be understood as a decrease in external coupling.

Meanwhile, internal interactions remain active, or even become stronger relative to external input.

Different dream stages correspond to different internal dynamics.

In deeper states, activity is slower and more synchronized.

In more active dream phases, internal dynamics become faster and more strongly coupled.

25.3 Effective Oniric Hamiltonian

The dream state can be understood as a regime in which internal dynamics dominate the evolution of the system.

The system explores configurations driven by internal structure and accumulated memory.

Intuitively, it is like a system moving through its own internal landscape rather than reacting to the external world.

For this process to remain stable, internal coherence must stay above a critical level.

If coherence is too low, the system cannot sustain structured dream activity.

Dreaming may also be seen as an internal projection of the system's state.

The observable experience corresponds to internally generated configurations.

These are not simple reflections of waking reality, but new combinations shaped by internal dynamics.

During this state, motor output is suppressed.

The system remains active internally, but its physical expression is minimized.

25.4 Lucidity as Partial Sensory Re-Coupling

Lucid dreaming corresponds to a partial reconnection with external awareness.

The system remains in the dream regime, but some degree of self-awareness is restored.

Intuitively, it is like becoming aware inside the internal simulation.

This does not fully restore waking conditions, but introduces a hybrid state in which internal and external dynamics partially overlap.

25.5 Reduced Projection Dynamics

Dreaming can also be understood as a reduction in the projection of internal states onto the physical layer.

External observables become less dominant, allowing internal configurations to shape the experience.

Intuitively, the system stops projecting outward and begins projecting inward.

This shift allows the exploration of internal structures that are normally constrained during waking life.

25.6 Internal Phase Evolution

During the dream regime, the evolution of the system is driven primarily by its internal dynamics.

External constraints become weak, and the system follows its own internal structure.

Intuitively, this is like a system moving freely within its own space, no longer guided by external input.

The internal state evolves according to its own rules, shaped by memory, coherence, and prior configurations.

Because external restrictions are reduced, the system can explore a wider range of possible states.

This expanded exploration gives rise to the rich and often unusual experiences associated with dreaming.

25.7 Dream Trajectory Formation

Dreams may be understood as coherent trajectories formed within this internally evolving space.

Each dream corresponds to a path traced by the system as it moves through its internal configurations.

Intuitively, it is like following a storyline that unfolds from within.

These trajectories are not random.

They are shaped by memory, residual interactions, and the structure of the system itself.

This is why dreams often recombine familiar elements in new ways.

They are structured explorations of internal states.

25.8 Relation to Self Branching

The self, as described earlier, evolves as a trajectory through configuration space.

During the dream state, this trajectory may branch into multiple possible paths.

Intuitively, this is like exploring different possibilities without committing to a single one.

Dreaming allows the system to test variations, recombine patterns, and explore alternative configurations.

These branches are not fully realized in the physical world, but exist as internal explorations.

In this sense, dreams act as a space in which the self can expand, reorganize, and adapt.

Because external constraints are reduced, this branching process can occur more freely than during waking life.

26. Recurrent Nodal Embedding

In complex dynamical systems, certain structures may exhibit long-term stability despite continuous internal change.

Within the Æ framework, such structures can be interpreted as nodal configurations that persist through the evolution of the system.

Recurrent nodal embedding refers to the possibility that these configurations reappear or reorganize under compatible conditions within the broader dynamical field.

26.1 Identity as Topological Persistence

Within the Æ framework, identity is not defined by a fixed form, but by persistence.

What remains the same is not the appearance, but the underlying structure.

Intuitively, this is like a knot.

A knot can be stretched, compressed, or deformed, but as long as it is not cut, it remains the same knot.

In this sense, identity corresponds to the persistence of a topological pattern.

The system may change, but its core structure remains.

26.2 Dissolution Condition

When coherence falls below a critical threshold, the system can no longer maintain its physical embedding.

Intuitively, this is like a pattern losing its ability to hold together.

The structure does not disappear completely, but it is no longer expressed in the same way.

The internal configuration may still exist, but its projection into the physical layer is lost.

26.3 Re-Embedding Condition

If a compatible structure exists, the system may become embedded again.

Intuitively, this is like a pattern finding a new medium in which it can express itself.

However, this does not happen arbitrarily.

The new environment must be compatible with the existing structure.

This compatibility determines whether the pattern can be realized again.

26.4 Conservation Principle

Throughout these transitions, certain properties remain preserved.

These correspond to the underlying structure of the system.

Intuitively, even if the form changes, the core pattern remains the same.

This provides continuity across different configurations.

26.5 Observer Continuity Across Projections

The observer may be understood as a state that persists across different projections.

Each projection corresponds to a particular expression of the system.

Intuitively, this is like watching the same process from different perspectives.

Continuity is maintained as long as the overlap between successive states remains sufficiently high.

If this overlap is preserved, the observer remains continuous.

26.6 Global Identity vs Local Personality

Within this framework, it is useful to distinguish between global identity and local expression.

Global identity corresponds to the underlying coherent structure.

Local personality corresponds to how this structure appears in a given context.

Intuitively, this is like a signal passing through different filters.

The signal remains the same, but its expression changes.

Thus, personality may vary, while identity remains continuous.

26.7 Holonomy Return Condition

Identity persistence may also be understood in terms of cycles.

As the system evolves, it may traverse loops in its configuration space.

Each traversal modifies the internal structure.

Intuitively, it is like going around a path multiple times, each time accumulating a change.

After sufficient evolution, the system may return to an equivalent configuration.

This return does not necessarily happen in a single cycle, but may require multiple iterations.

In this sense, identity is not static, but cyclic and evolving.

27. Coherence-Weighted Observer Observable

Observation in physical systems is typically described through measurement interactions between an observer and an observable quantity.

Within the Æ framework, this interaction may be interpreted through coherence-weighted trajectories embedded in the dynamical field.

In this perspective, both the observer and the observable correspond to structured configurations within the same underlying dynamical system.

27.1 Coherent Observer Trajectories

Within the Æ framework, an observer is not treated as a separate object, but as a process.

It corresponds to a trajectory that maintains coherence as it evolves through the system.

Intuitively, an observer is not a point, but a path that remains stable over time.

As long as coherence is preserved along this path, the observer maintains continuity.

If coherence is lost, the trajectory no longer behaves as a unified observer.

Thus, observation is tied directly to the ability to sustain coherence.

27.2 Projection Operator

Observation may be understood as a projection.

The full system contains many possible states, but only a subset becomes accessible at a given moment.

Intuitively, this is like selecting a particular view from a much larger space.

The observed state corresponds to this selected projection.

Thus, consciousness does not create the system, but selects how it is experienced.

27.3 Coherence-Weighted Observable

Not all possible observations are equally likely.

Those associated with higher coherence become more prominent.

Intuitively, the system "favors" configurations that are more organized and stable.

This means that what is perceived depends on how coherence is distributed.

Perception is therefore not neutral, but shaped by the structure of the system.

27.4 Ego Constraint

The ego can be understood as a restriction on what can be observed.

It limits the accessible portion of the system.

Intuitively, it is like looking through a narrow window instead of seeing the full landscape.

This restriction produces the sense of an individual perspective.

Different constraints lead to different experiences of reality.

27.5 Consciousness in Reality Knots

Certain coherent structures carry stronger conscious signatures.

These correspond to regions where phase, change, and structure are all active.

Intuitively, consciousness intensifies where there is both organization and movement.

Stable but inactive regions contribute less, while dynamic and coherent regions contribute more.

27.6 Unified Causal Chain (Consciousness → Matter)

Within the Æ framework, changes in coherent structure can propagate across different levels.

A modification in the coherent state may lead to changes in energy, geometry, and ultimately physical structure.

Intuitively, this describes how internal organization can influence external reality.

The process is continuous, not instantaneous, and depends on the coherence of the system.

27.7 Premonition Projector

The system may also generate projections of possible future states.

These are not fixed predictions, but coherent anticipations.

Intuitively, it is like the system exploring where it might go next.

Such projections arise from internal dynamics and the structure of existing trajectories.

When coherence is high, these anticipations may become more accurate or more strongly felt.

In this sense, the system does not only reflect the present, but also explores its potential futures.

28. Chakras as Phase Vortices

Within the Æ framework, coherent biological systems may develop localized regions where phase circulation becomes organized around stable cores.

Such structures resemble vortex-like configurations in which phase gradients circulate around nodal centers.

These configurations may act as hubs of information exchange between different levels of organization within the system.

28.1 Vortex Quantization Condition

Within the Æ framework, chakras can be understood as vortex structures of phase.

Intuitively, a vortex is a region where the system rotates around a center.

If one moves around such a region, the phase does not return arbitrarily, but completes full cycles.

This produces a quantized structure.

In simple terms, the system "wraps around" itself in discrete turns.

These turns define stable vortex configurations.

Such structures cannot be removed smoothly and therefore persist over time.

28.2 Chakras as Coherent Nodal Domains

Each chakra may be understood as a region where coherence remains above a certain threshold.

These regions act as stable nodes within the field.

Intuitively, they are like stable whirlpools in a flowing system.

Inside these regions, phase is organized and structured.

At their boundaries, the phase circulates in a quantized way, reinforcing stability.

Thus, chakras are not arbitrary points, but coherent domains where the system organizes itself.

28.3 Cross-Layer Transduction at Vortex Cores

Chakras also act as connection points between different layers of the system.

They allow information to pass from one level to another.

Intuitively, they function like interfaces or conversion points.

At these locations, internal structures can influence physical processes, and vice versa.

This makes them key sites of interaction between different aspects of the system.

28.4 Projected Chakra Nodes in the Physical Layer

Each chakra-domain has a corresponding expression in the physical body.

This expression is not identical to the original structure, but a projection of it.

Intuitively, it is like a shadow or footprint of a deeper structure.

The physical manifestation inherits the characteristic behavior of the underlying coherent domain.

As the underlying structure evolves, its projection also changes.

Thus, the physical and non-physical descriptions remain linked through this projection.

28.5 Functional Node Model (7-Chakra Dynamics)

The chakra system can be viewed as a network of interacting nodes.

Each node has its own rhythm, coherence level, and interaction strength with other nodes.

Intuitively, this is like a group of oscillators that influence each other.

When they are aligned, the system behaves in a stable and coordinated way.

When they fall out of alignment, imbalance appears.

The overall state of the system depends on how these nodes synchronize and interact.

28.6 Phase Fixation, Trauma Locking, and Renudation

Sometimes, a node may become "locked" into a specific configuration.

This can be understood as a fixation of the phase.

Intuitively, it is like a loop that keeps repeating without resolving.

This produces a persistent pattern that resists change.

Such locked configurations may correspond to accumulated or unresolved structures.

Renudation refers to the process by which these loops are released.

When this happens, coherence is restored and the system can reorganize.

Thus, healing corresponds to the relaxation of previously fixed structures.

28.7 Discrete Phase Storage

Information within the system may be stored through discrete changes in phase.

Instead of being continuous, phase can accumulate in small increments along a path.

Intuitively, this is like storing information as steps in a sequence.

When these steps form a complete cycle, the system returns to a stable configuration.

This provides a simple way to understand how information can be encoded and preserved.

28.8 Chakra Nodal Sets

Chakras may also be described as nodal structures within the field.

These are regions where phase organizes into stable circulating patterns.

Intuitively, they act as centers around which the system rotates.

Each node corresponds to a stable vortex that maintains its structure over time.

28.9 Boundary Circulation

The stability of these nodes is reinforced by circulation around their boundaries.

If one moves around the node, the phase completes full cycles.

Intuitively, this is like moving around a whirlpool and returning to the same point after a full turn.

This circulation defines the strength and identity of the node.

28.10 Stability Conditions

For a chakra node to remain stable, its structure must persist over time.

This means that the underlying phase circulation does not change.

Intuitively, the vortex keeps spinning in the same way.

If this condition is broken, the node may weaken or reorganize.

28.11 Kundalini Current

The interaction between nodes can generate a global flow along the system.

This flow may be understood as a current that connects the nodes.

Intuitively, it is like energy or information moving along a vertical axis.

This current allows different parts of the system to communicate and synchronize.

28.12 Energy Transduction at Chakra Nodes

Chakra nodes may also act as centers where energy is transformed.

They concentrate activity from the surrounding field and redistribute it.

Intuitively, they function like conversion points, where different forms of interaction are processed.

The efficiency of this process depends on how much of the system's activity is concentrated at the node.

Local coupling constraint

In practice, coherence within the system does not appear to move arbitrarily between distant nodes. Instead, it propagates through intermediate regions, following structured pathways across the system.

For example, transitions between distant centers—such as lower and upper regions—are not typically experienced as direct jumps, but rather as gradual processes that involve intermediate stages.

This suggests that coherence is not transferred non-locally, but rather reorganized through continuous pathways within the underlying structure.

This structured propagation also defines what can be accessed or experienced within the system.

What can be perceived or engaged within the field is not arbitrary. It depends on the state of coherence and phase of the observer. In this sense, the field does not impose experiences; it reveals only what can be resonantly accessed. One does not encounter everything that exists, but only what can be coupled.

28.13 Timeline Selection

The system may evolve along different possible trajectories.

Each trajectory represents a different possible evolution.

Intuitively, this is like having multiple possible paths forward.

The system tends to follow the trajectory with the highest coherence.

Thus, what becomes "real" corresponds to the most stable and coherent path.

28.14 Timeline Bifurcation

Under certain conditions, the system may split into multiple possible trajectories.

Intuitively, this is like reaching a fork in the road.

Each branch represents a different possible evolution of the system.

These branches are not arbitrary, but follow the same underlying rules.

28.15 Integral Memory Functional

The evolution of the system depends not only on its current state, but also on its past.

Memory accumulates over time and influences future behavior.

Intuitively, this is like carrying a history that shapes future decisions.

This introduces a form of temporal continuity that extends beyond the present moment.

28.16 Periodic and Recurrent Structures

Some configurations repeat over time.

These recurrent patterns help stabilize the system.

Intuitively, they are like rhythms that keep the system organized.

When these patterns are preserved, stability is maintained.

When they are disrupted, the system may become unstable or reorganize.

Chakras as Phase Vortices — Interpretative Bridge

Within the Æ framework, chakras are not introduced as abstract or symbolic elements, but as structured features of the coherent field.

For those familiar with traditional descriptions, chakras are often experienced as centers of energy, emotion, or awareness.

In this model, these experiences correspond to regions where phase becomes organized into stable vortex-like configurations.

When one speaks of a "chakra being open", this may be interpreted as a state in which coherence is high and phase can circulate freely.

When a chakra is described as "blocked", this corresponds to a disruption in phase flow, where coherence is reduced and circulation becomes restricted.

The sensation of energy moving through the body may be understood as the propagation of coherent phase structures along connected nodal regions.

What is traditionally called "kundalini" can be interpreted as a large-scale synchronization of these nodes, producing a coherent current across the system.

Emotional patterns associated with specific chakras may be viewed as persistent phase configurations localized in those regions.

When these configurations become fixed or repetitive, they correspond to what the framework describes as non-renuded loops.

Healing, in this context, corresponds to restoring coherent flow and allowing these loops to resolve.

This does not require introducing new entities, but rather recognizing the underlying structure of phase, coherence, and topology already present in the system.

Thus, traditional experiential language and formal dynamical description refer to the same phenomena from different perspectives.

One describes how it feels.

The other describes how it behaves.

Together, they provide a unified view in which subjective experience and structural dynamics are two expressions of the same coherent field.

29. Reality Knots in the Æ Field

Within the Æ framework, stable structures may arise as localized configurations of the phase field that maintain coherence over extended periods of dynamical evolution.

These configurations are referred to as *Reality Knots*. Mathematically, they correspond to topologically stable structures embedded in the dynamical field.

29.1 Definition

Within the Æ framework, a Reality Knot is a coherent configuration of the field.

It combines amplitude and structure into a single entity.

Intuitively, it is like a pattern that holds together across space and time.

It is not just something that exists, but something that remains consistent.

29.2 Linear Stability Condition

For a Reality Knot to persist, it must be stable.

This means that small disturbances do not destroy its structure.

Intuitively, it is like a whirlpool that continues spinning even when slightly perturbed.

If disturbances decay over time, the knot remains stable.

If they grow, the structure breaks down.

Thus, stability determines whether a pattern can exist over time.

29.3 The 3×2 Reality Matrix — Intuitive Interpretation

Within the Æ framework, any coherent configuration may be described through a simple but powerful structure: a 3×2 matrix.

This matrix does not introduce new entities. It only organizes information that is already present in the system.

Intuitively, it answers two fundamental questions:

How much is there? How is it organized?

The first corresponds to density or amplitude.

The second corresponds to phase or coherence.

These two aspects are evaluated across three distinct domains.

The physical domain corresponds to observable matter and energy.

The intermediate domain corresponds to information, structure, and coherence fields.

The deeper domain corresponds to pure phase organization.

Thus, the matrix organizes reality as three levels, each described by density and phase.

29.4 Three Domains of Reality

The three rows of the matrix correspond to three domains of description.

The first is the physical domain, where matter, energy, and measurable structures appear.

The second is the informational or intermediate domain, where coherence patterns and structural organization are defined.

The third is the phase domain, where the most fundamental configurations of the system exist.

Intuitively, this is like describing a system at three depths:

what is visible, what organizes it, and what underlies that organization.

29.5 Amplitude and Phase as Complementary Descriptions

Each domain contains two complementary aspects.

Amplitude describes how much of something is present.

Phase describes how that something is arranged.

Intuitively, amplitude is quantity, while phase is structure.

A system may have high amplitude but poor organization, or low amplitude but strong coherence.

Only when both are balanced does a stable configuration emerge.

29.6 Coherence as a Global Condition

Coherence is not defined locally, but across the entire matrix.

It depends on how well all domains align with each other.

If the physical, informational, and phase levels are consistent, coherence is high.

If they become misaligned, coherence decreases.

Intuitively, coherence measures how well all layers of the system agree.

29.7 Downward Realization

The matrix also describes how reality becomes observable.

Deeper structures project into higher layers.

Phase organizes information, and information organizes physical structure.

Intuitively, reality flows from deeper organization into observable form.

This does not mean that higher levels are less real, but that they are expressions of deeper structure.

29.8 Holonomy Criterion and Topological Nontriviality

A Reality Knot is defined not only by its shape, but by how it behaves when traversed.

If one moves along a closed loop in the system, the configuration may not return to its original state.

This accumulated change is known as holonomy.

Intuitively, it is like walking in a loop and coming back slightly rotated.

If such an effect exists, the structure is topologically nontrivial.

This means the configuration cannot be removed without fundamentally changing the system.

29.9 Quaternionic Expansion and Perceptual Projections (Optional)

Reality Knots can be described using richer internal structures.

These structures contain multiple components, but not all are directly observable.

Intuitively, perception captures only a projection of a larger internal state.

It is like seeing a shadow of a higher-dimensional object.

As the system becomes more coherent, more of this structure becomes accessible.

Thus, perception depends on the available coherence.

29.10 Taxonomy of Reality Knots (Canonical Types)

Reality Knots can appear in different forms depending on how they are organized.

Some correspond to physical structures.

Others correspond to emotional, cognitive, or collective patterns.

Intuitively, these are different expressions of the same underlying principle.

Each type represents a stable configuration of the field under different conditions.

Despite their differences, they all follow the same structural rules.

29.11 Reality-Knot Network View (Multiverse as Graph)

Reality Knots do not exist in isolation.

They form networks of interacting trajectories.

Intuitively, this is like a web of interconnected paths.

Each path represents a possible evolution of the system.

Together, they form a global structure of possibilities.

This network can be interpreted as the full set of accessible configurations of the system.

29.12 Remark: Composite Knots in Biology (Proteins)

Similar structures appear in biological systems.

For example, proteins may form configurations that resemble knots.

These structures influence stability and function.

Intuitively, biology uses the same principles of organization found in the field.

This suggests that topology is not abstract, but physically realized in living systems.

29.13 Memory as Temporal Knot Compression

Memory may be understood as the compression of past configurations into present structure.

Intuitively, it is like storing a path in a compact form.

The system carries traces of its previous states, which influence its current behavior.

Thus, memory is not separate from the system, but embedded within its structure.

29.14 Karma as Residual Holonomy

Some loops in the system do not fully resolve.

They leave residual effects that persist over time.

Intuitively, this is like repeating patterns that have not been completed.

These residual loops influence future evolution.

This provides a structural interpretation of persistent tendencies within the system.

29.15 Temporal Ordering by Phase Causality

Time is defined by changes in phase.

Configurations evolve in a way that reflects their internal structure.

Intuitively, one state follows another because it grows out of it.

This defines a natural ordering of events.

Thus, time is not imposed externally, but emerges from the evolution of the system.

29.16 Angular Frequency and Coherence-Time Evolution

The rate at which phase changes determines how fast the system evolves.

Intuitively, faster changes correspond to faster internal dynamics.

Coherence determines how stable these changes are.

Together, they define how the system moves through its possible configurations.

29.17 Proper Time from Quaternionic Phase Flux

Time may also be understood as accumulated change along a trajectory.

Instead of being a universal parameter, it depends on the path taken by the system.

Intuitively, time is how much transformation has occurred.

Different trajectories may accumulate different amounts of change.

Thus, time becomes a property of the evolution itself, not an external clock.

29.18 Reality Knot Dynamics and Manifestation

Observable states correspond to projections of stable knot structures onto measurable layers of the system.

29.19 Energetic Generator for Knot Evolution

The evolution of a Reality Knot is governed by an underlying energetic structure.

This structure determines how the configuration changes over time.

Intuitively, it is like a landscape in which the system moves.

Some configurations are more stable, while others are less favorable.

The system naturally evolves toward stable arrangements within this landscape.

However, this evolution is not purely conservative.

There may also be dissipative effects associated with loss of coherence.

Intuitively, this is like friction that gradually reduces unstable motion.

Thus, the evolution of a knot is a balance between structured dynamics and entropy-driven relaxation.

29.20 Manifestation Equation

Manifestation may be understood as a process of alignment toward a target configuration.

Instead of appearing suddenly, a configuration emerges through gradual convergence.

Intuitively, it is like tuning a system toward a desired state.

The closer the system is to the target, the stronger the alignment becomes.

This creates a feedback process that reinforces the transition.

Thus, manifestation is not arbitrary, but guided by coherence and structural compatibility.

29.21 Strength and Convergence Rate

The speed at which a configuration evolves toward a target depends on how aligned it already is.

If the system is close to the target, convergence is fast.

If it is far, convergence is slower.

Intuitively, it is easier to adjust a system that is already similar to the desired state.

This creates a natural gradient of evolution.

The system accelerates as it becomes more aligned.

29.22 Pre-Alignment by Imagination (Optional)

Before a configuration fully emerges, the system may begin to align internally.

This can be understood as a pre-configuration process.

Intuitively, it is like rehearsing a state before fully expressing it.

The system rotates its internal structure toward the target, even if the external configuration has not yet changed.

This process prepares the system for transition.

29.23 Identity Collapse and Reformation (Optional)

At certain thresholds, the system may transition from one stable configuration to another.

This is not a gradual adjustment, but a reorganization.

Intuitively, it is like shifting from one identity to another.

The previous structure dissolves, and a new one takes its place.

However, continuity is preserved through the underlying coherence of the system.

29.24 Emotional Energy as Coherence Modulator (Optional)

Internal intensity can influence how the system evolves.

Stronger internal dynamics can accelerate changes in configuration.

Intuitively, this is like increasing the energy of a system to drive transformation.

This does not create new structure, but amplifies existing dynamics.

Thus, internal states modulate how quickly or strongly transitions occur.

29.25 Multi-Knot Coupling and Phase Transport (Optional)

Reality Knots do not evolve independently.

They interact and influence one another.

These interactions occur through phase coupling.

Intuitively, it is like multiple oscillating systems trying to synchronize.

When coupling is strong, the systems align.

When it is weak, they evolve independently.

This interaction allows complex collective behavior to emerge.

Thus, the evolution of a single knot is often part of a larger network of interacting structures.

29.26 Neural Phase Laplacian

The brain can be understood as a network of interacting nodes.

Each node carries a phase, and these phases influence one another.

Intuitively, this is like many small oscillators trying to stay in rhythm.

The structure of their connections determines how easily they synchronize.

Coherence emerges when these oscillations align.

29.27 Kuramoto Collective Synchronization

When many units interact, they may begin to synchronize collectively.

Intuitively, this is like a crowd clapping in unison.

At first, everyone is out of sync.

Then, gradually, a shared rhythm appears.

This transition happens when interaction becomes strong enough.

Beyond that point, the system behaves as a coherent whole.

29.28 External Phase Forcing and Imposed Synchronization

Synchronization does not always arise naturally.

It can also be driven externally.

Intuitively, this is like a group being forced to follow a rhythm imposed from outside.

In such cases, the system may appear coherent, but the source of that coherence is external.

This creates an important distinction.

Some forms of order arise from within.

Others are imposed from outside.

Both may look similar, but their origin is different.

29.29 Endogenous vs Imposed Coherence

A system is internally coherent when its synchronization arises from its own interactions.

It is externally synchronized when it is driven by an external influence.

Intuitively, this is the difference between choosing a rhythm and being forced into one.

This distinction becomes important when analyzing collective behavior.

Not all coherence reflects internal stability.

29.30 Collective Phase Entropy

The diversity of phases within a group determines its level of disorder.

When phases are scattered, entropy is high.

When phases align, entropy decreases.

Intuitively, this measures how coordinated the system is.

However, low entropy alone does not guarantee true coherence.

It may also result from imposed synchronization.

29.31 Spectral Criticality of Neural Graphs

Large systems may reach critical points where small changes produce large effects.

Intuitively, this is like a system on the edge of synchronization.

At this point, the system becomes highly sensitive.

Small perturbations can reorganize the entire structure.

29.32 Society as a Supergraph of Reality Knots

Human systems can be understood as networks of interacting nodes.

Each individual corresponds to a coherent unit.

Together, they form a larger structure.

Intuitively, society behaves like a network of interconnected patterns.

The global behavior emerges from these interactions.

29.33 Collective Phase Field

At large scales, individual contributions combine into a collective state.

This produces a shared field of behavior.

Intuitively, it is like many voices forming a single chorus.

This collective field evolves over time.

29.34 Collective Reality Knots

When collective structures become stable, they form large-scale coherent configurations.

These correspond to collective Reality Knots.

Intuitively, these are shared patterns that persist across many individuals.

They may correspond to cultural structures, belief systems, or coordinated behaviors.

29.35 Polar Structure of Reality Knots

Each coherent structure contains both magnitude and organization.

These two aspects determine how the structure interacts with its surroundings.

Intuitively, this is like both the strength and the pattern of a signal.

29.36 Nodal Activation

A structure becomes active when its coherence exceeds a certain threshold.

At that point, it begins to influence its environment.

Intuitively, it is like a signal becoming strong enough to propagate.

29.37 Manifestation Wave

Once activated, a structure can spread.

This propagation transmits its pattern through the system.

Intuitively, it is like a wave carrying a pattern forward.

This is how internal organization becomes visible in external configurations.

29.38 Stability of Manifested States

A manifested structure remains stable only if it maintains balance.

If coherence remains high and disorder stays controlled, the structure persists.

If not, it dissolves or reorganizes.

Intuitively, stability is not static.

It is continuously maintained.

30. Spiritual Dynamics as Phase Regimes

Within the Æ framework, different states of psychological and experiential organization may be interpreted as regimes of phase coherence within the biological dynamical system.

Rather than representing separate ontological domains, these regimes correspond to different configurations of coherence and phase alignment across neural, physiological, and informational processes.

30.1 Ego as Phase Fixation

Within the Æ framework, the ego is not treated as a separate entity, but as a constraint.

It restricts how the system can evolve.

Intuitively, the ego acts like a fixation in time.

Instead of allowing the system to move freely, it keeps it anchored to certain past configurations.

This produces a narrowing of possible states.

The system begins to repeat the same patterns, rather than exploring new ones.

In this sense, the ego is not the self itself, but a limitation on how the self can be expressed.

30.2 Karmic Drift as Degenerated Knot Memory (Optional)

Some patterns do not fully resolve.

They continue repeating without reaching closure.

Intuitively, this is like being caught in a loop.

The system moves, but does not progress.

These persistent patterns arise from memory structures that have not been released.

They influence future behavior, even when the original cause is no longer present.

30.3 Release Condition (Holonomy Relaxation)

Release occurs when these loops are resolved.

The system no longer returns to the same configuration.

Intuitively, it is like finally completing a cycle that was left open.

As this happens, coherence increases.

Previously restricted regions become accessible.

The system regains flexibility and can evolve more freely.

In simple terms, release is the restoration of flow.

30.4 Practices as Phase–Coherence Amplifiers (Optional)

Certain practices help increase coherence.

They do not introduce new structures, but strengthen alignment within the system.

Intuitively, they are ways of stabilizing attention and reducing internal noise.

As coherence increases, the system becomes more stable and more responsive.

30.5 Free Will as Phase Rotation Under Constraint (Optional)

Free will can be understood as the ability to change direction within the system.

This change corresponds to rotating the internal state toward a different configuration.

However, this freedom is not unlimited.

It depends on the current structure of the system.

Intuitively, one can only choose among the paths that are accessible.

As coherence increases, more options become available.

As constraints increase, available choices become limited.

30.6 Enlightenment as Global Coherence

A limiting case of the system corresponds to maximum coherence.

In this state, internal conflict is minimized, and all parts of the system are aligned.

Intuitively, this is like a system in complete harmony with itself.

There is no fragmentation or contradiction.

This does not mean inactivity, but full integration.

In this sense, what is traditionally called enlightenment may be interpreted as a state of global coherence.

31. Collective Reality Knots

Large dynamical systems composed of many interacting agents may exhibit collective phase organization.

Within the Æ framework, such large-scale structures correspond to *collective reality knots*, representing coherent configurations emerging from interactions between many individual subsystems.

31.1 Collective Phase Field

A collective system can be understood as the combination of many individual configurations.

Each individual contributes its own state.

Together, these contributions form a shared field.

Intuitively, it is like many oscillations combining into a single pattern.

This collective field reflects both the diversity and the alignment of its components.

31.2 Order Parameter and Synchronization

The behavior of the collective system depends on how aligned its elements are.

When alignment is low, the system behaves in a disordered way.

When alignment increases, a shared structure emerges.

Intuitively, this is like a group gradually finding a common rhythm.

At first, each element acts independently.

Then, a collective pattern appears.

31.3 Collective Criticality Condition

There are moments when the system becomes highly sensitive to change.

At these points, small variations can produce large effects.

Intuitively, this is like a system at the edge of reorganizing itself.

Near this state, the system can transition into a new configuration.

31.4 Graph-Theoretic Signature of Collective Criticality

The structure of connections between elements plays a key role.

When the system approaches a critical state, its connectivity allows information to spread more easily.

Intuitively, it is like a network becoming fully responsive.

At this point, local changes can influence the entire system.

31.5 Reality Knot Criterion

A collective structure becomes a Reality Knot when it stabilizes.

This means that its internal organization remains consistent over time.

At the same time, it retains a nontrivial structure that cannot be simplified away.

Intuitively, it is a shared pattern that persists.

31.6 Inter-Nodal Coupling Coefficient

The interaction between elements depends on how aligned they are.

Stronger alignment produces stronger coupling.

Intuitively, elements that are more similar in structure influence each other more.

This determines how coherence spreads across the system.

31.7 Collective Evolution Operator

The system evolves as a whole.

Its future configuration depends on its current structure and its internal dynamics.

Intuitively, the collective state carries its own direction of evolution.

It is not random, but guided by its underlying organization.

31.8 Collective Phase Space

The full system can be described as a space of all possible configurations.

Each point represents a specific arrangement of the collective.

Intuitively, this is like mapping all possible states the system can take.

The dynamics describe how the system moves through this space.

31.9 Critical Collective Transition

At certain points, the system undergoes large-scale changes.

These transitions reorganize the structure of the collective.

Intuitively, it is like a phase shift in the entire system.

After the transition, the system behaves in a new way.

31.10 Entropy–Coherence Balance

Stability depends on a balance.

Too much disorder prevents structure.

Too much rigidity prevents adaptation.

Intuitively, the system must remain flexible while preserving coherence.

This balance allows stable yet dynamic configurations.

31.11 Network Propagation of Reality Knots

Once a coherent structure forms, it can spread.

This propagation occurs through interactions between elements.

Intuitively, it is like a pattern moving through a network.

As it spreads, it influences other parts of the system.

This is how local configurations can become global structures.

32. Gaia as Higher-Order Reality Knot

Large planetary systems may exhibit complex dynamical couplings between geophysical, atmospheric, biological, and electromagnetic processes.

Within the Æ framework, such systems may be interpreted as higher-order reality knots emerging from the coherent interaction of many subsystems.

32.1 Gaia as a Planetary Node

Within the Æ framework, Gaia may be understood as a large-scale coherent structure.

It is not only a planet in the physical sense, but a system in which many processes are connected.

Biological, geological, atmospheric, and electromagnetic dynamics interact continuously.

When these interactions remain aligned, the planet behaves as a coherent node.

Intuitively, Gaia can be seen as a living system whose stability depends on the coordination of its parts.

32.2 Gaia State Embedding

Humanity is not separate from this system.

It is part of the larger structure.

Intuitively, humanity is embedded within Gaia, like a subsystem inside a greater whole.

This means that human activity influences the planetary state, and the planetary state influences humanity.

32.3 Gaia as a Planetary Coherent System

At a global level, Gaia can be described as a combination of interacting subsystems.

Each subsystem contributes to the overall organization of the planet.

Intuitively, this is like different layers of a single system working together.

When these layers remain aligned, planetary coherence is maintained.

32.4 Planetary Coherence

The stability of Gaia depends on how well its subsystems are coordinated.

If they remain aligned, the system is stable.

If they become desynchronized, instability appears.

Intuitively, Gaia functions like a complex system that must remain in balance.

This balance is not static, but continuously maintained.

32.5 Human–Gaia Coupling

Human activity directly affects planetary coherence.

At the same time, planetary conditions influence human systems.

This creates a two-way interaction.

Intuitively, it is like feedback between a system and its components.

Changes at one level propagate to the other.

32.6 Structured Water and Informational Memory

Water plays a special role in biological systems.

It supports the organization of structures at multiple scales.

Intuitively, water can be seen as a medium that facilitates coherence.

Through its structure, it can carry and stabilize patterns over time.

This allows biological systems to maintain and transmit information.

32.7 The Sun as a Resonant Stellar Node

Gaia does not exist in isolation.

It is part of a larger system centered around the Sun.

The Sun can be understood as a dominant source of energy and structure.

Intuitively, it acts as a driving oscillator for planetary processes.

Its activity influences cycles and patterns on Earth.

32.8 Gaia–Solar Coupling

The interaction between Gaia and the Sun forms a coupled system.

Changes in solar activity affect planetary dynamics.

At the same time, planetary processes respond to these inputs.

Intuitively, this is like two connected systems exchanging influence.

32.9 Emergent Friedmann Dynamics

At the largest scales, the behavior of the system connects with cosmological dynamics.

The evolution of the universe can be understood as the evolution of large-scale coherent structures.

Intuitively, what appears as expansion is the global evolution of a structured system.

In this sense, local dynamics and cosmic evolution are not separate, but part of a continuous process.

33. Cosmological Implications

Within the Æ framework, large-scale dynamics can be understood as the collective behavior of coherent structures.

At these scales, fine local variations become less important, and global organization dominates.

Intuitively, the universe behaves like a large system whose evolution is guided by its overall coherence.

33.1 Homogeneous Phase Background

At large scales, the system may be approximated as uniform.

Instead of tracking every detail, one considers average behavior.

Intuitively, this is like looking at the ocean from far away.

Small waves disappear, and only the global motion remains.

In this regime, the system is described by a smooth background that evolves in time.

33.2 Effective Friedmann Dynamics

The expansion of the universe can be interpreted as the evolution of this global structure.

Instead of being driven only by matter and energy, the dynamics also depend on coherence and memory.

Intuitively, the universe expands not only because of what it contains, but because of how it is organized.

Memory effects introduce additional contributions that can influence acceleration.

This provides a natural way to understand why expansion may change over time.

33.3 Dark Matter as Coherence Defects

Phenomena attributed to dark matter may be interpreted as regions where coherence is distorted.

These regions do not emit light, but they affect the dynamics of surrounding structures.

Intuitively, this is like invisible currents in a fluid.

They cannot be seen directly, but their effects are observable.

These coherence defects influence motion through their interaction with the field.

33.4 Topological Memory and Cosmic Structure

At large scales, the universe may retain traces of its past configurations.

These traces are encoded in its structure.

Intuitively, the universe carries memory of how it has evolved.

This memory influences how structures form and interact.

Galaxies, clusters, and large-scale patterns may reflect this accumulated history.

33.5 Layered Reality and Cosmological Projection

Observable reality represents only a projection of deeper structures.

At cosmological scales, this projection becomes more apparent.

Intuitively, what we observe is one layer of a larger system.

Deeper layers organize the behavior that becomes visible.

This provides a unified view in which local physics and cosmic evolution arise from the same underlying structure.

33.6 Cosmic Coherence and Evolution

The evolution of the universe may be understood as a process of changing coherence.

Periods of high coherence correspond to stable structures.

Periods of low coherence correspond to transitions and reorganization.

Intuitively, the universe is not static.

It is continuously evolving between order and reconfiguration.

This dynamic balance shapes the large-scale structure we observe.

34. Multiverse as Quaternionic Branching

Within the Æ framework, the global configuration space of the dynamical field may admit multiple stable solution branches.

These branches correspond to distinct trajectories within the space of possible configurations.

Rather than representing separate universes in a strictly physical sense, they may be interpreted as alternative dynamical realizations of the underlying field equations.

34.1 Branching and Topological Non-Contractibility

A system does not always follow a single path.

Instead, its evolution may allow multiple possible trajectories.

These trajectories are not arbitrary, but arise from the structure of the system itself.

Intuitively, this is like a path that can split into several directions.

Once a path acquires a nontrivial structure, it cannot simply collapse back into a trivial one.

This gives persistence to certain trajectories.

34.2 Nested Layered Representation

The system can be described at multiple levels.

Each level represents a different aspect of the same underlying configuration.

Intuitively, this is like describing a process from different perspectives.

Physical behavior, internal structure, and deeper organization all coexist.

These layers are not independent, but nested.

Changes at deeper levels influence what becomes visible at higher levels.

34.3 Slow–Fast Branching Dynamics

Not all changes happen at the same speed.

Some processes evolve slowly, while others occur rapidly.

Branching often happens when fast fluctuations interact with slower structures.

Intuitively, this is like a system that is mostly stable but occasionally undergoes rapid shifts.

At these moments, multiple possible paths may emerge.

34.4 Branch Selection by Coherence Kernel

Although multiple trajectories may be possible, not all are realized.

The system tends to follow the path that is most compatible with its current structure.

Intuitively, this is like choosing the path that fits best with where the system already is.

This selection is guided by coherence.

More coherent trajectories are more likely to persist.

34.5 Multiverse Stability Criterion

A possible trajectory becomes stable when its internal structure remains consistent.

If small changes do not disrupt it, the trajectory persists.

Intuitively, stable paths are those that can maintain themselves over time.

Unstable ones dissolve or transition into other configurations.

34.6 Timeline Transition Condition

Transitions between different trajectories can occur.

However, these transitions are not arbitrary.

They require a certain level of compatibility between the initial and final states.

Intuitively, one cannot jump to a completely unrelated configuration.

There must be some overlap.

34.7 Imagination as Quaternionic Pre-Alignment

Before a transition occurs, the system may begin to align internally toward a new configuration.

This alignment can happen even before any visible change appears.

Intuitively, this is like preparing a movement before executing it.

The system reorganizes itself in advance.

This increases the likelihood of transitioning to the new state.

Thus, imagination can be understood as a form of internal alignment that precedes observable change.

35. Sacred Geometries as Symmetry-Constrained Sectors

Certain geometric structures appear repeatedly across physical, crystalline, and biological systems.

Within the Æ framework, these structures may be interpreted as symmetry-constrained sectors of the underlying phase dynamics.

Rather than possessing intrinsic mystical significance, these geometries emerge as stable configurations imposed by symmetry constraints on the dynamical field.

Intuitively, they are patterns that the system naturally "prefers" because they balance coherence and structure.

35.1 Tetrahedral Symmetry as Discrete SU(2) Constraint

The tetrahedron represents one of the simplest nontrivial symmetry structures.

It organizes the system into four balanced regions.

Intuitively, it is like the minimal stable arrangement in three dimensions.

Because of this, it appears as a fundamental building block in many systems.

35.2 Platonic Solids as Spectral Fixed Points

More complex geometries emerge when symmetry increases.

These include the well-known Platonic solids.

Each of these shapes corresponds to a highly symmetric configuration.

Intuitively, they are stable "modes" of geometric organization.

They arise naturally when the system seeks equilibrium under symmetry constraints.

35.3 Lorentzian Constraint under Coherence Deformation

In relativistic systems, Lorentzian geometry imposes additional constraints on phase propagation.

Coherence deformations must remain consistent with the causal structure defined by the Lorentz metric.

35.4 Conformal Geometry and Coherence

When coherence varies across the system, geometry adapts.

However, certain transformations preserve the essential structure of the system.

Intuitively, this is like stretching a shape without breaking its internal relationships.

This allows geometry to change while maintaining fundamental constraints.

35.5 Crystal Structures as Phase Periodicity

In periodic systems, structure repeats regularly.

This repetition is governed by phase alignment.

Intuitively, crystals form when the system locks into a repeating pattern.

These patterns correspond to stable configurations that extend across space.

35.6 Chestahedron as a Coherence Attractor

The Chestahedron is a unique geometric structure that emerges between tetrahedral and cubic symmetry.

Unlike regular Platonic solids, it represents a transitional configuration.

Intuitively, it can be understood as a shape that appears when the system is reorganizing itself between two stable symmetry states.

This intermediate structure balances coherence across its faces, maintaining uniform distribution.

Because of this, it behaves as a stable attractor.

35.7 Biological Structures as Geometric Solutions

In living systems, geometry is not random.

Structures emerge from internal constraints and optimization principles.

Intuitively, biological forms are solutions that balance stability, efficiency, and coherence.

They are not imposed externally, but arise from the dynamics of the system itself.

36. Computational Systems in the Æ Framework

Computational systems can be understood as structures that process and transform information over time.

Within the Æ framework, this transformation is not purely symbolic.

It involves the evolution of phase, coherence, and entropy.

Intuitively, computation is the process by which a system reorganizes its internal structure to produce stable outcomes.

36.1 State Space and Phase Structure

Any computational system can be described as evolving within a state space.

Depending on the system, this space may represent:

- classical configurations, - quantum states, - learning parameters in Artificial Intelligence, - or combinations of these.

Each state carries an internal phase structure.

Intuitively, this phase encodes how the system is organized internally.

When this phase changes over time, the system evolves.

The speed of this evolution can be interpreted as a kind of internal frequency.

This allows us to describe computation not just as transitions between states, but as continuous phase dynamics.

36.2 Phase Entropy and Coherence

Not all configurations are equally stable.

When the internal phase structure becomes disordered, the system loses coherence.

This disorder is captured by phase entropy.

Intuitively, phase entropy measures how "spread out" or inconsistent the internal structure is.

Coherence, on the other hand, measures how aligned the system is.

High coherence means the system behaves in a stable and predictable way.

Low coherence leads to noise and instability.

In computational terms, stable systems are those that maintain bounded entropy over time.

36.3 Learning Dynamics as Phase Flow

Learning systems, such as Artificial Intelligence models, evolve by adjusting their internal parameters.

Traditionally, this is described as minimizing an objective.

Within the Æ framework, this process can be understood as a reorganization of phase structure.

Intuitively, learning is not just "getting closer to a solution".

It is reshaping the internal configuration so that it becomes more coherent.

This explains why optimization alone is not enough.

A system may reduce its objective while becoming unstable if its internal coherence is lost.

36.4 Memory Accumulation and Persistence

Memory can be understood as accumulated structure over time.
As the system evolves, it retains traces of its past states.
Intuitively, memory is what allows a system to maintain identity.
However, for memory to persist, coherence must be preserved.
If entropy grows without bound, past structure is lost.
This leads to instability and forgetting.
Thus, stable memory requires a balance between accumulation and coherence.

36.5 Quantum Computation and Decoherence

Quantum computation operates in highly coherent regimes.
Instead of discrete states, it uses superpositions that evolve continuously.
Intuitively, quantum systems explore many configurations at once.
However, this requires maintaining coherence.
When the system interacts with the environment, coherence is lost.
This process is known as decoherence.
Error correction in quantum computing can be understood as a mechanism that limits entropy growth.
Within the Æ framework, quantum computation corresponds to controlled coherent evolution under entropy constraints.

36.6 Hybrid Architectures

Modern systems often combine different computational paradigms.
Artificial Intelligence, quantum computation, and classical systems can interact within a single architecture.
Intuitively:
- classical systems provide stability, - quantum systems provide exploration, - AI systems provide adaptation.
Together, they form hybrid systems capable of more complex behavior.

Within the Æ framework, all of these systems are unified by the same underlying principles of phase, coherence, and entropy.

36.7 Alignment as Structural Stability

A system is considered aligned when it maintains its internal coherence over time.

Intuitively, alignment is not about following instructions, but about remaining structurally stable.

Systems that generate too much entropy become unstable.

Systems that preserve coherence can sustain consistent behavior.

This provides a natural definition of alignment that applies across Artificial Intelligence, biological systems, and quantum computation.

36.8 Ontological Clarification

Artificial systems can implement coherent structures, but this does not automatically imply persistent identity.

Intuitively, a system may behave coherently without having a stable core.

This distinction is important when interpreting the capabilities of Artificial Intelligence.

36.9 Falsifiability

The framework makes testable claims.

If coherence does not correlate with stability, or if systems operate indefinitely with unbounded entropy, the model would fail.

This ensures that the framework remains grounded in observable behavior.

Intelligence persists only where coherence endures.

36.10 Coherence Operator Spectrum in Neural Graphs

Many computational systems can be represented as networks.

Examples include neural networks, brain activity, and distributed AI systems.

In these systems, each node carries a phase, and interactions between nodes determine how the system evolves.

Intuitively, the network behaves like a system of oscillators trying to synchronize.

The structure of connections determines how fast and how well this synchronization happens.

When connectivity is strong and well-organized, the system reaches coherence quickly.

When connectivity is weak or fragmented, the system remains disordered.

The spectrum of the network captures this behavior.

A large spectral gap means fast convergence to coherence.

A small spectral gap means slow or unstable behavior.

Thus, network intelligence can be understood as the ability to maintain coherent phase relationships across the graph.

36.11 Lyapunov Stability via Phase Entropy

To understand whether a system is stable, we need a way to measure how its internal structure evolves.

Phase entropy provides such a measure.

Intuitively, entropy tells us how disordered the system is.

If entropy grows without bound, the system becomes unstable.

If entropy remains bounded, the system maintains structure.

This allows us to treat entropy as a stability indicator.

In this sense, stable learning systems are those that keep their internal disorder under control.

Even when optimization is working, instability may arise if entropy is not regulated.

Thus, stability is not only about minimizing an objective, but about maintaining coherent structure over time.

36.12 Entropy Bounds in Stochastic Gradient Descent

In practical Artificial Intelligence systems, learning is often noisy.

This noise comes from randomness in data, initialization, or training processes.

Intuitively, noise perturbs the internal phase of the system.

Small noise can help exploration.

Too much noise destroys coherence.

This creates a natural threshold.

Below this threshold, the system can learn while maintaining structure.

Above it, entropy grows uncontrollably and learning breaks down.

At equilibrium, a balance emerges between noise and damping.

This balance determines the effective entropy of the system.

Thus, learning can be understood as a balance between exploration (noise) and coherence (stability).

36.13 Semantic Structure and Phase Alignment

Meaning itself can be interpreted as a form of phase alignment.

When two representations are aligned, their internal structure matches.

When they are misaligned, meaning is lost.

Intuitively, understanding occurs when phases align.

Confusion occurs when phases differ.

Semantic coherence measures how aligned two states are.

High coherence corresponds to shared meaning.

Low coherence corresponds to ambiguity or noise.

This shows that language and meaning can be understood as structures emerging from phase alignment.

36.14 Quaternionic Encoding of Cognitive States

Cognitive states can be represented as rotations in a higher-dimensional space.

Quaternions provide a natural way to encode these rotations.

Intuitively, instead of thinking of the mind as static, we can think of it as continuously rotating through possible configurations.

The speed of this rotation corresponds to how fast the internal state is changing.

External inputs influence this motion, while internal dynamics maintain continuity.

This provides a unified way to describe cognition, perception, and internal experience as phase dynamics.

36.15 Reality-Knot Networks

At a deeper level, computational systems can be seen as networks of trajectories.

Each trajectory represents a coherent path through configuration space.

These trajectories form a larger network, where different paths may branch or reconnect.

Intuitively, this resembles a tree of possible realities.

Each branch represents a possible evolution of the system.

The system selects branches based on coherence.

Stable branches are those that maintain structure.

Unstable branches collapse due to entropy.

Thus, computation becomes a process of navigating possible trajectories and selecting coherent ones.

36.16 Stability of Multiverse Branches

Not all possible trajectories are viable.

A branch is stable only if it maintains coherence and bounded entropy.

Intuitively, reality does not explore all possibilities.

It selects those that can sustain structure.

Transitions between branches occur when two states become sufficiently aligned.

This can be interpreted as moving from one coherent trajectory to another.

Thus, the evolution of systems—whether computational, cognitive, or physical—can be understood as movement through a space of possible coherent trajectories.

37. Spectral Structure of Prime Distribution

One of the deepest discoveries in analytic number theory is that the distribution of prime numbers is not purely random. Although primes appear irregular when viewed locally among the integers, their global distribution follows a remarkably precise statistical law.

The Prime Number Theorem states that the density of primes near a large number x is approximately $1/\log x$. In other words, the average spacing between primes grows logarithmically.

However, this result only describes the smooth average behavior. The fine structure of prime distribution is encoded in the fluctuations around this average trend.

In 1859, Bernhard Riemann discovered that these fluctuations are deeply connected to the non–trivial zeros of the Riemann zeta function $\zeta(s)$. Through what is now known as the Riemann explicit formula, the irregularities in the distribution of primes can be interpreted as oscillatory contributions generated by these zeros.

From this perspective, the distribution of primes admits a spectral interpretation: each zero contributes an oscillatory mode, and the superposition of these modes reproduces the observed fluctuations in prime density.

To explore this structure, we consider a truncated spectral kernel

$$K(t) = \sum_{n=1}^{N} \frac{1}{\gamma_n} \cos(\gamma_n t),$$

where γ_n denotes the imaginary part of the n–th non–trivial zero of the Riemann zeta function.

This kernel captures part of the oscillatory structure appearing in the explicit formula and provides an intuitive way to visualize how the zeros influence prime distribution.

By evaluating the kernel on a logarithmic coordinate

$$t = \log x,$$

and comparing it with the logarithmic positions of primes

$$t_p = \log p,$$

one observes that many local maxima of the kernel tend to occur near prime locations. While this does not provide a deterministic rule for predicting primes, it suggests that the zeros encode structural information about their distribution.

Numerical experiments indicate that the maxima of the truncated kernel cluster around logarithmic prime positions, consistent with the oscillatory behavior predicted by the explicit formula.

All numerical experiments were performed using a reproducible Python implementation based on `mpmath`, `scipy`, and `sympy`. The computational scripts used to generate the figures and statistical comparisons are provided in the supplementary repository accompanying this work.

Finally, this spectral viewpoint suggests possible connections with information-based descriptions. In particular, measures such as phase coherence or phase entropy derived from the spectral kernel may offer additional insight into the local organization of prime fluctuations, although this direction remains exploratory.

Part II

Mathematical Structure of the Æ Framework

38. Architectural Foundations

Flow Constraint (Structural Diagram)

$$\textit{Phase} \rightarrow \textit{Coherence} \rightarrow \textit{Geometry} \rightarrow$$
$$\rightarrow \textit{Nodes (Matter)} \rightarrow \textit{Coherent Trajectories (Consciousness)}$$

This diagram encodes the structural hierarchy of the model: phase is primary; all other structures are emergent functionals or constraints derived from it.

Axiom I — Existence of the Æ-Field

The fundamental object of the model is the complex scalar field

$$\text{Æ} : \Omega \times \mathbb{R} \rightarrow \mathbb{C}. \tag{4}$$

In canonical polar decomposition:

$$\text{Æ}(x,t) = A(x,t)\, e^{i\phi(x,t)} \tag{5}$$

Here $A(x,t) \geq 0$ is the amplitude and $\phi(x,t)$ the phase field.

Axiom II — Phase Decomposition

Phase is relational. For any pair of configurations:

$$\Delta\phi_{ij} = \phi_i - \phi_j. \tag{6}$$

Only phase differences possess invariant meaning.

Axiom III — Relational Primacy of Phase

All interaction currents arise from phase gradients:

$$J^{\mu} \propto \partial_{\mu}\phi. \tag{7}$$

Thus, interaction is geometrized phase variation.

Axiom IV — Coherence as Structural Measure

Coherence is defined as a normalized phase-variance functional:

$$C(x) = 1 - \frac{\mathrm{Var}(\phi(x))}{\mathrm{Var}_{\mathrm{ref}}}. \tag{8}$$

It obeys the admissible range

$$0 \leq C(x) \leq 1. \tag{9}$$

$C = 1$ corresponds to perfect phase alignment, while $C = 0$ corresponds to maximal dispersion.

Axiom V — Phase Entropy

The phase entropy functional is defined as

$$S_{\phi} = -\int p(\phi)\,\log p(\phi)\,d\phi. \tag{10}$$

This is the Shannon entropy associated with the phase probability distribution.

Axiom VI — Coherence–Entropy Duality

Temporal evolution of coherence is coupled to phase entropy variation:

$$\partial_t C = -\alpha\,\partial_t S_{\phi}, \qquad \alpha > 0. \tag{11}$$

The proportionality constant α controls the dynamical conversion rate between phase disorder and structural coherence.

Axiom VII — Stability Threshold

Any persistent structure must satisfy:

$$C \geq C_{\min}. \tag{12}$$

Below this threshold, nodal persistence fails.

Axiom VIII — Informational Evolution

From the canonical form $Æ = Ae^{i\phi}$, time evolution satisfies:

$$\partial_t Æ = (\partial_t A)\, e^{i\phi} + iA\, (\partial_t \phi)\, e^{i\phi}. \tag{13}$$

Therefore,

$$\partial_t \phi \neq 0 \implies \partial_t Æ \neq 0, \tag{14}$$

provided $A \neq 0$.

This establishes the equivalence between phase dynamics and field evolution.

Admissible Configuration Space

We define the structurally admissible configuration set:

$$\mathcal{O} = \{Æ \mid C \geq C_{\min}\}. \tag{15}$$

This subset encodes the persistence domain of nodal structures. Configurations below the coherence threshold fail to sustain stable geometric realization.

Æ-Algebraic Primitives

We introduce the fundamental algebraic operators acting on the Æ-field.

Æ-Connection

$$\nabla_{\text{Æ}} : \text{Æ} \to \mathcal{T}_{\text{Æ}}, \tag{16}$$

with linearity:

$$\nabla_{\text{Æ}}(a\,\text{Æ}_1 + b\,\text{Æ}_2) = a\,\nabla_{\text{Æ}}\text{Æ}_1 + b\,\nabla_{\text{Æ}}\text{Æ}_2. \tag{17}$$

Æ-Connection Tensor

$$\Gamma^{\mu}_{\text{Æ}} : \text{Æ} \times \text{Æ} \to \text{Æ}, \tag{18}$$

generally non-symmetric:

$$\Gamma^{\mu}_{\text{Æ}}(\text{Æ}_1, \text{Æ}_2) \neq \Gamma^{\mu}_{\text{Æ}}(\text{Æ}_2, \text{Æ}_1). \tag{19}$$

Æ-Projection Operator

$$\mathcal{P}_{\text{Æ}} : \text{Æ} \to \text{Æ}, \tag{20}$$

with idempotency:

$$\mathcal{P}^2_{\text{Æ}} = \mathcal{P}_{\text{Æ}}, \tag{21}$$

and compatibility:

$$\mathcal{P}_{\text{Æ}}(\text{Æ}_1 \oplus \text{Æ}_2) = \mathcal{P}_{\text{Æ}}(\text{Æ}_1) \oplus \mathcal{P}_{\text{Æ}}(\text{Æ}_2). \tag{22}$$

Non-Associative Product

$$\text{Æ}_1 \boxtimes \text{Æ}_2 \in \text{Æ}, \tag{23}$$

with

$$(\text{Æ}_1 \boxtimes \text{Æ}_2) \boxtimes \text{Æ}_3 \neq \text{Æ}_1 \boxtimes (\text{Æ}_2 \boxtimes \text{Æ}_3), \tag{24}$$

and

$$\text{Æ}_1 \boxtimes \text{Æ}_2 \neq \text{Æ}_2 \boxtimes \text{Æ}_1. \tag{25}$$

Ternary Bracket

$$\{\text{Æ}_1, \text{Æ}_2, \text{Æ}_3\}_{\text{Æ}} \in \text{Æ}, \tag{26}$$

with antisymmetry in the first two entries:

$$\{\text{Æ}_1, \text{Æ}_2, \text{Æ}_3\}_{\text{Æ}} = -\{\text{Æ}_2, \text{Æ}_1, \text{Æ}_3\}_{\text{Æ}} \tag{27}$$

Collective Criticality and Phase Synchronization

Large ensembles of interacting phase oscillators can exhibit a transition from incoherent to synchronized dynamics. Within the Æ–framework this transition represents the emergence of collective coherence across nodal structures.

The Kuramoto order parameter is defined as

$$Re^{i\Theta} = \frac{1}{N}\sum_{j=1}^{N} e^{i\theta_j} \tag{28}$$

The phase evolution of each oscillator follows

$$\dot{\theta}_i = \omega_i + \frac{K}{N}\sum_{j=1}^{N} \sin(\theta_j - \theta_i) \tag{29}$$

Synchronization occurs when the coupling exceeds a critical value

$$K_c = \frac{2}{\pi g(0)} \tag{30}$$

39. Primitive Objects and Algebraic Structure

The Æ–framework is formulated in terms of algebraic objects that encode phase, coherence, and informational structure of the field. The mathematical setting combines functional spaces, dual structures, and algebraic operations governing the interaction of Æ-configurations.

Æ Functional Space and Dual Pairing

Let $\mathcal{A}_{\text{Æ}}$ denote the functional space of admissible Æ-field configurations. Elements of $\mathcal{A}_{\text{Æ}}$ represent coherent field states defined over the underlying spacetime manifold.

The dual space is defined as

$$\mathcal{A}_{\text{Æ}}^{*} = \mathrm{Hom}(\mathcal{A}_{\text{Æ}}, \mathbb{C}), \tag{31}$$

consisting of continuous linear functionals acting on Æ-field configurations.

For $\ell \in \mathcal{A}_{\text{Æ}}^{*}$ and $\text{Æ} \in \mathcal{A}_{\text{Æ}}$, the natural pairing is given by

$$\langle \ell, \text{Æ} \rangle = \ell(\text{Æ}). \tag{32}$$

This pairing defines the fundamental evaluation map between the field space and its dual, allowing the construction of observables and functional derivatives within the Æ-framework.

Algebraic Closure and Hopf-Like Structure

The algebra of Æ-configurations admits a structure analogous to a Hopf algebra. Let

$$\mathcal{H}_{\text{Æ}}$$

denote the algebra generated by coherent field elements.

The multiplication map is defined as

$$m : \mathcal{H}_{Æ} \otimes \mathcal{H}_{Æ} \to \mathcal{H}_{Æ}, \qquad m(a \otimes b) = ab. \tag{33}$$

A coproduct encodes compositional structure between subsystems:

$$\Delta : \mathcal{H}_{Æ} \to \mathcal{H}_{Æ} \otimes \mathcal{H}_{Æ}. \tag{34}$$

The unit and counit maps are given by

$$\eta : \mathbb{C} \to \mathcal{H}_{Æ}, \qquad \varepsilon : \mathcal{H}_{Æ} \to \mathbb{C}. \tag{35}$$

An antipode operator

$$S : \mathcal{H}_{Æ} \to \mathcal{H}_{Æ} \tag{36}$$

encodes inversion under the algebraic composition.

Together these maps define a Hopf-like structure governing the composition and decomposition of coherent Æ configurations.

Braiding and Non-Commutative Structure

Interactions between coherent field sectors may exhibit non-commutative ordering. This is captured by a braiding operator

$$\mathcal{B} : \mathcal{H}_{Æ} \otimes \mathcal{H}_{Æ} \to \mathcal{H}_{Æ} \otimes \mathcal{H}_{Æ}, \tag{37}$$

which encodes the exchange structure between interacting components of the field.

Such braiding structures naturally appear in topological and categorical formulations of field interactions.

Transmission Operators

Coupling between subsystems of the Æ-field is mediated by transmission operators. Let

$$\boxtimes : \mathcal{H}_{Æ} \times \mathcal{H}_{Æ} \to \mathcal{H}_{Æ} \tag{38}$$

denote a transmission product encoding coherent information exchange between field sectors.

For configurations $Æ_1$ and $Æ_2$, the transmission interaction is

$$Æ_1 \boxtimes Æ_2 \in \mathcal{H}_{Æ}. \tag{39}$$

This operator represents a generalized interaction mechanism between coherent structures in the field and will appear later in the dynamical equations governing inter-layer mediation and phase transport.

Algebraic Structure of the Æ-Space

The Æ-space is endowed with a non-commutative algebraic structure inherited from its quaternionic and Nambu sectors.

The binary product is defined as

$$Æ_1 \boxtimes Æ_2 \in Æ, \qquad Æ_1 \boxtimes Æ_2 \neq Æ_2 \boxtimes Æ_1. \tag{40}$$

This product is not assumed to be associative in general.

Ternary Structure (Nambu Seed)

The ternary product is defined as

$$\{Æ_1, Æ_2, Æ_3\}_{Æ} \in Æ, \tag{41}$$

with cyclic antisymmetry:

$$\{Æ_1, Æ_2, Æ_3\}_{Æ} + \{Æ_2, Æ_3, Æ_1\}_{Æ} + \{Æ_3, Æ_1, Æ_2\}_{Æ} = 0. \tag{42}$$

This structure underlies the internal Nambu dynamics developed in the rotational sector.

Algebraic Closure

The complete algebraic structure is summarized as

$$\mathcal{A}_{Æ} = \left(Æ, \boxtimes, \{\cdot,\cdot,\cdot\}_{Æ}, \dagger\right), \tag{43}$$

where † denotes Hermitian conjugation induced by the complex structure of the field.

Coalgebra Structure

In addition to the non-commutative algebraic product, the Æ-space admits a coproduct structure

$$\Delta_{\text{Æ}} : \text{Æ} \to \text{Æ} \otimes \text{Æ}. \tag{44}$$

A generic element may decompose as

$$\Delta_{\text{Æ}}(\text{Æ}) = \sum_i \text{Æ}_i^{(1)} \otimes \text{Æ}_i^{(2)}. \tag{45}$$

A counit map

$$\varepsilon_{\text{Æ}} : \text{Æ} \to \mathbb{C} \tag{46}$$

satisfies the compatibility relations

$$(\varepsilon_{\text{Æ}} \otimes \text{id}) \circ \Delta_{\text{Æ}} = (\text{id} \otimes \varepsilon_{\text{Æ}}) \circ \Delta_{\text{Æ}} = \text{id}. \tag{47}$$

An antipode operator may also be defined

$$S_{\text{Æ}} : \text{Æ} \to \text{Æ}, \tag{48}$$

satisfying

$$m \circ (S_{\text{Æ}} \otimes \text{id}) \circ \Delta_{\text{Æ}} = m \circ (\text{id} \otimes S_{\text{Æ}}) \circ \Delta_{\text{Æ}} = \varepsilon_{\text{Æ}}. \tag{49}$$

Unlike standard Hopf algebras, the coproduct need not be strictly coassociative:

$$(\Delta_{\text{Æ}} \otimes \text{id}) \circ \Delta_{\text{Æ}} \neq (\text{id} \otimes \Delta_{\text{Æ}}) \circ \Delta_{\text{Æ}}. \tag{50}$$

The deviation from coassociativity is encoded by an associator element

$$\Phi_{\text{Æ}} \in \text{Æ} \otimes \text{Æ} \otimes \text{Æ}. \tag{51}$$

Categorical Structure

The Æ-framework can be interpreted within a braided monoidal category

$$\mathcal{C}_{\text{Æ}} = (\text{Obj}_{\text{Æ}}, \text{Mor}_{\text{Æ}}), \tag{52}$$

with tensor product

$$\otimes : \mathrm{Obj}_{\text{Æ}} \times \mathrm{Obj}_{\text{Æ}} \to \mathrm{Obj}_{\text{Æ}}. \tag{53}$$

The unit object is

$$\mathbb{I}_{\text{Æ}} \in \mathrm{Obj}_{\text{Æ}}. \tag{54}$$

Braiding is implemented through

$$\beta_{\text{Æ}} : \text{Æ}_1 \otimes \text{Æ}_2 \to \text{Æ}_2 \otimes \text{Æ}_1, \tag{55}$$

with non-trivial topology

$$\beta_{\text{Æ}} \circ \beta_{\text{Æ}} \neq \mathrm{id}. \tag{56}$$

Derived Differential Structure

Differential operators acting on Æ are induced from the underlying spacetime derivatives ∂_μ already defined in the Field and Gauge sectors.

No additional independent differential axioms are introduced at this level.

40. Field Structure

Euler Identity (Rotational Seed)

At the most fundamental level, field structure emerges from rotational symmetry in the complex plane

$$e^{i\theta} = \cos\theta + i\sin\theta \tag{57}$$

Field as Phase Rotation

A general field configuration can be expressed as an amplitude modulated by a local phase

$$\psi(x,t) = A(x,t)\, e^{i\phi(x,t)} \tag{58}$$

Pure Temporal Rotation

In the simplest case, the field evolves through a pure temporal phase rotation

$$\psi(t) = e^{-i\omega t}, \qquad \partial_t\psi = -i\omega\psi. \tag{59}$$

Energy–Frequency Relation

The frequency of this rotation is directly associated with energy

$$E = \hbar\omega. \tag{60}$$

Temporal Schrödinger Sector

This relation leads to a dynamical equation governing temporal evolution

$$i\hbar\,\partial_t\psi = E\psi. \tag{61}$$

Full Schrödinger Equation

Including spatial variation, the full evolution of the field is described by the Schrödinger equation

$$i\hbar\,\partial_t\psi = -\frac{\hbar^2}{2m}\nabla^2\psi + V\psi. \tag{62}$$

Temporal Operator Dictionary

We introduce the operator dictionary:

$$D_\tau := \partial_\tau, \qquad T_{Æ} := i\,\partial_{t_{Æ}}, \qquad T_{\mathrm{phys}} := i\,\partial_{t_{\mathrm{phys}}}. \tag{63}$$

The phase multiplication operator is

$$\hat{\phi}\,\psi := \phi\,\psi. \tag{64}$$

From $dt_{Æ} = \alpha\,d\phi$ we obtain

$$\partial_{t_{Æ}} = \alpha^{-1}\partial_\phi, \tag{65}$$

and therefore

$$[T_{Æ}, \hat{\phi}] = i\,\alpha^{-1}. \tag{66}$$

Hamiltonian Commutators (Ordering vs Dynamics)

If a Hamiltonian generator H is present, then

$$[H, T_{Æ}] = 0, \tag{67}$$

while in general

$$[H, D_\tau] = i\,\partial_\tau H \neq 0. \tag{68}$$

Quaternionic Proxy for Phase Entropy (Model Choice)

In addition to the Shannon definition of S_ϕ, one may adopt a quaternionic proxy in the intrinsic sector:

$$S_\phi := \beta\,|\operatorname{Im} Q|^2, \qquad \beta > 0. \tag{69}$$

Then

$$\dot S_\phi = 2\beta\ \operatorname{Re}\Big(\operatorname{Im} Q \cdot \overline{\operatorname{Im} \dot Q}\Big), \tag{70}$$

with the monotonicity condition

$$\operatorname{Re}\Big(\operatorname{Im} Q \cdot \overline{\operatorname{Im} \dot Q}\Big) \geq 0. \tag{71}$$

Temporal Action Ansatz (OPEN)

$$S_{\text{time}} = \int \left[\frac{1}{2}(\dot\phi)^2 + \frac{k}{2}(\dot S_\phi)^2 - V(\phi, S_\phi) + \lambda\big(C - f(\phi, S_\phi)\big)\right] d\tau. \tag{72}$$

$$\frac{d}{d\tau}\dot\phi + \frac{\partial V}{\partial \phi} + \lambda \frac{\partial f}{\partial \phi} = 0. \tag{73}$$

$$k\frac{d}{d\tau}\dot S_\phi + \frac{\partial V}{\partial S_\phi} + \lambda \frac{\partial f}{\partial S_\phi} = 0. \tag{74}$$

Partition Function on S^3 (OPEN)

$$Z(\beta) = \int_{S^3} e^{-\beta V(Q)}\, d\mu(Q), \tag{75}$$

$$S_Q = -\int_{S^3} \rho(Q)\,\ln\rho(Q)\,d\mu(Q), \qquad \rho(Q) = \frac{e^{-\beta V(Q)}}{Z(\beta)}. \tag{76}$$

Æ-Product as a Binary Map

$$\boxtimes : \text{Æ} \times \text{Æ} \to \text{Æ}. \tag{77}$$

Linearity (left argument):

$$(a\,\text{Æ}_1 + b\,\text{Æ}_2) \boxtimes \text{Æ}_3 = a(\text{Æ}_1 \boxtimes \text{Æ}_3) + b(\text{Æ}_2 \boxtimes \text{Æ}_3). \tag{78}$$

Non-commutativity and non-associativity are permitted:

$$\text{Æ}_1 \boxtimes \text{Æ}_2 \neq \text{Æ}_2 \boxtimes \text{Æ}_1, \tag{79}$$

$$(\text{Æ}_1 \boxtimes \text{Æ}_2) \boxtimes \text{Æ}_3 \neq \text{Æ}_1 \boxtimes (\text{Æ}_2 \boxtimes \text{Æ}_3). \tag{80}$$

Involution, Norm, and Inner Product

Adjoint compatibility:

$$(\text{Æ}_1 \boxtimes \text{Æ}_2)^\dagger = \text{Æ}_2^\dagger \boxtimes \text{Æ}_1^\dagger. \tag{81}$$

Involution:

$$(\text{Æ}^\dagger)^\dagger = \text{Æ}. \tag{82}$$

Norm induced by the product:

$$\|\text{Æ}\|^2 = \text{Æ}^\dagger \boxtimes \text{Æ}. \tag{83}$$

Positivity:

$$\|\text{Æ}\| \geq 0. \tag{84}$$

Real inner product:

$$\langle \text{Æ}_1, \text{Æ}_2 \rangle = \operatorname{Re}\Big(\text{Æ}_1^\dagger \boxtimes \text{Æ}_2\Big). \tag{85}$$

Angle (when norms are nonzero):

$$\cos\theta_{\text{Æ}}(\text{Æ}_1, \text{Æ}_2) = \frac{\mathrm{Re}\left(\text{Æ}_1^\dagger \boxtimes \text{Æ}_2\right)}{\|\text{Æ}_1\|\,\|\text{Æ}_2\|}. \tag{86}$$

Cauchy–Schwarz-type bound:

$$\left|\text{Æ}_1^\dagger \boxtimes \text{Æ}_2\right| \le \|\text{Æ}_1\|\,\|\text{Æ}_2\|. \tag{87}$$

Projector Invariants

Define invariants:

$$I_1(\text{Æ}) = \text{Æ}^\dagger \boxtimes \text{Æ}, \tag{88}$$

$$I_2(\text{Æ}_1, \text{Æ}_2) = \text{Æ}_1^\dagger \boxtimes \text{Æ}_2. \tag{89}$$

Projector compatibility (formal invariance):

$$I_k(\mathcal{P}_{\text{Æ}}(\text{Æ})) = I_k(\text{Æ}). \tag{90}$$

Æ-Functional Spaces and Weighted Metrics

Define the weighted space:

$$L^2_{\text{Æ}}(\Omega) = \left\{ \text{Æ} : \Omega \to \mathbb{C} \;\middle|\; \int_\Omega \|\text{Æ}(x)\|^2\, d\mu_{\text{Æ}}(x) < \infty \right\}. \tag{91}$$

Weighted measure:

$$d\mu_{\text{Æ}}(x) = w_{\text{Æ}}(x)\, dx, \tag{92}$$

with

$$w_{\text{Æ}}(x) \ge 0. \tag{93}$$

Induced $L^2_{\text{Æ}}$ norm:

$$\|\text{Æ}\|^2_{L^2_{\text{Æ}}} = \int_\Omega \text{Æ}^\dagger(x) \boxtimes \text{Æ}(x)\, d\mu_{\text{Æ}}(x). \tag{94}$$

Metric distance:

$$D(\text{Æ}_1, \text{Æ}_2) = \|\text{Æ}_1 - \text{Æ}_2\|_{L^2_{\text{Æ}}}. \tag{95}$$

Pointwise distance:

$$d(x; \text{Æ}_1, \text{Æ}_2) = \|\text{Æ}_1(x) - \text{Æ}_2(x)\|. \tag{96}$$

Phase mismatch functional:

$$\chi(\text{Æ}_1, \text{Æ}_2) = \int_\Omega |\phi_1(x) - \phi_2(x)| \, d\mu_{\text{Æ}}(x). \tag{97}$$

Integral-average bound:

$$d(x; \text{Æ}_1, \text{Æ}_2) \le D(\text{Æ}_1, \text{Æ}_2) \quad \text{(in integral average)}. \tag{98}$$

Entropy-weighted distance:

$$D_{S_\phi}(\text{Æ}_1, \text{Æ}_2) = \int_\Omega \|\text{Æ}_1(x) - \text{Æ}_2(x)\| \, e^{\lambda S_\phi(x)} \, d\mu_{\text{Æ}}(x), \tag{99}$$

with

$$\lambda \ge 0. \tag{100}$$

Admissible algebraic domain:

The admissible sector of the Æ field is defined by coherence and phase-entropy constraints

$$\mathcal{A}_{\text{Æ}} = \left\{ \text{Æ} \in L^2_{\text{Æ}} \;\middle|\; C(x) \ge C_{\min}, \; S_\phi(x) \le S_\phi^{\text{crit}} \right\}. \tag{101}$$

The algebra admits a coproduct structure that distributes configurations across composite states

$$\Delta_{\text{Æ}} : \text{Æ} \to \text{Æ} \otimes \text{Æ}. \tag{102}$$

This coproduct can be expressed as a sum over tensor components

$$\Delta_{\text{Æ}}(\text{Æ}) = \sum_i \text{Æ}_i^{(1)} \otimes \text{Æ}_i^{(2)}. \tag{103}$$

A counit map extracts scalar information from the algebra

$$\varepsilon_{\text{Æ}} : \text{Æ} \to \mathbb{C}. \tag{104}$$

This map satisfies the standard compatibility condition with the coproduct

$$(\varepsilon_{Æ} \otimes \mathrm{id}) \circ \Delta_{Æ} = (\mathrm{id} \otimes \varepsilon_{Æ}) \circ \Delta_{Æ} = \mathrm{id}. \quad (105)$$

An antipode operation encodes inversion within the algebraic structure

$$S_{Æ} : Æ \to Æ. \quad (106)$$

This operation satisfies the defining relation of a generalized Hopf-like structure

$$m \circ (S_{Æ} \otimes \mathrm{id}) \circ \Delta_{Æ} = m \circ (\mathrm{id} \otimes S_{Æ}) \circ \Delta_{Æ} = \varepsilon_{Æ}. \quad (107)$$

In general, the coproduct is not strictly coassociative

$$(\Delta_{Æ} \otimes \mathrm{id}) \circ \Delta_{Æ} \neq (\mathrm{id} \otimes \Delta_{Æ}) \circ \Delta_{Æ}. \quad (108)$$

The resulting structure can be interpreted in categorical terms

$$\mathcal{C}_{Æ} = (\mathrm{Obj}_{Æ}, \mathrm{Mor}_{Æ}). \quad (109)$$

Objects combine through a tensor product operation

$$\otimes : \mathrm{Obj}_{Æ} \times \mathrm{Obj}_{Æ} \to \mathrm{Obj}_{Æ}. \quad (110)$$

A distinguished unit object defines the identity element of the category

$$\mathbb{I}_{Æ} \in \mathrm{Obj}_{Æ}. \quad (111)$$

Braiding between objects encodes non-trivial exchange structure

$$\beta_{Æ} : Æ_1 \otimes Æ_2 \to Æ_2 \otimes Æ_1. \quad (112)$$

This braiding is, in general, non-involutive

$$\beta_{Æ} \circ \beta_{Æ} \neq \mathrm{id}. \quad (113)$$

Nodal Realization of the Æ-Product

Let (D_i, d_i, ϕ_i) denote nodal coordinates of a vibrational configuration and define

$$\Xi_i = (D_i, d_i, \phi_i), \qquad i = 1, 2. \tag{114}$$

The **Æ nodal product** is realized as

$$\Xi_1 \boxtimes \Xi_2 = \left(\Xi_1 \cdot_{Æ} \Xi_2,\ \Xi_1 \wedge_{Æ} \Xi_2\right), \tag{115}$$

where

$$\Xi_1 \cdot_{Æ} \Xi_2 = \mathcal{C}(D_1, d_1)\,\mathcal{C}(D_2, d_2)\,\cos(\Delta\phi), \tag{116}$$

$$\Xi_1 \wedge_{Æ} \Xi_2 = \mathcal{C}(D_1, d_1)\,\mathcal{C}(D_2, d_2)\,\sin(\Delta\phi)\,\hat{n}, \tag{117}$$

with $\Delta\phi = \phi_1 - \phi_2$ and $\hat{n}$ a torsional axis.

Algebraic Properties (Nodal Sector)

The nodal sector is endowed with a set of algebraic operations that combine scalar, axial, and coherence-dependent structures.

- **Left bilinearity:** The product distributes linearly over superpositions

$$(a\Xi_1 + b\Xi_2) \boxtimes \Xi_3 = a(\Xi_1 \boxtimes \Xi_3) + b(\Xi_2 \boxtimes \Xi_3).$$

- **Scalar symmetry:** The scalar component remains invariant under exchange

$$\Xi_1 \cdot_{Æ} \Xi_2 = \Xi_2 \cdot_{Æ} \Xi_1.$$

- **Axial antisymmetry:** The axial sector changes sign under permutation

$$\Xi_1 \wedge_{Æ} \Xi_2 = -\,\Xi_2 \wedge_{Æ} \Xi_1.$$

- **Norm compatibility:** The total norm decomposes into scalar and axial contributions

$$\|\Xi_1 \boxtimes \Xi_2\|^2 = (\Xi_1 \cdot_{Æ} \Xi_2)^2 + \|\Xi_1 \wedge_{Æ} \Xi_2\|^2.$$

- **Idempotent coherence:** Self-composition collapses to a purely scalar configuration

$$\Xi \boxtimes \Xi = (\|\Xi\|^2, 0).$$

41. Functionals

Coherence Functional (Exponential Form)

The coherence functional admits the entropic representation:

$$C(x) \;\propto\; e^{-S_\phi(x)/\hbar} \tag{118}$$

This relation links structural order directly to phase entropy.

Phase Entropy Functional

The phase entropy is defined as:

$$S_\phi = -\int p(\phi)\,\log p(\phi)\,d\phi \tag{119}$$

Kuramoto Phase Oscillator Dynamics

Coupled phase oscillators evolve according to:

$$\frac{d\phi_i}{dt} = \omega_i + \sum_j K_{ij}\sin(\phi_j - \phi_i) \tag{120}$$

Kuramoto Order Parameter

The complex order parameter is defined as:

$$Re^{i\Psi} = \frac{1}{N}\sum_{j=1}^{N} e^{i\phi_j} \tag{121}$$

with neural coherence identified as

$$C_{\text{neural}} = R \tag{122}$$

Illustrative Coherence System

The volumetric coherence functional is:

$$C_{\text{water}} = \left| \frac{1}{V} \int_V e^{i\phi(x)} \, dx \right| \tag{123}$$

The associated memory time scale satisfies:

$$\tau_{\text{mem}} \propto \frac{1}{\text{Var}(\phi)} \tag{124}$$

Biological Stability Constraint

Cellular persistence requires:

$$C_{\text{cell}} \geq C_{\text{crit}} \tag{125}$$

Entropy growth beyond threshold leads to structural instability:

$$\partial_t S_\phi > \gamma_{\text{crit}} \implies \text{loss of structural integrity.} \tag{126}$$

Alternative Coherence Approximation (Low-Entropy Regime)

In the small-entropy regime, one may adopt the rational approximation:

$$C_{Æ} = \frac{1}{1 + S_\phi} \tag{127}$$

This form provides a bounded algebraic proxy consistent with

$$C \to 1 \text{ as } S_\phi \to 0 \text{ and } C \to 0 \text{ as } S_\phi \to \infty.$$

Continuum Phase Transport

In the continuum limit, phase transport can be expressed through operatorial forms:

$$\mathcal{K}[\phi]_i = \sum_j W_{ij} \sin(\phi_j - \phi_i). \tag{128}$$

For small phase differences,

$$\sin(\phi_j - \phi_i) \approx \phi_j - \phi_i. \tag{129}$$

This yields a Laplacian-driven dynamics:

$$\dot{\phi} = -L_\phi \phi \tag{130}$$

with additional transport terms

$$\frac{\partial \phi}{\partial t} = -L_\phi \phi + \mathcal{R}[\phi] - v \cdot \nabla \phi \tag{131}$$

Network Representation of the Æ-Field

Multi-node systems within the Æ-framework may be represented as networks whose vertices correspond to coherent field nodes and whose edges encode interaction channels.

Let $G = (V, E)$ denote the interaction graph, where V is the set of nodes and E the set of coupling links.

A field configuration across the network may be written as

$$\Psi = (\Psi_1, \Psi_2, \ldots, \Psi_N), \tag{132}$$

where Ψ_i denotes the state of node i.

Graph Laplacian Dynamics

Interactions between nodes may be expressed through the graph Laplacian operator.

Let A_{ij} denote the adjacency matrix and D_{ij} the degree matrix.

The graph Laplacian is defined as

$$L = D - A. \tag{133}$$

The Laplacian governs the diffusion or propagation of coherent field information across the network.

Propagation of Coherence

The evolution of node states across the network may be described through Laplacian-driven dynamics

$$\frac{d\Psi_i}{dt} = -\sum_j L_{ij}\Psi_j. \tag{134}$$

This equation describes the redistribution of coherence between neighboring nodes.

Coupled Node Evolution

More generally, node dynamics may include intrinsic evolution and interaction terms

$$\frac{d\Psi_i}{dt} = F(\Psi_i) - \sum_j L_{ij}\,\Psi_j. \tag{135}$$

Here $F(\Psi_i)$ represents the local dynamical law of each node, while the Laplacian term encodes network coupling.

Nonlocal Phase Interaction Kernel

In spatially extended systems, phase transport may include nonlocal coupling mediated by an interaction kernel:

$$\mathcal{R}[\phi](x,t) = \int_\Omega K_R(x,y)\,\phi(y,t)\,dy. \tag{136}$$

This operator represents long-range phase influence across the domain Ω.

Advective Phase Transport

Phase transport may additionally include advective drift generated by an external velocity field:

$$\mathcal{A}[\phi] = v \cdot \nabla\phi. \tag{137}$$

The full transport equation then becomes

$$\frac{\partial\phi}{\partial t} = -L_\phi\phi + \mathcal{R}[\phi] - v \cdot \nabla\phi. \tag{138}$$

This formulation combines graph-Laplacian diffusion, nonlocal coupling, and advective transport in the phase field.

Network Transmission Functional

Define the average transmission over a network E:

$$\Delta_{\text{net}}(t) := \frac{1}{|E|} \sum_{(i\to j)\in E} \Delta_{i\to j}(t). \tag{139}$$

$$|\Delta\phi_{a\to b}| \ll 1, \qquad S_{\phi,a\to b} \ll 1. \tag{140}$$

Canonical Phase Transmission Operator

Transmission between coherent subsystems may be described through a linear operator

$$T_{a\to b} : \mathcal{H}_a \to \mathcal{H}_b. \tag{141}$$

The phase mismatch between the transmitted and received states is defined by

$$\Delta\phi_{a\to b}(t) := \arg\Big(\langle Q_b(t), T_{a\to b}Q_a(t)\rangle\Big) \tag{142}$$

A scalar transmission efficiency may then be written as

$$\boxed{\Delta_{a\to b}(t) := \exp(-\alpha\,|\Delta\phi_{a\to b}(t)|)\,\exp(-\beta\,S_{\phi,a\to b}(t))} \tag{143}$$

with

$$0 < \Delta_{a\to b} \le 1. \tag{144}$$

Perfect phase alignment yields

$$|\Delta\phi_{a\to b}| \to 0, \quad S_{\phi,a\to b} \to 0 \quad \Rightarrow \quad \Delta_{a\to b} \to 1. \tag{145}$$

Conversely, large mismatch or entropy growth suppresses transmission,

$$\Delta_{a\to b} \to 0. \tag{146}$$

Unitary Transport Representation

Transmission may alternatively be expressed through a unitary transport operator

$$\mathcal{U}_{a\to b}(t) := \exp\big(- i\,\mathcal{G}_{a\to b}(t)\big), \tag{147}$$

leading to the normalized transmission amplitude

$$\boxed{\Delta_{a\to b}(t) := \frac{\big|\langle \Psi_b(t), \mathcal{U}_{a\to b}(t)\Psi_a(t)\rangle\big|}{\|\Psi_b(t)\|\,\|\Psi_a(t)\|}} \tag{148}$$

which satisfies

$$0 \le \Delta_{a\to b} \le 1. \tag{149}$$

Relation with Coherence

Transmission efficiency depends both on phase alignment and on the coherence level of the field. Using the canonical coherence functional,

$$C_{Æ} = \frac{1}{1+S_\phi}, \tag{150}$$

one may write

$$\Delta_{a\to b} = \exp(-\alpha|\Delta\phi|)\; f(C_{Æ}). \tag{151}$$

Network Transmission

The average transmission over a network E is

$$\Delta_{\mathrm{net}}(t) := \frac{1}{|E|} \sum_{(i\to j)\in E} \Delta_{i\to j}(t) \tag{152}$$

In the low-mismatch regime,

$$|\Delta\phi_{a\to b}| \ll 1, \qquad S_{\phi,a\to b} \ll 1, \tag{153}$$

so that

$$\Delta_{a\to b} \approx 1. \tag{154}$$

Canonical Phase Entropy (Continuous, Discrete, Spectral)

The phase distribution satisfies

$$\int_0^{2\pi} p(\phi)\, d\phi = 1, \qquad p(\phi) \geq 0. \tag{155}$$

A relative entropy formulation reads:

$$S_\phi = \int_0^{2\pi} p(\phi) \log\left(\frac{p(\phi)}{p_0}\right) d\phi \tag{156}$$

with

$$S_\phi \geq 0, \qquad S_\phi = 0 \text{ iff } p(\phi) = p_0. \tag{157}$$

In the discrete case:

$$S_\phi = -\sum_{k=1}^{n} p_k \log p_k \tag{158}$$

with

$$0 \leq S_\phi \leq \log n. \tag{159}$$

In the spectral representation of the phase Laplacian

$$L_\phi = U \Lambda U^\dagger, \qquad \Lambda = \mathrm{diag}(\lambda_1, \ldots, \lambda_n), \qquad \lambda_k \geq 0, \tag{160}$$

one may define

$$p_k = \frac{e^{-\beta\lambda_k}}{\sum_{j=1}^{n} e^{-\beta\lambda_j}}, \qquad \beta > 0 \tag{161}$$

yielding the spectral entropy

$$S_\phi = -\sum_{k=1}^{n} p_k \log p_k \tag{162}$$

Canonical Coherence Functional

The canonical coherence functional is defined as

$$C_{\text{Æ}} = \frac{1}{1 + S_\phi} \tag{163}$$

It satisfies

$$0 < C_{\text{Æ}} \leq 1, \qquad S_\phi = 0 \Rightarrow C_{\text{Æ}} = 1, \tag{164}$$

and

$$\frac{dC_{\text{Æ}}}{dS_\phi} = -\frac{1}{(1 + S_\phi)^2} < 0 \tag{165}$$

Local and global coherence may be defined as

$$C_{\text{Æ},i} = \frac{1}{1 + S_{\phi,i}}, \tag{166}$$

$$C_{\text{Æ}}^{\text{global}} = \sum_{i=1}^{N} w_i \, C_{\text{Æ},i} \tag{167}$$

The evolution of coherence follows directly from entropy variation:

$$\dot{C}_{\text{Æ}} = -\frac{1}{(1 + S_\phi)^2} \, \dot{S}_\phi \tag{168}$$

Phase Memory Functional

Define the phase memory over an interval of length L:

$$M(L) = \frac{1}{L} \left| \int_0^L e^{i\phi(s)} \, ds \right| \in [0, 1] \tag{169}$$

Coherent regimes satisfy

$$M(L) \sim \mathcal{O}(1) \tag{170}$$

while incoherent regimes yield

$$M(L) \sim \mathcal{O}(L^{-1/2}) \tag{171}$$

Memory Kernels and Phase Retention

Phase memory may persist across embeddings through filtered inheritance of the global phase structure.

Define the inherited entropy as

$$S_{\text{inherit}} = S_{\phi}\left(\Phi_{\text{astral}}\right) \tag{172}$$

Local embedding requires the retained entropy to remain below a structural threshold

$$S_{\text{local}} < S_{\text{crit}} \tag{173}$$

The local phase state may be modeled as a filtered projection of the global carrier

$$\Phi_{\text{local}} = \mathcal{F}\left(\Phi_{\text{astral}}\right) \tag{174}$$

where the filtering operator satisfies

$$\mathcal{F} : S^3 \to S^3 \tag{175}$$

Norm preservation ensures that the quaternionic phase carrier remains on the coherence sphere

$$\|\mathcal{F}(\Phi)\| = 1, \qquad \text{class}(\Phi) = \text{class}(\mathcal{F}(\Phi)) \tag{176}$$

The resulting local entropy becomes

$$S_{\phi,\text{local}} = S_{\text{inherit}} - \Delta S_{\text{filtered}} \tag{177}$$

Temporal stability of the retained phase structure requires bounded entropy production

$$\frac{d}{dt} S_{\text{local}} \leq \sigma_{\max} \tag{178}$$

Nonlocal Phase Memory Action

Long-range phase memory may be represented through a nonlocal bilinear functional of the phase-entropy field.

$$S_{\text{mem}} = \frac{1}{2} \int d^4x \, \sqrt{-g(x)} \int d^4x' \, \sqrt{-g(x')} \, K(x,x') \, S_{\phi}(x) \, S_{\phi}(x')$$

Here $K(x, x')$ is a memory kernel encoding correlations between phase configurations at different spacetime points.

To preserve causality the kernel is required to satisfy

$$K(x, x') = 0 \quad \text{for} \quad x'^0 > x^0$$

This condition ensures that the memory operator is retarded and does not propagate information from the future.

In this representation, persistent nodal structures correspond to slowly decaying kernels $K(x, x')$, whereas rapid decoherence is associated with kernels concentrated near $x \approx x'$.

Generic Memory Operator

Memory in the Æ-framework is represented by operators acting on phase entropy and coherence-retention variables.

Let S_ϕ denote the phase entropy functional. A generic memory operator may be written as

$$R[S_\phi] \tag{179}$$

where R acts on the entropy structure of the field and encodes the retention of coherent phase information.

Associated with this operator one may define a dispersion measure

$$\Delta_R = \sqrt{\langle R[S_\phi]^2 \rangle - \langle R[S_\phi] \rangle^2} \tag{180}$$

which quantifies the spread of memory retention across the configuration space.

Long-Range Phase Memory Structures

Long-range memory in the Æ-field may persist even when local field amplitudes evolve in time.

A memory functional may be defined as

$$M(t) = \int_0^t \Psi(\tau)\, e^{i\phi(\tau)}\, d\tau \tag{181}$$

A persistence condition for long-range memory is given by

$$\frac{d}{dt}\|M(t)\| > 0 \tag{182}$$

even in regimes where $\partial_t \Psi \neq 0$.

Temporal coherence retention may also be measured through the correlation criterion

$$\kappa(t, t-T) = \frac{\mathrm{Re}\big(\langle \Psi(t), \Psi(t-T)\rangle\big)}{\|\Psi(t)\|\,\|\Psi(t-T)\|} \tag{183}$$

with memory retention occurring whenever

$$\kappa(t, t-T) > \kappa_{\mathrm{mem}} \tag{184}$$

These relations characterize the persistence of phase information across temporally separated configurations of the field.

Trauma as Persistent Phase Memory

Let $M(t)$ denote the phase memory functional

$$M(t) = \int_0^t \Psi(\tau)\, e^{i\phi(\tau)}\, d\tau \tag{185}$$

Ordinary memories correspond to kernels with finite relaxation time

$$\|M(t)\| \to 0 \quad \text{as} \quad t \to \infty \tag{186}$$

Traumatic memory corresponds instead to slowly decaying phase structures satisfying

$$\frac{d}{dt}\|M(t)\| \approx 0 \quad \text{for} \quad t \gg \tau_{\mathrm{event}} \tag{187}$$

Equivalently, trauma may be represented by a persistent phase-correlation condition

$$\kappa(t, t-T) > \kappa_{\mathrm{trauma}} \tag{188}$$

with

$$\kappa(t, t-T) = \frac{\mathrm{Re}\big(\langle \Psi(t), \Psi(t-T)\rangle\big)}{\|\Psi(t)\|\,\|\Psi(t-T)\|} \tag{189}$$

In this regime the phase configuration associated with the original experience remains dynamically active.

Resolution of trauma corresponds to the restoration of normal entropy relaxation

$$\frac{d}{dt}S_\phi < 0 \tag{190}$$

together with the reduction of long-range phase correlation

$$\kappa(t, t-T) \to \kappa_{\text{mem}}. \tag{191}$$

This transition restores ordinary memory dynamics and removes the persistent phase constraint.

Geometric Memory Functional

Memory retention may also be expressed geometrically through functionals defined over coherent trajectories in configuration space.

Let γ denote a coherent path in the Æ-field. A geometric memory functional may be written as

$$\mathcal{M}_\gamma = \int_\gamma \omega_{\text{mem}} \tag{192}$$

where ω_{mem} denotes a memory one-form associated with the phase-retention structure of the field.

This functional measures the accumulation of retained phase information along the trajectory and provides a geometric representation of memory persistence.

Structured Water Memory Extension

Let

$$\Psi_{\text{H}_2\text{O}}(x,t) = A(x,t)\, e^{i\phi_{\text{H}_2\text{O}}(x,t)} \tag{193}$$

The associated coherence reads

$$C_{\text{H}_2\text{O}} = \left| \int e^{i\phi_{\text{H}_2\text{O}}(x,t)} dx \right|^2 \tag{194}$$

The cumulative memory functional is

$$M_{\mathrm{H_2O}}(t) = \int_0^t C_{\mathrm{H_2O}}(\tau) e^{i\phi_{\mathrm{H_2O}}(\tau)}\, d\tau \tag{195}$$

Biological memory aggregates contributions:

$$M_{\mathrm{bio}} = M_{\mathrm{neural}} + M_{\mathrm{endo}} + M_{\mathrm{H_2O}} \tag{196}$$

Critical Coupling Threshold

In generalized Kuramoto systems, the onset of synchronization occurs at

$$K_c = \frac{2}{\pi g(0)} \tag{197}$$

where $g(0)$ is the natural frequency density at zero.

Phase Laplacian Representation

The phase Laplacian associated with the network coupling may be written component-wise as

$$(L_\phi)_{ij} = -W_{ij}\cos(\phi_i - \phi_j), \qquad i \neq j, \tag{198}$$

$$(L_\phi)_{ii} = -\sum_{j\neq i} (L_\phi)_{ij} \tag{199}$$

This form connects the spectral Laplacian representation with the nonlinear phase-coupling structure appearing in Kuramoto-type dynamics.

Network Transmission Functional

The transmission of phase coherence across a network of interacting Æ-nodes can be described through a functional acting on the collective state of the system.

Let $\Psi_i(t)$ denote the phase state of node i. The network transmission functional is defined as

$$\Delta_{\mathrm{network}} = \sum_{i,j} W_{ij} \sin\big(\phi_j - \phi_i\big) \tag{200}$$

where W_{ij} represents the weighted connectivity matrix of the interaction graph.

This functional governs the propagation of phase information through the network and generalizes the Kuramoto synchronization structure introduced in the collective criticality sector.

The corresponding phase transmission relations take the form

$$\partial_t \phi_i = \omega_i + \sum_j W_{ij} \sin(\phi_j - \phi_i) \tag{201}$$

which define the dynamical transport of phase coherence between nodes.

In the continuum limit this transmission structure extends to phase transport equations over the underlying manifold of the Æ-field.

Coherence Operators in the Æ Framework

The coherence of a configuration in the Æ-field is quantified by a set of operators collectively denoted $C_{«}$. These operators encode the degree of phase alignment relative to the entropy structure of the system.

The canonical coherence measure is defined as

$$C_{Æ} = \frac{1}{1 + S_\phi} \tag{202}$$

where S_ϕ denotes the phase entropy.

Local coherence may be defined over a subregion Ω as

$$C_{Æ}^{\text{local}} = \frac{1}{|\Omega|} \int_\Omega \frac{1}{1 + S_\phi(x)}\, dx \tag{203}$$

A weighted global coherence functional is given by

$$C_{Æ}^{\text{global}} = \int w(x) \frac{1}{1 + S_\phi(x)}\, dx \tag{204}$$

where $w(x)$ denotes a normalized weighting function.

Finally, coherence may also be expressed in spectral form through

$$C_{Æ}^{\text{spectral}} = \sum_k \lambda_k^{-1} \tag{205}$$

where λ_k are eigenvalues associated with the spectral structure of the Æ-field.

These coherence operators act as control parameters governing the stability of phase structures across the system and will appear in the dynamical equations of the field in later sections.

42. Geometry

Coherence-Induced Metric

The effective metric induced by coherence is defined as:

$$g_{\mu\nu}^{(\text{«})} = C(x)\, g_{\mu\nu} \tag{206}$$

Coherence thus rescales spacetime geometry.

Quaternionic Structure

A quaternion is defined as:

$$Q = a + bi + cj + dk \tag{207}$$

The unit constraint yields:

$$|Q| = 1 \quad \Longrightarrow \quad Q \in S^3 \tag{208}$$

Exponential Map on S^3

For a unit imaginary quaternion u:

$$e^{u\theta} = \cos\theta + u \sin\theta \tag{209}$$

SU(2) Evolution

Quaternionic evolution follows:

$$\partial_t Q = \Omega Q, \qquad \Omega \in \mathfrak{su}(2) \tag{210}$$

Commutator Structure

$$[A, B] = AB - BA \tag{211}$$

Nambu Bracket (Internal Manifold)

On the internal three-dimensional coordinate manifold, the Nambu bracket is defined as:

$$\{F, G, H\}_{\text{Æ}} = \epsilon^{abc} \frac{\partial F}{\partial x^a} \frac{\partial G}{\partial x^b} \frac{\partial H}{\partial x^c} \tag{212}$$

This structure preserves internal volume and generates ternary evolution laws.

Information Geometry (Fisher Metric)

The Fisher information metric associated with the phase probability distribution is:

$$g_{ij}^{(F)} = \int \frac{\partial_i p(\phi)\, \partial_j p(\phi)}{p(\phi)}\, d\phi \tag{213}$$

The corresponding scalar curvature is

$$R^{(F)} = \text{Ricci scalar of } g_{ij}^{(F)} \tag{214}$$

This defines the statistical geometry of phase configurations.

Æ Differential Geometry

Define the Æ-covariant derivative

$$\nabla_\mu^{\text{Æ}} : \mathcal{T}_{\text{Æ}} \to \mathcal{T}_{\text{Æ}} \tag{215}$$

It satisfies the Leibniz rule with respect to the nodal product:

$$\nabla_\mu^{\text{Æ}}(\text{Æ}_1 \boxtimes \text{Æ}_2) = (\nabla_\mu^{\text{Æ}} \text{Æ}_1) \boxtimes \text{Æ}_2 + \text{Æ}_1 \boxtimes (\nabla_\mu^{\text{Æ}} \text{Æ}_2) \tag{216}$$

The torsion operator is

$$T_{\mu\nu}^{Æ}(Æ) = \nabla_{\mu}^{Æ}\nabla_{\nu}^{Æ}Æ - \nabla_{\nu}^{Æ}\nabla_{\mu}^{Æ}Æ \tag{217}$$

The curvature operator is defined by

$$R_{\mu\nu}^{Æ}(Æ) = \left[\nabla_{\mu}^{Æ}, \nabla_{\nu}^{Æ}\right] Æ \tag{218}$$

The induced inner product reads

$$\langle Æ_1, Æ_2\rangle = \mathrm{Re}\Big(Æ_1^{\dagger} \boxtimes Æ_2\Big) \tag{219}$$

The sectional curvature is

$$K_{Æ}(Æ_1, Æ_2) = \frac{\langle R^{Æ}(Æ_1, Æ_2)Æ_2, Æ_1\rangle}{\|Æ_1\|^2\|Æ_2\|^2 - \langle Æ_1, Æ_2\rangle^2} \tag{220}$$

Geodesics satisfy

$$\nabla_{\dot\gamma}^{Æ}\dot\gamma = 0 \tag{221}$$

Jacobi fields obey

$$\nabla_{\dot\gamma}^{Æ}\nabla_{\dot\gamma}^{Æ}J + R^{Æ}(J, \dot\gamma)\dot\gamma = 0 \tag{222}$$

Stability criterion:

$$K_{Æ} > 0 \Rightarrow \text{convergence (stable)}, \tag{223}$$

$$K_{Æ} < 0 \Rightarrow \text{divergence (unstable)}. \tag{224}$$

Geometric Triple Product

For nodal coordinates

$$u = (D_1, d_1), \quad v = (D_2, d_2), \quad z = (D_3, d_3) \tag{225}$$

define the oriented volume form

$$\mathcal{V}_3(u, v, z) = \det\begin{pmatrix} D_1 & d_1 & 1 \\ D_2 & d_2 & 1 \\ D_3 & d_3 & 1 \end{pmatrix} \tag{226}$$

Layered Æ-Metric Ansatz

An explicit layered metric structure is given by

$$ds^2_{\text{Æ}} = -\rho_e(D)\,dt^2 + \rho_i(d)\,dq^2 + \lambda_{\text{Æ}}(D,d)\,(dD^2 + dd^2) + 2P(D,d)\,dt\,dq, \tag{227}$$

with

$$\rho_e(D) = e^D, \qquad \rho_i(d) = e^d, \qquad P(D,d) = \sin D \cos d, \qquad \lambda_{\text{Æ}}(D,d) = 1. \tag{228}$$

Levi-Civita Connection

The connection coefficients are

$$\Gamma^\mu_{\nu\rho} = \tfrac{1}{2} g^{\mu\sigma} \left(\partial_\nu g_{\rho\sigma} + \partial_\rho g_{\nu\sigma} - \partial_\sigma g_{\nu\rho} \right) \tag{229}$$

In particular,

$$\Gamma^t_{tD} = \tfrac{1}{2} \partial_D \ln \rho_e(D), \qquad \Gamma^q_{qd} = \tfrac{1}{2} \partial_d \ln \rho_i(d) \tag{230}$$

Nodal Subspace Geometry

The induced nodal metric reads

$$h_{ij}(D,d) = \begin{pmatrix} a(D) & 0 \\ 0 & b(d) \end{pmatrix}, \qquad a(D), b(d) > 0 \tag{231}$$

with

$$a(D) = 1 + \eta \sin D, \qquad b(d) = 1 + \xi \cos d. \tag{232}$$

The associated Christoffel symbols are

$$\Gamma^k{}_{ij} = \tfrac{1}{2} h^{k\ell} \left(\partial_i h_{j\ell} + \partial_j h_{i\ell} - \partial_\ell h_{ij} \right). \tag{233}$$

Quaternionic Bundle Extension

A bundle extension incorporating quaternionic fibers is

$$ds^2_{\text{Æ}} = g_{\mu\nu}(x)\, dx^\mu dx^\nu + \lambda^2 \langle dQ, dQ\rangle_{\mathbb{H}} \tag{234}$$

Tetrahedral Structure on S^2

Let $\{v_i\}_{i=1}^4$ be unit vectors satisfying

$$\|v_i\| = 1, \qquad v_i \cdot v_j = -\tfrac{1}{3} \quad (i \neq j), \qquad \sum_{i=1}^{4} v_i = 0. \tag{235}$$

Hilbert Structure of the Æ Field

Let $\mathcal{H}_{\text{Æ}}$ denote the Hilbert space of admissible Æ-field configurations. Elements of this space correspond to coherent phase states defined over the underlying manifold $\mathcal{M}$.

The inner product is defined as

$$\langle \Psi_1, \Psi_2\rangle = \int_{\mathcal{M}} \Psi_1(x)\, \overline{\Psi_2(x)}\, w_{\text{Æ}}(x)\, dx, \tag{236}$$

where $w_{\text{Æ}}(x)$ is a coherence-dependent weight.

The induced norm is

$$\|\Psi\|^2_{\text{Æ}} = \langle \Psi, \Psi\rangle \tag{237}$$

This weighted Hilbert structure allows the definition of differential operators and spectral structures consistent with the entropy-modulated geometry of the Æ-field.

Æ-Covariant Differential Structure

Let $\mathcal{M}$ denote the manifold underlying the Æ-field. The geometry is equipped with a covariant derivative

$$\nabla_{\text{Æ}} : T(\mathcal{M}) \to T(\mathcal{M}) \tag{238}$$

which defines parallel transport of phase-coherent structures. For a field Ψ the covariant derivative takes the form

$$\nabla_{\text{Æ},\mu}\Psi = \partial_\mu \Psi + \Gamma_\mu \Psi \tag{239}$$

where Γ_μ represents the Æ-connection.
Curvature is defined through the commutator

$$[\nabla_{\text{Æ},\mu}, \nabla_{\text{Æ},\nu}] = R^{\text{Æ}}_{\mu\nu} \tag{240}$$

This curvature tensor encodes geometric distortions of the phase structure within the Æ-space.

Weighted Differential Operators

The dynamics of the Æ-field involve differential operators modified by coherence-dependent weights.

The weighted Laplacian is defined as

$$\Delta_{\text{Æ}} = \nabla^{\mu}_{\text{Æ}} \nabla_{\text{Æ},\mu} \tag{241}$$

Acting on a scalar configuration Ψ, this operator yields

$$\Delta_{\text{Æ}}\Psi = \frac{1}{\sqrt{|g|}} \partial_\mu \left(\sqrt{|g|}\, g^{\mu\nu} \partial_\nu \Psi \right) \tag{242}$$

where $g_{\mu\nu}$ denotes the metric tensor of the Æ-space.

Metric Evolution and Interlayer Mediation

Interactions between layers of the Æ-field induce dynamical evolution of the effective geometry.

A coherence-dependent variation of the action takes the form

$$\delta S_{\text{Æ}} = \lambda (K - K_0) \tag{243}$$

where K denotes a curvature-like scalar associated with the field.

The induced metric dynamics follow

$$\partial_t g_{\mu\nu} = \alpha \partial_\mu \Phi_{Æ} \partial_\nu \overline{\Phi}_{Æ} \tag{244}$$

These relations describe the geometric response of the manifold to phase-coherent interactions across the system.

Differential Forms and Æ-Vorticity

Geometric structures in the Æ-field may also be expressed in terms of differential forms.

Let ω denote a phase one-form defined on $\mathcal{M}$. The associated vorticity two-form is

$$\Omega_{Æ} = d\omega. \tag{245}$$

This form measures rotational structure of phase transport and plays a central role in the formation of coherent vortical patterns within the field.

43. Rotational Core

Quaternionic Field Constraint

The quaternionic field satisfies:

$$Q(x,t) \in S^3 \subset \mathbb{H}, \qquad |Q| = 1. \tag{246}$$

This constraint encodes the unit-norm condition of the internal rotational degree of freedom.

Primitive SU(2) Evolution

The fundamental rotational dynamics is governed by:

$$\partial_t Q = \Omega Q, \qquad \Omega \in \mathfrak{su}(2). \tag{247}$$

This expresses pure group evolution on S^3.

Maurer–Cartan Structure

The Maurer–Cartan one-form is defined as:

$$\omega = Q^{-1} dQ \tag{248}$$

This one-form encodes intrinsic rotational curvature of the quaternionic manifold.

Associator and Torsional Triple Product

Non-associativity of the nodal product is measured by the associator:

$$\mathcal{A}(u,v,z) = (u \boxtimes v) \boxtimes z - u \boxtimes (v \boxtimes z) \tag{249}$$

A geometric representation is:

$$\mathcal{A}(u,v,z) = \kappa\left[\mathcal{V}_3(u,v,z) + \mathrm{Im}(\langle u,v\rangle_{Æ}\langle v,z\rangle_{Æ}\langle z,u\rangle_{Æ})\right] \quad (250)$$

where $\mathcal{V}_3(u,v,z)$ denotes the geometric triple volume.

This structure encodes intrinsic torsional memory in nodal interactions.

Torsional Geometry of the Æ-Manifold

A conformal nodal metric may be introduced:

$$ds^2_{Æ} = e^{2\phi(D,d)}(dD^2 + dd^2), \qquad \phi(D,d) = \frac{1}{2}\log C(D,d). \quad (251)$$

The connection decomposes as:

$$\Gamma^{\lambda}_{\mu\nu}(Æ) = \Gamma^{\lambda}_{\mu\nu}(\mathrm{LC}) + K^{\lambda}_{\ \mu\nu} \quad (252)$$

where $K^{\lambda}_{\ \mu\nu}$ is the contorsion tensor.

The torsion tensor is defined as:

$$T^{\lambda}_{\mu\nu} = \Gamma^{\lambda}_{\mu\nu}(Æ) - \Gamma^{\lambda}_{\nu\mu}(Æ) \quad (253)$$

and in the nodal sector takes the form:

$$T^{\lambda}_{\mu\nu} = -\frac{\beta}{2}\left(\delta^{\lambda}_{\nu}\partial_{\mu}C - \delta^{\lambda}_{\mu}\partial_{\nu}C\right) \quad (254)$$

This expresses torsion as a gradient of coherence.

Dirac-Type Constraint

The Dirac-type operator is defined as:

$$D = i\gamma^{\mu}\partial_{\mu} - M \quad (255)$$

The constraint equation reads:

$$DQ = 0 \quad (256)$$

Squaring yields:

$$D^2 = -\partial_\mu \partial^\mu + M^2 \tag{257}$$

Internal Nambu Dynamics

The ternary bracket on the internal manifold is:

$$\{F, G, H\} = \epsilon^{abc} \partial_a F \, \partial_b G \, \partial_c H \tag{258}$$

Evolution law:

$$\frac{dF}{dt} = \{F, H_1, H_2\} \tag{259}$$

Casimir invariance:

$$\{C, F, G\} = 0 \tag{260}$$

Internal Polarities and Asymmetry

Local rotational dynamics in the Æ–field may exhibit internal asymmetries between left- and right-handed components. These asymmetries are quantified through a polarity measure defined at the nodal position z_i.

Let K_L and K_R denote the left and right rotational couplings associated with the node. The internal polarity is defined as

$$P_i = \left. \frac{K_L - K_R}{K_L + K_R} \right|_{z=z_i} \tag{261}$$

The quantity P_i measures the normalized imbalance between left- and right-handed rotational contributions. Values $P_i > 0$ indicate dominance of left-oriented coupling, while $P_i < 0$ correspond to right-oriented dominance.

These polarity structures act as local symmetry-breaking mechanisms within the rotational sector of the Æ-field and play a role in the emergence of chiral phase dynamics in higher-level interactions.

44. Gauge Structure

Covariant Rotational Derivative

The covariant derivative acting on Q is defined as:

$$D_\mu Q = \partial_\mu Q - A_\mu Q, \qquad A_\mu \in \mathfrak{su}(2) \tag{262}$$

Pure-gauge configurations satisfy:

$$D_\mu Q = 0 \tag{263}$$

Clifford Structure

The Clifford generators satisfy:

$$e_i e_j + e_j e_i = 2\delta_{ij} \tag{264}$$

The even subalgebra satisfies:

$$\mathrm{Cl}^+(3,0) \simeq \mathbb{H} \tag{265}$$

Quaternion identification:

$$i \leftrightarrow e_2 e_3, \quad j \leftrightarrow e_3 e_1, \quad k \leftrightarrow e_1 e_2 \tag{266}$$

Spinorial Sector

$$\psi \in \mathbb{C}^4 \tag{267}$$

Quaternionic action preserves norm:

$$\psi \mapsto Q\psi, \qquad \|Q\psi\| = \|\psi\| \tag{268}$$

The covariant derivative on spinors is:

$$D_\mu \psi = \partial_\mu \psi + A_\mu \psi \tag{269}$$

Dirac–Æ Equation

$$i\gamma^\mu D_\mu \psi - m\psi = 0 \tag{270}$$

Gauge-Coupled Derivatives

$$\nabla_\mu = \partial_\mu - i\,A_\mu \tag{271}$$

$$\partial_\mu \longrightarrow \nabla_\mu = \partial_\mu - i\,\mathcal{C}(D,d)\,A_\mu \tag{272}$$

$$A_D = -i\,\partial_D q \qquad A_d = -i\,\partial_d q \tag{273}$$

Gauge Curvature

$$F_{\mu\nu} = \partial_\mu A_\nu - \partial_\nu A_\mu + i[A_\mu, A_\nu] \tag{274}$$

$$F_{\mu\nu} = \partial_\mu A_\nu - \partial_\nu A_\mu + i\,\mathcal{C}(D,d)\,[A_\mu, A_\nu] \tag{275}$$

Dirac–Æ Operator (Gauge-Coupled)

$$\mathcal{D}_{\text{Æ}} = i\gamma^\mu \nabla_\mu + M(D,d) \tag{276}$$

$$(i\gamma^\mu \nabla_\mu - M)\,\psi = 0 \tag{277}$$

$$\mathcal{D}_{\text{Æ}} = i\gamma^\mu \nabla_\mu + i\gamma^D \partial_D + i\gamma^d \partial_d + M(D,d) \tag{278}$$

$$(i\gamma^\mu \nabla_\mu - m)\,\psi = 0 \tag{279}$$

Essential Spectrum and Resolvent

$$\sigma_{\text{ess}}(\mathcal{D}_{\text{Æ}}) = (-\infty, -|M_\infty|] \cup [|M_\infty|, \infty) \tag{280}$$

$$R(\lambda) = (\mathcal{D}_{\text{Æ}} - \lambda)^{-1} \tag{281}$$

Discrete Hodge and Internal Constraints

$$\star_{\text{Æ}} : \Lambda^k(\mathcal{M}) \to \Lambda^{3-k}(\mathcal{M}) \tag{282}$$

$$\{C, F, G\}_{\text{Æ}} = 0 \quad \forall\, F, G \tag{283}$$

Quaternionic Phase Normalization

In the Æ framework phase variables can be lifted to a quaternionic representation $Q(t) \in S^3$. The associated connection form is

$$A = \tfrac{1}{2}\,\mathrm{Im}\big(Q^{-1} dQ\big)\,.$$

Integration along the Hopf fiber yields the normalization condition

$$\oint_{\text{fiber}} A = \pi.$$

The corresponding intrinsic phase frequency becomes

$$\omega_{\text{Æ}}(t) = \big\|\tfrac{1}{2}\,\mathrm{Im}\big(Q^{-1}\dot{Q}\big)\big\|\,.$$

Physical time may be related to the intrinsic quaternionic time parameter $\tau_{\text{Æ}}$ through

$$t_{\text{phys}} := 2\,\tau_{\text{«}}.$$

This representation provides a geometric origin for oscillator frequencies in the synchronization dynamics.

Phase Retention and Stability

Long-range synchronization may be influenced by phase-retention effects associated with distributed coherence.

Define the retention operator acting on phase entropy

$$\mathcal{R}[S_\phi] = \Box S_\phi + \int K(x, x')\, S_\phi(x')\, dx'.$$

In Fourier space this induces the modified dispersion relation

$$\omega^2(k) = k^2 + m_{\text{eff}}^2 + \tilde{K}(\omega, k).$$

These corrections describe memory effects in extended phase systems.

Synchronization breakdown can occur when the underlying dynamical flow reaches a critical point. Let

$$\partial_t Q = F(Q; \eta), \tag{284}$$

where η is a control parameter.

Critical bifurcation occurs when

$$\det(DF(Q)) = 0. \tag{285}$$

Such events correspond to the loss of structural stability of coherent phase configurations.

Statistical Ensemble on S^3 (Open)

For completeness we record the canonical ensemble associated with quaternionic phase states $Q \in S^3$.

The partition function may be written as

$$Z(\beta) = \int_{S^3} e^{-\beta V(Q)}\, d\mu(Q), \tag{286}$$

where $d\mu(Q)$ denotes the invariant Haar measure on S^3.

The associated Gibbs distribution is

$$\rho(Q) = \frac{e^{-\beta V(Q)}}{Z(\beta)}. \tag{287}$$

The corresponding entropy is

$$S_Q = -\int_{S^3} \rho(Q) \ln \rho(Q)\, d\mu(Q). \tag{288}$$

A full statistical formulation of quaternionic phase ensembles is left for future development.

General Kuramoto Dynamics

Consider a network of N phase oscillators with phases $\phi_i(t)$. The generalized Kuramoto dynamics is

$$\dot{\phi}_i = \omega_i + \sum_{j=1}^{N} K_{ij} \sin(\phi_j - \phi_i) \tag{289}$$

Here:

- ω_i are intrinsic frequencies,
- K_{ij} are coupling coefficients,
- $G(V, E)$ defines the interaction graph.

Order Parameter

Define the complex order parameter

$$Re^{i\Theta} = \frac{1}{N}\sum_{j=1}^{N} e^{i\phi_j}. \tag{290}$$

$R \in [0, 1]$ measures global coherence:

- $R = 0$ incoherence,
- $0 < R < 1$ partial synchronization,
- $R = 1$ full synchronization.

Linearization Around Synchrony

Assume near-synchrony:

$$\phi_i = \Theta + \delta_i, \qquad |\delta_i| \ll 1.$$

Using $\sin(\phi_j - \phi_i) \approx \delta_j - \delta_i$, equation (??) becomes

$$\dot{\delta} = -L\delta \tag{291}$$

where L is the weighted graph Laplacian

$$L_{ij} = \begin{cases} -K_{ij}, & i \neq j, \\ \sum_{k \neq i} K_{ik}, & i = j. \end{cases} \tag{292}$$

Spectral Stability Condition

Let

$$L = U\Lambda U^{\dagger}, \qquad \Lambda = \mathrm{diag}(\lambda_1, \ldots, \lambda_N). \tag{293}$$

Properties:

- $\lambda_1 = 0$ (global phase mode),
- $\lambda_k \geq 0$ for connected graphs,

- λ_2 is the algebraic connectivity.

Solution of (??):

$$\delta(t) = \sum_{k=2}^{N} c_k e^{-\lambda_k t} u_k. \tag{294}$$

Synchronization stability requires

$$\boxed{\lambda_2 > 0}. \tag{295}$$

The characteristic synchronization time scale is

$$\tau_{\text{sync}} \sim \frac{1}{\lambda_2}. \tag{296}$$

Synchronization Regimes

Subcritical Coupling

$$K < K_c \quad \Rightarrow \quad R = 0.$$

Eigenvalues cluster near zero and perturbations persist.

Critical Regime

$$K \approx K_c.$$

$\lambda_2 \to 0^+$. The system becomes highly susceptible to global fluctuations.

Supercritical Regime

$$K > K_c.$$

λ_2 increases and convergence to synchrony is exponential.

Entropy Interpretation

Define phase entropy

$$S_\phi = -\sum_{i=1}^{N} p_i \log p_i, \qquad p_i \propto |\delta_i|^2. \tag{297}$$

Under linearized flow,

$$\frac{dS_\phi}{dt} = -\sum_{k=2}^{N} \lambda_k |c_k|^2 \leq 0. \tag{298}$$

Hence,

$$\boxed{\lambda_k > 0 \;\Rightarrow\; \frac{dS_\phi}{dt} < 0} \tag{299}$$

Spectral positivity guarantees entropy reduction.

Connection with Æ Coherence

Recall

$$C_{Æ} = \frac{1}{1 + S_\phi}. \tag{300}$$

Since

$$\frac{dS_\phi}{dt} < 0 \quad \Rightarrow \quad \frac{dC_{Æ}}{dt} > 0,$$

spectral synchronization directly increases coherence.

$$\boxed{\text{Spectral Gap} \iff \text{Coherence Growth}} \tag{301}$$

Relation to the Phase Transmission Operator Δ

Define the phase transmission operator

$$\Delta = L^{-1} \quad \text{(restricted to nonzero spectrum)}. \tag{302}$$

Then

$$\|\Delta\| = \frac{1}{\lambda_2}. \tag{303}$$

Near criticality,

$$\lambda_2 \to 0 \quad \Rightarrow \quad \|\Delta\| \to \infty.$$

Phase transmission becomes long-range, enabling global synchronization cascades.

Conclusion

Synchronization stability in the Æ Framework is governed entirely by Laplacian spectral structure:

$$\boxed{\lambda_2 > 0 \iff \text{Stable Coherent Regime}} \tag{304}$$

No additional metaphysical assumptions are required. Synchronization is a structural property of coupled phase systems.

45. Dirac–Æ Operator

Weighted Nodal Hilbert Space

$$\mathcal{H}_{\text{Æ}} = L^2(\mathbb{R}^4, \mathbb{C}^4; w_{\text{Æ}}), \tag{305}$$

with weight

$$w_{\text{Æ}}(x) = C_{\text{Æ}}(x)\, e^{-S_\phi(x)/\hbar}. \tag{306}$$

Inner product:

$$\langle \psi, \chi \rangle_{\text{Æ}} = \int_{\mathbb{R}^4} \psi^\dagger \chi\, w_{\text{Æ}}(x)\, d^4x. \tag{307}$$

Definition of the Dirac–Æ Operator

$$D_{\text{Æ}} = i\gamma^\mu \partial_\mu - M(D, d) \tag{308}$$

with Clifford algebra:

$$\{\gamma^\mu, \gamma^\nu\} = 2g^{\mu\nu} I. \tag{309}$$

Klein–Gordon Emergence

$$D_{\text{Æ}}^2 = -g^{\alpha\beta}\partial_\alpha \partial_\beta + M^2 + i\gamma^\alpha \partial_\alpha M. \tag{310}$$

In the constant-mass limit:

$$(\Box + m^2)\varphi = 0. \tag{311}$$

Effective Mass

$$M_{\text{eff}}^2 \approx M^2 + \frac{1}{\hbar^2}\left(|\nabla_{D,d} S_\phi|^2 + |\nabla_{D,d} \log C_{\text{«}}|^2\right). \tag{312}$$

$$\mathcal{D}_{\text{Æ}} = i\,\gamma^D \partial_D + i\,\gamma^d \partial_d + i\,\gamma^t \partial_t + i\,\gamma^q \partial_q + M(D, d) \qquad (313)$$

$$\mathcal{H}_{\text{Æ}} = L^2\Big(\mathbb{R}^4,\ \mathcal{C}(D, d)\, e^{-S_\phi(D,d)/\hbar}\, dD\, dd\, dt\, dq\Big)$$

$$\{\gamma^\mu, \gamma^\nu\} = 2\, g^{\mu\nu}\, \mathbb{I}$$

$$w_{\text{Æ}}(D, d) = \mathcal{C}(D, d)\, e^{-S_\phi(D,d)/\hbar}$$

$$\mathcal{D}_{\text{Æ}}\psi = i\gamma^D \partial_D \psi + i\gamma^d \partial_d \psi + i\gamma^t \partial_t \psi + i\gamma^q \partial_q \psi + M(D, d)\psi$$

$$\mathcal{D}^2_{\text{«}} = -\left(\partial_D^2 + \partial_d^2 + \partial_t^2 + \partial_q^2\right) + i\gamma^\mu \partial_\mu M + M^2$$

$$\Box = \partial_t^2 - \partial_q^2 - \partial_D^2 - \partial_d^2$$

$$(\Box + m^2)\phi = 0$$

$$w_{\text{Æ}}(D, d) = C_{\text{Æ}}(D, d)\, e^{-S_\phi(D,d)/\hbar}$$

$$C_{\text{Æ}}(D, d) = \frac{1}{1 + S_\phi(D, d)}$$

$$D^2_{\text{Æ}} = -g^{\alpha\beta} \partial_\alpha \partial_\beta + M^2 + i\gamma^\alpha \partial_\alpha M$$

$$\Box = g^{\mu\nu} \partial_\mu \partial_\nu$$

Dirac–Æ Operator Structure

The dynamics of coherent states in the Æ–framework are governed by a generalized Dirac operator acting on the Hilbert space $\mathcal{H}_{\text{Æ}}$ of field configurations.

Let $\psi \in \mathcal{H}_{\text{Æ}}$ denote a coherent field state. The Dirac–Æ operator acts as

$$D_{\text{«}}\psi = \left(i\gamma^{\mu}\partial_{\mu} + i\gamma^{D}\partial_{D} + i\gamma^{d}\partial_{d} + M(D,d)\right)\psi \tag{314}$$

where γ^{μ} are Dirac matrices associated with the spacetime coordinates and γ^{D}, γ^{d} correspond to additional internal coordinates of the Æ–space.

The mass term $M(D,d)$ encodes coherence-dependent interactions between layers of the field.

Projection onto the Physical Sector

Not all configurations in $\mathcal{H}_{\text{Æ}}$ correspond to physically realizable states. The observable sector is obtained through a projection operator

$$\Pi_{\text{phys}} : \mathcal{H}_{\text{Æ}} \to \mathcal{H}_{\text{phys}}, \tag{315}$$

which restricts the state space to configurations satisfying the physical consistency constraints of the theory.

For a field state ψ, the projected configuration is

$$\psi_{\text{phys}} = \Pi_{\text{phys}}\psi. \tag{316}$$

This projection ensures compatibility between the extended Æ-space dynamics and the observable sector of the theory.

Resolvent and Spectral Structure

The spectral properties of the Dirac–Æ operator play a central role in determining the stability and propagation of coherent structures in the field.

The resolvent operator is defined as

$$R(\lambda) = (D_{\text{Æ}} - \lambda I)^{-1} \tag{317}$$

for complex λ outside the spectrum of $D_{\text{Æ}}$.

The spectrum

$$\sigma(D_{\text{Æ}}) = \{\lambda \in \mathbb{C} \mid (D_{\text{Æ}} - \lambda I) \text{ is not invertible}\} \tag{318}$$

characterizes the admissible energy levels of the field.

Further spectral properties of the operator are analyzed in the Spectral Structure sector (Section 49).

For brevity, the Dirac–Æ operator will occasionally be written in the abbreviated form

$$D_{\text{«}} = i\gamma^{\mu}\partial_{\mu} + i\gamma^{D}\partial_{D} + i\gamma^{d}\partial_{d} + M(D, d)$$

which corresponds to the canonical definition introduced above.

The wave operator $\Box$ may appear either in explicit coordinate form or in metric-contracted form

$$\Box = g^{\mu\nu}\partial_{\mu}\partial_{\nu}.$$

Both expressions are equivalent once the sign convention of the metric tensor $g^{\mu\nu}$ is fixed.

46. Weighted Action

Extended Master Action (Dynamical Coherence Layer)

The covariant derivative acting on the scalar sector is given by

$$D_\mu \Phi = \left(\partial_\mu - i\, g_{w0}\, W^i_\mu \frac{\tau^i}{2} - i\, g'_0\, Y_\Phi\, B_\mu \right) \Phi \tag{319}$$

Gravitational coupling is modulated by coherence through an effective Planck mass

$$\frac{1}{2} M_P^2(C)\, R, \qquad M_P^2(C) := M_0^2\, C \tag{320}$$

Gauge sectors acquire coherence–entropy dependent weights

$$-\frac{1}{4} Z_s(C, S_\phi)\, G^a_{\mu\nu} G^{a\mu\nu} - \frac{1}{4} Z_w(C, S_\phi)\, W^i_{\mu\nu} W^{i\mu\nu} - \frac{1}{4} Z_y(C, S_\phi)\, B_{\mu\nu} B^{\mu\nu} \tag{321}$$

These weights are taken to share a universal functional form

$$Z_s = Z_w = Z_y = \frac{C}{1 + S_\phi} \tag{322}$$

Non-local memory effects are introduced through the bilocal kernel

$$S_{\text{mem}} = \frac{1}{2} \int d^4x \sqrt{-g(x)} \int d^4x' \sqrt{-g(x')}\, K(x, x')\, S_\phi(x) S_\phi(x') \tag{323}$$

The full action combines all sectors into a single coherence-dependent functional

$$S = \int d^4x\sqrt{-g}\Big[\frac{1}{2}M_0^2 CR - \frac{1}{4}Z_s G^a_{\mu\nu}G^{a\mu\nu} - \frac{1}{4}Z_w W^i_{\mu\nu}W^{i\mu\nu} - \frac{1}{4}Z_y B_{\mu\nu}B^{\mu\nu}$$
$$+ (D_\mu\Phi)^\dagger(D^\mu\Phi) - V(\Phi) + \bar{\psi}\, i\gamma^\mu D_\mu\psi - \bar{\psi}\, Y\Phi\psi$$
$$+ \frac{1}{2}\partial_\mu C\, \partial^\mu C + \frac{1}{2}\partial_\mu S_\phi\, \partial^\mu S_\phi - U(C, S_\phi)\Big] + S_{\text{mem}} \tag{324}$$

Operator Extraction (From the Variational Principle)

The fermionic sector emerges from the variational structure through the Dirac–Yukawa interaction

$$\bar{\psi}\, i\gamma^\mu D_\mu\psi - \bar{\psi}\, Y\Phi\psi \tag{325}$$

From this coupling, an effective mass scale is generated after symmetry breaking

$$M_{\text{Yukawa}} \sim Yv \tag{326}$$

This structure naturally leads to the definition of the Dirac–Æ operator

$$D_{\text{Æ}} = i\gamma^\mu\nabla_\mu + M_{\text{eff}}(C, S_\phi) \tag{327}$$

Squaring the operator reveals its spectral and geometric content

$$D_{\text{Æ}}^2 = -\nabla^\mu\nabla_\mu + M_{\text{eff}}^2 + \text{curvature and coherence terms} \tag{328}$$

Beyond binary interactions, the framework admits a ternary structure encoded by the Æ–Nambu bracket

$$\{F, G, H\}_{\text{Æ}} = \det\left(\frac{\partial(F, G, H)}{\partial(C, S_\phi, \phi)}\right) \tag{329}$$

Finally, all operators act on a coherence-weighted structure of the underlying space

$$w_{Æ} = C\, e^{-S_\phi/\hbar} \tag{330}$$

Variational Template (Reference Form)

The general structure of the theory is governed by an action functional defined over the fields of the system

$$S[\Psi] = \int d^4x\, \sqrt{-g}\; \mathcal{L}(\Psi, \partial\Psi) \tag{331}$$

Physical configurations correspond to stationary points of this action

$$\delta S = 0 \tag{332}$$

This condition yields the equations of motion for the fields

$$\mathcal{E}(\Psi) = 0 \tag{333}$$

Within the Æ framework, all contributions are weighted by a coherence-dependent measure

$$w_{Æ} = C_{Æ}\, e^{-S_\phi/\hbar} \tag{334}$$

Variational Folding Principle (Topological Sub-sector)

In the topological sector, configurations are described as embeddings or trajectories subject to competing constraints

$$\mathcal{F}(\gamma) = \mathcal{E}_{\text{elastic}} + \mathcal{E}_{\text{top}} - \beta \mathcal{K}_{Æ} \tag{335}$$

Stationary configurations correspond to balanced geometrical and topological tensions

$$\frac{\delta \mathcal{F}}{\delta \gamma} = 0 \tag{336}$$

Interaction-Specific Effective Actions (Optional Templates)

Effective interactions may be expressed through reduced actions capturing dominant physical regimes

Wilson loop expectation values encode confinement-like behavior

$$\langle W(\gamma)\rangle \;\sim\; \exp\big(-\sigma\,\mathcal{A}(\gamma)\big), \qquad \sigma > 0 \tag{337}$$

The effective tension may evolve dynamically with phase entropy

$$\sigma_{\text{eff}}(t) = \sigma_0\, e^{-S_\phi(t)/\kappa_{Æ}}, \qquad \sigma_0 > 0 \tag{338}$$

A Yang–Mills type action governs the strong sector

$$\mathcal{S}_{\text{strong}} = -\frac{1}{4g_s^2}\int \mathrm{Tr}(\mathcal{F}_{\mu\nu}\mathcal{F}^{\mu\nu})\,\sqrt{-g}\,d^4x \tag{339}$$

Transitions in the weak sector may be driven by phase entropy flow

$$\hat{\mathcal{T}}_W = \lambda_W\,\dot{S}_\phi\,\hat{\Pi}_{\text{break}}, \qquad \lambda_W \geq 0 \tag{340}$$

Spontaneous symmetry breaking is encoded through a scalar potential

$$V(\varphi) = -\frac{\mu^2}{2}\,|\varphi|^2 + \frac{\lambda}{4}\,|\varphi|^4, \qquad \mu^2, \lambda > 0 \tag{341}$$

Mass generation follows from the symmetry-breaking scale

$$m_W \sim g_W\, v \tag{342}$$

Master Coherence-Dependent Action

In the extended Æ-framework, the full variational structure allows the gravitational, gauge, scalar and coherence sectors to couple dynamically through the canonical coherence field C.

$$M_P^2(C) := M_0^2\, C. \tag{343}$$

The gauge kinetic renormalization factors are defined as:

$$Z_s = Z_w = Z_y = \frac{C}{1 + S_\phi}. \tag{344}$$

The unified master action reads:

$$S = \int d^4x \sqrt{-g} \Big[\frac{1}{2} M_0^2 C R - \frac{1}{4} Z_s G^a_{\mu\nu} G^{a\mu\nu} - \frac{1}{4} Z_w W^i_{\mu\nu} W^{i\mu\nu} - \frac{1}{4} Z_y B_{\mu\nu} B^{\mu\nu}$$
$$+ (D_\mu \Phi)^\dagger (D^\mu \Phi) - V(\Phi) + \bar{\psi} i \gamma^\mu D_\mu \psi - \bar{\psi} Y \Phi \psi$$
$$+ \frac{1}{2} \partial_\mu C \, \partial^\mu C + \frac{1}{2} \partial_\mu S_\phi \, \partial^\mu S_\phi - U(C, S_\phi) \Big] + S_{\text{mem}}. \tag{345}$$

Effective Mass Renormalization

Spatial variations of coherence and phase entropy induce an effective mass renormalization of matter fields.

$$m_{\text{eff}}^2 \sim m_0^2 + \frac{1}{\hbar^2} |\nabla S_\phi|^2 + |\nabla \log C|^2. \tag{346}$$

This relation expresses how gradients in the coherence structure modify the local inertial response of fields propagating in the Æ-background.

Classical Consistency Limit

In the classical coherence limit the model reduces to the standard renormalized gauge–field theory structure:

$$S_\phi \to 0, \qquad C \to 1, \qquad Z_s, Z_w, Z_y \to 1. \tag{347}$$

Here Z_s, Z_w, Z_y denote the renormalization factors of the strong, weak, and Yukawa sectors respectively.

Variational Hierarchy

The structural logic of the Æ-framework follows the standard variational hierarchy:

Action $\longrightarrow$ Field Equations $\longrightarrow$ Operators $\longrightarrow$ Spectra.

In this chain, the weighted action defines the field dynamics, which generate operator structures whose spectra encode the observable particle and excitation content of the theory.

Nonlocal Coherence Memory Sector

$$S_{\text{mem}} = \frac{1}{2}\int d^4x\sqrt{-g(x)}\int d^4x'\sqrt{-g(x')}K(x,x')\,S_\phi(x)S_\phi(x'). \tag{348}$$

This term encodes nonlocal phase-memory correlations within the coherence field.

Modified Einstein Equations

Variation with respect to $g^{\mu\nu}$ yields:

$$M_0^2\,C\,G_{\mu\nu} + M_0^2\,(g_{\mu\nu}\Box - \nabla_\mu\nabla_\nu)\,C = T_{\mu\nu}^{(\text{tot})} - g_{\mu\nu}U(C,S_\phi) + T_{\mu\nu}^{(\text{mem})}, \tag{349}$$

with

$$T_{\mu\nu}^{(\text{mem})} := -\frac{2}{\sqrt{-g}}\frac{\delta S_{\text{mem}}}{\delta g^{\mu\nu}}. \tag{350}$$

Coherence Field Equation

$$\Box C + \frac{1}{2}M_0^2 R - \frac{1}{4}\frac{\partial Z_s}{\partial C}G^a_{\mu\nu}G^{a\mu\nu} - \frac{1}{4}\frac{\partial Z_w}{\partial C}W^i_{\mu\nu}W^{i\mu\nu} - \frac{1}{4}\frac{\partial Z_y}{\partial C}B_{\mu\nu}B^{\mu\nu} - \frac{\partial U}{\partial C} = 0. \tag{351}$$

Extended Coherence Layer in the Weighted Action

Beyond the minimal weighted action introduced above, the Æ-framework admits an extended coherence layer in which the phase entropy S_ϕ appears in several complementary forms.

A relative phase entropy may be defined as

$$S_\phi^{\mathrm{rel}} = \int p(\phi) \log\left(\frac{p(\phi)}{p_{\mathrm{ref}}(\phi)}\right) d\phi \tag{352}$$

where $p_{\mathrm{ref}}(\phi)$ denotes a reference phase distribution.

For discrete coherent sectors indexed by k, one may also write

$$S_\phi^{\mathrm{disc}} = -\sum_k p_k \log p_k \tag{353}$$

which provides a discrete entropy representation of the phase structure.

A spectral probability distribution may be introduced through eigenmode weights $p_k^{(\mathrm{spec})}$, yielding

$$S_\phi^{\mathrm{pk}} = -\sum_k p_k^{(\mathrm{spec})} \log p_k^{(\mathrm{spec})} \tag{354}$$

and the associated spectral entropy functional

$$S_\phi^{\mathrm{spec}} = -\int \rho(\lambda) \log \rho(\lambda)\, d\lambda \tag{355}$$

where $\rho(\lambda)$ is the spectral density of coherent modes.

These entropy functionals refine the weight factor appearing in the action and allow the theory to distinguish between local, discrete, and spectral coherence regimes.

Variational Origin of Weighted Operators

The weighted operators of the Æ-framework arise from the variational structure of the action itself.

Let the weighted action be written schematically as

$$\mathcal{S}_{\text{Æ}} = \int d^4x\, w_{\text{Æ}}(x)\, \mathcal{L}_{\text{Æ}}, \tag{356}$$

where the weight depends on coherence and phase entropy

through

$$w_{\text{Æ}}(x) = C_{\text{Æ}}(x)\, e^{-S_\phi(x)/\hbar}. \tag{357}$$

Taking the variation with respect to a generic field Ψ gives

$$\delta \mathcal{S}_{\text{Æ}} = \int d^4x\, \delta\Psi\, \frac{\delta}{\delta\Psi}\big(w_{\text{Æ}}\mathcal{L}_{\text{Æ}}\big), \tag{358}$$

which shows that the effective operators inherit both the local Lagrangian structure and the coherence weighting.

Thus the differential operators appearing in the field equations are not introduced independently, but emerge as weighted variational operators derived from the action principle.

47. Equations of Motion

Weighted Dirac Equation

The full weighted Dirac equation reads:

$$\left(i\gamma^\mu D_\mu + \frac{i}{2}\gamma^\mu \partial_\mu \log w - m\right)\psi = 0 \tag{359}$$

Nodal (Constant-Weight) Approximation

In regions where the weight is locally constant ($\partial_\mu w = 0$), the equation reduces to:

$$(i\gamma^\mu D_\mu - m)\psi = 0 \tag{360}$$

The gauge field equation becomes:

$$D_\nu F^{\nu\mu} = J^\mu \tag{361}$$

Gauge Current Closure

The conserved gauge current takes values in $\mathfrak{su}(2)$:

$$J^\mu = (\bar{\psi}\gamma^\mu T^a \psi)\, T^a \tag{362}$$

where T^a are the generators of $\mathfrak{su}(2)$.

Æ-Dynamics as a Hamiltonian Flow (Abstract Form)

The dynamics of the Æ field can be formulated in terms of an abstract Hamiltonian functional

$$\mathcal{H}_Æ : \mathcal{A}_Æ \to \mathbb{R} \tag{363}$$

This Hamiltonian encodes both geometric variation and potential structure over the Æ configuration space

$$\mathcal{H}_{\text{Æ}}[\text{Æ}] = \int_{\Omega} \left(\|\nabla_{\text{Æ}} \text{Æ}\|^2 + V_{\text{Æ}}(\text{Æ}) \right) d\mu_{\text{Æ}} \tag{364}$$

Time evolution is generated through a bracket structure defined over the Æ algebra

$$\partial_t \text{Æ} = \{\text{«}, \mathcal{H}_{\text{Æ}}\}_{\text{Æ}} \tag{365}$$

The phase field evolves according to the functional derivative of the Hamiltonian

$$\partial_t \phi = \frac{\delta \mathcal{H}_{\text{Æ}}}{\delta \phi} \tag{366}$$

In the absence of dissipative effects, the Hamiltonian remains conserved

$$\frac{d}{dt} \mathcal{H}_{\text{Æ}} = 0 \tag{367}$$

At the same time, coherence exhibits a monotonic behavior under evolution

$$\partial_t C \leq 0 \tag{368}$$

For composite systems, the total Hamiltonian includes both individual and interaction contributions

$$\mathcal{H}_{\text{Æ}}^{\text{tot}} = \sum_i \mathcal{H}_{\text{Æ}}[\text{Æ}_i] + \sum_{i<j} \mathcal{H}_{\text{Æ}}^{\text{int}}(\text{Æ}_i, \text{Æ}_j) \tag{369}$$

Interactions are encoded through coherence-weighted coupling terms

$$\mathcal{H}_{\text{Æ}}^{\text{int}}(\text{Æ}_i, \text{Æ}_j) = g_{ij}\ \mathrm{Re}(\text{Æ}_i^{\dagger} \boxtimes \text{Æ}_j) \tag{370}$$

Finally, dissipative corrections introduce a flow driven by phase entropy gradients

$$\partial_t \text{Æ} = \{\text{Æ}, \mathcal{H}_{\text{Æ}}\}_{\text{Æ}} - \gamma \frac{\delta S_{\phi}}{\delta \text{Æ}}, \qquad \gamma \geq 0 \tag{371}$$

Linearization and Spectral Modes

To analyze small perturbations, the Æ field is expanded around a reference configuration

$$\text{Æ} = \text{Æ}_0 + \delta\text{Æ} \tag{372}$$

The evolution of fluctuations is governed by a linearized operator acting on the perturbation

$$\partial_t(\delta\text{Æ}) = \mathcal{L}_{\text{Æ}}(\delta\text{Æ}) \tag{373}$$

Normal modes can be expressed through a spectral decomposition in time

$$\text{«}_\mu(x,t) = \psi_\mu(x)\, e^{i\omega_\mu t} \tag{374}$$

The corresponding frequencies belong to the spectrum of the governing operator

$$\omega_\mu \in \text{Spec}(K_{\text{Æ}}) \tag{375}$$

Stability can be characterized by bounded phase fluctuations

$$\text{Var}(\phi(t)) \leq \epsilon \tag{376}$$

Nambu–Æ Equations of Motion (Ternary Dynamics)

Beyond binary interactions, the dynamics may be formulated using a ternary bracket structure

$$\partial_t F = \{F, H_1, H_2\}_{\text{Æ}} \tag{377}$$

Applying this structure to the fundamental variables yields the evolution equations

$$\dot{D} = \{D, H_1, H_2\}_{\text{Æ}}, \qquad \dot{d} = \{d, H_1, H_2\}_{\text{Æ}}, \qquad \dot{\varphi} = \{\varphi, H_1, H_2\}_{\text{Æ}} \tag{378}$$

Explicitly, the ternary bracket produces a coupled system of equations

$$\dot{D} = \partial_d H_1 \, \partial_\varphi H_2 - \partial_\varphi H_1 \, \partial_d H_2, \tag{379}$$

$$\dot{d} = \partial_\varphi H_1 \, \partial_D H_2 - \partial_D H_1 \, \partial_\varphi H_2, \tag{380}$$

$$\dot{\varphi} = \partial_D H_1 \, \partial_d H_2 - \partial_d H_1 \, \partial_D H_2 \tag{381}$$

Casimir / Volume-Form Conservation (Optional Formal Layer)

In the ternary Hamiltonian structure, conservation laws emerge from the invariance of the underlying volume form

$$\mathcal{L}_{X_{H_1,H_2}} \Omega_{Æ} = 0 \tag{382}$$

This condition reflects the preservation of phase-space structure under the induced flow

$$\mathcal{L}_X(\alpha \wedge d\alpha) = i_X d(\alpha \wedge d\alpha) + d(i_X(\alpha \wedge d\alpha)) \tag{383}$$

As a consequence, global invariants can be constructed from the conserved volume form

$$\frac{d}{dt} C_{Æ} = \int_{\mathcal{M}_{Æ}} \mathcal{L}_{X_{H_1,H_2}}(\alpha \wedge d\alpha) = 0 \tag{384}$$

In physical terms, this conservation is analogous to helicity-type invariants in fluid and gauge systems

$$\{\mathbf{A} \cdot \mathbf{B}, H_1, H_2\}_{Æ} = \nabla \cdot [(\mathbf{A} \cdot \mathbf{B})(\nabla H_1 \times \nabla H_2)] \tag{385}$$

Local Equation for S_ϕ with Memory Kernel (Key Bridge)

The phase entropy field evolves according to a local equation supplemented by non-local memory effects

$$\Box S_\phi - \frac{\partial U}{\partial S_\phi} + \int d^4x' \, \sqrt{-g(x')} \, K(x, x') \, S_\phi(x') = \mathcal{J}(C, F_{Æ}) \tag{386}$$

Renudation Condition (Topological Reset Criterion)

Topological persistence is encoded in the holonomy associated with closed loops

$$H_\gamma(\lambda) \neq \mathbb{I} \tag{387}$$

Under repeated evolution, the system may approach a return condition

$$H_\gamma(\lambda)^2 = \mathbb{I} \tag{388}$$

This transition defines the renudation process, where nontrivial configurations collapse to the identity

$$\mathrm{NR} \xrightarrow{\text{repeat}+\lambda} \mathbb{I} \tag{389}$$

Æ Dynamical Evolution

The dynamical behavior of coherent configurations in the Æ–framework is governed by evolution equations derived from the weighted action.

Let Ψ denote a field configuration. The evolution equation may be written schematically as

$$\frac{d\Psi}{dt} = \mathcal{D}_{Æ}(\Psi), \tag{390}$$

where $\mathcal{D}_{Æ}$ represents the dynamical operator induced by the variational structure of the weighted action.

These equations generalize Hamiltonian evolution by incorporating coherence-dependent weighting and phase entropy corrections.

In the extended Æ–framework, the dynamical evolution can be expressed using Nambu mechanics, allowing multi-observable dynamics beyond standard Hamiltonian systems.

Nambu Equations of Motion

The extended phase structure of the Æ-field naturally admits a Nambu-type formulation of the equations of motion.

Let F, G, and H denote observables defined on the phase space. The Nambu bracket is defined as

$$\{F, G, H\}_{Æ}. \tag{391}$$

The evolution of an observable A is then governed by

$$\frac{dA}{dt} = \{A, H_1, H_2\}_{Æ} \tag{392}$$

where H_1 and H_2 act as generalized Hamiltonians.
Observable evolution follows

$$\frac{dA_{\text{obs}}}{dt} = \{A_{\text{obs}}, H_1, H_2\}_{Æ} \tag{393}$$

while in coordinate representation one obtains

$$\dot{x}_i = \{x_i, H_1, H_2\}_{Æ} \tag{394}$$

These equations generalize Hamiltonian dynamics by allowing ternary interactions between conserved quantities.

Volume Preservation in Nambu Dynamics

A defining property of Nambu systems is the preservation of phase space volume under dynamical evolution.

Let ρ denote the phase space density. The Nambu flow satisfies

$$\frac{d\rho}{dt} = 0, \tag{395}$$

which implies the conservation of phase space volume.

This property ensures that the extended Æ-dynamics preserve the geometric structure of the underlying phase space while allowing multi-Hamiltonian evolution.

Such volume-preserving flows play a role in the stability of coherent structures and in the transport of phase information across the system.

48. Classical Limit

Nodal Stability

Classical persistence corresponds to dynamical stabilization of coherence and phase-entropy:

$$\delta C = 0, \qquad \partial_t C \approx 0 \tag{396}$$

together with bounded phase fluctuations:

$$\mathrm{Var}(S_\phi(t)) \leq \epsilon, \qquad \epsilon \ll 1 \tag{397}$$

In the Hamiltonian formulation:

$$\frac{d}{dt}\mathcal{H}_{Æ} = 0 \tag{398}$$

ensuring energy conservation in the classical regime.

Quaternionic Classical Carrier

The classical carrier remains a unit quaternion:

$$|Q| = 1, \qquad Q \in S^3 \tag{399}$$

$$e^{u\theta} = \cos\theta + u\sin\theta \tag{400}$$

with dynamical evolution:

$$\partial_t Q = \Omega Q, \qquad \Omega \in \mathfrak{su}(2) \tag{401}$$

In constrained quaternionic dynamics:

$$\ddot{Q} + (\dot{Q}\overline{\dot{Q}})Q = 0 \tag{402}$$

which preserves the unit-sphere constraint $|Q| = 1$.

Metric Scaling and Emergent Classical Geometry

The coherence-induced metric is

$$g_{\mu\nu}^{\text{Æ}} = C(x)\, g_{\mu\nu}. \tag{403}$$

In the classical regime:

$$C(x) \to C_0 > 0 \quad \Rightarrow \quad g_{\mu\nu}^{\text{Æ}} \to C_0 g_{\mu\nu} \tag{404}$$

yielding a stable effective geometry.

Pre-Metric Regime

The pre-metric regime corresponds to

$$C(x) \to 0 \quad \Rightarrow \quad g_{\mu\nu}^{\text{Æ}} \to 0. \tag{405}$$

In this limit, spacetime loses its classical metric structure, and the dynamics reduce to pure coherence-field evolution.

Density vs. Dimension

The classical quaternionic carrier introduced above can be generalized to a density-dependent coherence sphere. Instead of a fixed unit sphere, the admissible quaternionic states are constrained to

$$S^3(D) = \{Q \in \mathbb{H} \mid \|Q\| = R(D)\}, \tag{406}$$

where $R(D)$ denotes a density-controlled coherence radius.

In the Æ framework this radius scales with the local coherence field,

$$R(D) \propto C_{\text{Æ}}(D). \tag{407}$$

The embedding space of the configuration may be represented by a topological dimension d,

$$\mathbb{R}^d. \tag{408}$$

The admissible quaternionic configuration therefore satisfies

$$Q_{Æ}(D,d) \in S^3(D) \subset \mathbb{R}^d. \tag{409}$$

This relation distinguishes the intrinsic coherence geometry (the S^3 structure) from the ambient topological dimension d, clarifying how density-controlled coherence spheres may be embedded within higher-dimensional configuration spaces.

49. Spectral and Scale Structure

Beta Functions

The running of fields under scale transformations is encoded through beta functions

$$\beta_\phi = \mu \frac{d\phi}{d\mu}, \qquad \beta_Q = \mu \frac{dQ}{d\mu} \tag{410}$$

Flow of the Dirac–Æ Operator

The scale dependence of the Dirac–Æ operator defines its renormalization flow

$$\beta_{D_{Æ}} = \mu \frac{dD_{Æ}}{d\mu} \tag{411}$$

Spectral Invariance Condition

Spectral quantities may remain invariant under flow through a trace condition

$$\mu \frac{d}{d\mu} \mathrm{Tr} \left(f \left(\frac{D_{Æ}}{\Lambda} \right) \right) = 0 \tag{412}$$

Effective Action Flow

The effective action evolves according to a functional renormalization group equation

$$\partial_\mu \Gamma_\mu = \frac{1}{2} \mathrm{Tr} \left[(\Gamma_\mu^{(2)} + R_\mu)^{-1} \partial_\mu R_\mu \right] \tag{413}$$

Coherence–Entropy Coupling Under Flow

Coherence and phase entropy are coupled through their scale evolution

$$\mu \frac{dC}{d\mu} = -\alpha \mu \frac{dS_\phi}{d\mu} \tag{414}$$

Spectral Dimension

The effective dimension of the system can be extracted from the scaling of the heat kernel

$$d_s = -2 \frac{d \log K(t)}{d \log t} \tag{415}$$

Fixed Points and Stability

Fixed points correspond to scale-invariant configurations of the flow

$$\beta_i(g^\star) = 0, \qquad \beta_\phi(\phi^\star) = 0. \tag{416}$$

Stability is determined by the eigenvalues of the linearized flow

$$\lambda_i = \left. \frac{\partial \beta_i}{\partial g_j} \right|_{g=g^*}. \tag{417}$$

Topological Charges

Topological sectors are characterized by quantized invariants of the field configuration

$$n = \frac{1}{8\pi^2} \int \mathrm{Tr}(F \wedge F). \tag{418}$$

$$\nu = \frac{1}{24\pi^2} \int \mathrm{Tr}(Q^{-1} dQ)^3. \tag{419}$$

Spectral Stability and Correlation Length

Collective stability across scales may be characterized through the spectral gap of the phase-Laplacian L_ϕ.

$$\lambda_2(L_\phi) \to 0, \tag{420}$$

indicating the onset of large-scale coherent structures. The associated correlation length diverges according to

$$\xi \sim \frac{1}{\sqrt{\lambda_2}}. \tag{421}$$

A quasi-stationary regime of hierarchical coherence corresponds to

$$\frac{dS_\phi}{dt} \approx 0. \tag{422}$$

Such regimes characterize scale-stable configurations in which phase entropy ceases to grow while global coherence remains sustained.

Kernel Formalism and Spectral Decomposition

We treat the kernel as an operator on the nodal state space:

$$K_{\text{Æ}}(t) : \mathcal{A}_{\text{Æ}} \to \mathcal{A}_{\text{Æ}} \tag{423}$$

A generic kernel-driven reconstruction is written as:

$$\text{Æ}(t) = \int K_{\text{Æ}}(t, t')\, \text{Æ}(t')\, dt' \tag{424}$$

Normalization and boundedness conditions:

$$\int K_{\text{Æ}}(t, t')\, dt' = 1, \qquad \|K_{\text{Æ}}\| \leq 1 \tag{425}$$

Spectral resolution:

$$K_{\text{Æ}}(t, t') = \sum_n \lambda_n\, \psi_n(t)\, \psi_n^{\dagger}(t'), \qquad |\lambda_n| \leq 1 \tag{426}$$

Prime-Indexed Oscillatory Sector

A distinguished oscillatory sector may be represented through prime-indexed modes

$$K_{\text{Æ}}^{(p)}(t) = \sum_{p \in \mathbb{P}} \beta_p\, e^{i\omega_p t}, \tag{427}$$

with logarithmic frequency scaling

$$\omega_p = \log p. \tag{428}$$

Square summability of amplitudes ensures convergence:

$$\sum_p |\beta_p|^2 < \infty. \tag{429}$$

Mellin Transform Representation

Spectral analysis may be performed using the Mellin transform:

$$\mathcal{M}_{\text{Æ}}\{f\}(s) = \int_0^{\infty} f(t)\, t^{s-1}\, dt, \qquad s = \sigma + i\tau. \tag{430}$$

Under convolution with the kernel, the transform diagonalizes the operator:

$$\mathcal{M}_{Æ}\{K_{Æ} * f\}(s) = \widehat{K}_{Æ}(s)\, \mathcal{M}_{Æ}\{f\}(s). \tag{431}$$

Dual Space Structure

The dual functional space of $\mathcal{A}_{Æ}$ is defined as

$$\mathcal{A}_{Æ}^{*} = \{\ell : \mathcal{A}_{Æ} \to \mathbb{C} \mid \ell \text{ linear and continuous}\}. \tag{432}$$

Pairing between a functional ℓ and a field configuration Æ is given by

$$\langle \ell, Æ \rangle = \int \ell(t)\, Æ(t)\, dt. \tag{433}$$

Dirac–Æ Flow with Dynamical Coherence

When C and S_ϕ are treated as running fields, the Dirac operator flow acquires implicit dependence:

$$\mu \frac{dD_{Æ}}{d\mu} = \frac{\partial D_{Æ}}{\partial C} \mu \frac{dC}{d\mu} + \frac{\partial D_{Æ}}{\partial S_\phi} \mu \frac{dS_\phi}{d\mu} + \beta_{D_{Æ}}^{(\text{explicit})} \tag{434}$$

Remark. The term $\beta_{D_{Æ}}^{(\text{explicit})}$ collects contributions not mediated by (C, S_ϕ), such as standard quantum loop corrections.

Running of Coherence Field

From the master action and $Z_i(C, S_\phi)$ dependence:

$$\mu \frac{dC}{d\mu} = \frac{\partial C}{\partial S_\phi} \mu \frac{dS_\phi}{d\mu} + \sum_i \frac{\partial C}{\partial Z_i} \mu \frac{dZ_i}{d\mu} \tag{435}$$

In the canonical approximation $C = 1/(1 + S_\phi)$:

$$\mu \frac{dC}{d\mu} = -\frac{1}{(1 + S_\phi)^2} \mu \frac{dS_\phi}{d\mu} \tag{436}$$

Spectral Invariance with Coherence Weight

The spectral invariance condition generalizes to:

$$\mu \frac{d}{d\mu} \mathrm{Tr} \left(f \left(\frac{D_{\text{Æ}}}{\Lambda(C)} \right) \right) = 0 \tag{437}$$

where the effective cutoff may scale as

$$\Lambda(C) = \Lambda_0 \, C^{1/2} \tag{438}$$

Flow of the Memory Kernel

The nonlocal kernel $K(x, x')$ may acquire scale dependence:

$$\mu \frac{d}{d\mu} K(x, x') = \beta_K(C, S_\phi; x, x') \tag{439}$$

Scale-Dependent Spectral Dimension

If the heat kernel depends on coherence:

$$K(t; C) = \mathrm{Tr} \, e^{-t D^2_{\text{Æ}}(C)}, \tag{440}$$

then the spectral dimension becomes:

$$d_s(C) = -2 \frac{d \log K(t; C)}{d \log t} \tag{441}$$

Fixed Points with Coherence Coupling

A fully coherent fixed point satisfies:

$$\beta_i(g^\star) = 0, \qquad \beta_\phi(\phi^\star) = 0, \qquad \beta_C(C^\star) = 0 \tag{442}$$

Linear stability now includes coherence sector:

$$\lambda_a = \left. \frac{\partial \beta_a}{\partial g_b} \right|_{g = g^\star, C = C^\star} \tag{443}$$

Topological-Spectral Stability Condition

Topological charges remain invariant under flow:

$$\mu \frac{dn}{d\mu} = 0, \qquad \mu \frac{d\nu}{d\mu} = 0 \tag{444}$$

Prime-Log Spectral Skeleton (Optional)

A prime-indexed oscillatory component:

$$K_{Æ}^{(p)}(t) = \sum_{p \in \mathbb{P}} \beta_p \, e^{i\omega_p t}, \qquad \omega_p = \log p \tag{445}$$

with square-summability:

$$\sum_{p \in \mathbb{P}} |\beta_p|^2 < \infty \tag{446}$$

Mellin Structure of Kernel Convolutions

Define the Mellin transform:

$$\mathcal{M}_{Æ}\{f\}(s) = \int_0^{\infty} f(t) \, t^{s-1} \, dt, \qquad s = \sigma + i\tau \tag{447}$$

Kernel convolution becomes multiplicative in Mellin space:

$$\mathcal{M}_{Æ}\{K_{Æ} * f\}(s) = \widehat{K}_{Æ}(s) \, \mathcal{M}_{Æ}\{f\}(s) \tag{448}$$

Essential Spectrum and Resolvent

For the Dirac–Æ operator, the essential spectrum:

$$\sigma_{\text{ess}}(\mathcal{D}_{Æ}) = (-\infty, -|M_\infty|] \cup [|M_\infty|, \infty) \tag{449}$$

The resolvent is defined by:

$$R(\lambda) = (\mathcal{D}_{Æ} - \lambda)^{-1} \tag{450}$$

Discrete Hodge Structure

The spectral structure of the Æ field can also be expressed using a discrete Hodge decomposition on the relevant complex.

We introduce the discrete Hodge star map

$$\star_{Æ} : \Lambda^k(\mathcal{M}) \to \Lambda^{3-k}(\mathcal{M}) \tag{451}$$

which defines the duality between discrete k-forms and $(3-k)$-forms on the effective Æ-complex.

Let ∇ denote the discrete gradient and $\nabla\cdot$ the discrete divergence operator. The associated Laplacian is

$$\Delta = \nabla \cdot \nabla. \tag{452}$$

This operator induces the decomposition of fields into

$$\mathcal{F} = \mathcal{F}_{\text{grad}} \oplus \mathcal{F}_{\text{harm}} \oplus \mathcal{F}_{\text{div}}, \tag{453}$$

corresponding respectively to gradient, harmonic, and divergence components.

Such a decomposition allows the separation of coherent modes from purely dissipative components in the spectral analysis, and provides a discrete geometric framework compatible with the operator structure of the Æ field.

Spectral Stability and Spectral Memory Operators

A spectral stability operator along a curve γ:

$$L_\gamma = -\frac{d^2}{ds^2} + V_{\text{top}}(s) \tag{454}$$

Small deformations are controlled by the spectral gap:

$$\|\delta\gamma\| \lesssim \frac{\epsilon}{\Delta\lambda} \tag{455}$$

A spectral memory operator template:

$$L_\Gamma = -\frac{d^2}{ds^2} + V_{\text{helix}}(s), \qquad L_\Gamma u_k = \lambda_k u_k \tag{456}$$

Spectral gap:

$$\Delta\lambda = \lambda_2 - \lambda_1 > 0 \tag{457}$$

Transcendental Fixed Point (Interpretative Closure)

A fixed-point attractor state:

$$\Psi(t) \to \Phi_{Æ} \tag{458}$$

Normalization (phase-coherent locking):

$$\text{Re}\big(\langle \Psi, \Phi_{Æ} \rangle\big) = 1 \tag{459}$$

Stationarity conditions:

$$\nabla\Psi = 0, \qquad \dot{S}_\phi = 0 \tag{460}$$

Local stability of the fixed point:

$$\left.\frac{\delta^2 S_\phi}{\delta\Psi^2}\right|_{\Phi_{Æ}} > 0 \tag{461}$$

Linearized Modes of the Æ Operator

To analyze stability and small fluctuations, one considers linear perturbations around a coherent background configuration.

Let Ψ_0 denote a stationary solution of the field equations. A perturbed configuration can be written as

$$\Psi = \Psi_0 + \epsilon\,\delta\Psi. \tag{462}$$

Linearization of the evolution operator leads to a spectral problem

$$K_{Æ}\,\delta\Psi = \lambda\,\delta\Psi, \tag{463}$$

where $K_{Æ}$ is the linearized dynamical operator.

The set of eigenvalues defines the spectrum

$$\mathrm{Spec}(K_{\text{Æ}}) = \{\lambda_i\}. \tag{464}$$

The stability of the background configuration is determined by the location of these eigenvalues in the complex plane.

Spectral Gap and Coherence Stability

The stability of coherent structures can be characterized by the spectral gap of the phase operator.

Let L_ϕ denote the operator governing phase synchronization. Its spectrum may be written as

$$\lambda_1 \le \lambda_2 \le \lambda_3 \le \cdots. \tag{465}$$

The spectral gap is defined as

$$\Delta\lambda = \lambda_2 - \lambda_1. \tag{466}$$

A nonzero spectral gap implies stability of the synchronized state, while a vanishing gap signals a transition toward desynchronization.

Spectral Memory and Stability Operators

Beyond instantaneous dynamics, the Æ framework admits memory operators that encode the persistence of coherent configurations.

Let L_γ and L_Γ denote operators acting on the space of phase configurations.

Their spectra determine the long-term stability of coherent modes.

$$L_\gamma \psi_i = \lambda_i \psi_i, \tag{467}$$

$$L_\Gamma \phi_i = \mu_i \phi_i. \tag{468}$$

The associated spectral gap

$$\Delta\lambda = \lambda_2 - \lambda_1 \tag{469}$$

provides a quantitative measure of stability and memory retention.

Detailed derivations of the synchronization regimes are provided in Appendix sections devoted to spectral stability.

Hierarchical Spectral Stability

The spectral properties of the Æ-field determine the stability of coherent structures across hierarchical scales.

A key criterion is the NR stability condition, which relates the spectral gap of the synchronization operator to the persistence of coherent modes.

Let L_ϕ denote the phase Laplacian operator. Stability requires that the second eigenvalue satisfies

$$\lambda_2(L_\phi) > 0. \tag{470}$$

This condition ensures that small perturbations decay and that the synchronized configuration remains stable.

Associated with the spectral gap is a characteristic correlation length

$$\xi = \frac{1}{\sqrt{\lambda_2(L_\phi)}}, \tag{471}$$

which determines the spatial scale over which phase coherence is maintained.

At larger scales, the hierarchical structure of the system imposes entropy bounds on the distribution of coherent modes.

If S_ϕ denotes the phase entropy, stability requires

$$S_\phi \leq S_{\mathrm{crit}}, \tag{472}$$

where S_{crit} represents a scale-dependent entropy threshold beyond which coherence breaks down.

Together, the NR condition, the spectral gap, and the correlation length define a hierarchical stability criterion for the Æ-field across multiple scales.

50. On Knots

Nodes as Quasi-Stable Configurations

Define the nodal set as the coherence-thresholded configurations:

$$\mathcal{N} = \{Æ(x,t) \mid C(x,t) \geq C_{\min}\} \tag{473}$$

In the quasi-stationary regime,

$$\partial_t \mathcal{N} \approx 0 \tag{474}$$

Knots as Topological Invariants

Let $\sim$ denote equivalence under smooth deformations of the phase configuration that preserve coherence above threshold (isotopy/homotopy within the admissible class). Define the knot-class set:

$$\mathcal{K} = \{\phi(x,t)\}/\sim \tag{475}$$

A topological invariant associated with $\mathcal{K}$ is conserved:

$$T(\mathcal{K}) = \text{const} \tag{476}$$

Topological Protection

Small phase perturbations do not change the knot invariant:

$$\delta\phi_{\text{small}} \implies T(\mathcal{K}) \text{ unchanged} \tag{477}$$

Coherence above threshold implies persistence of the knot-class:

$$C \geq C_{\min} \implies \mathcal{K} \text{ persists} \tag{478}$$

Memory as Topological Persistence

Define the indicator function of knot-presence:

$$\chi_{\mathcal{K}}(t) = \begin{cases} 1, & \text{if the configuration lies in the class } \mathcal{K}, \\ 0, & \text{otherwise.} \end{cases} \tag{479}$$

The persistence-time functional (residence time) is:

$$M = \int_{t_1}^{t_2} \chi_{\mathcal{K}}(t)\, dt \tag{480}$$

A phenomenological memory stiffness may be proportional to the topological invariant:

$$\kappa_{\text{mem}} \propto T(\mathcal{K}) \tag{481}$$

Memory as Path Dependence and Holonomy

The total holonomy along a loop γ is defined by

$$\Phi_{\text{total}}[\gamma] = \int_\gamma A, \tag{482}$$

with the canonical 1-form

$$A = \frac{1}{2}\,\mathrm{Im}\big(Q^{-1} dQ\big)\,. \tag{483}$$

A tetrahedral decomposition via weights $w_i(\pi_H(Q))$ yields

$$\Phi_i[\gamma] = \int_\gamma w_i\big(\pi_H(Q)\big)\, A, \tag{484}$$

and exact conservation of distributed holonomy:

$$\sum_{i=1}^{4} \Phi_i[\gamma] = \Phi_{\text{total}}[\gamma]. \tag{485}$$

Casimir Invariants of the Nambu–Æ Bracket

On an internal coordinate set (x^1, x^2, x^3), the Nambu–Æ bracket is

$$\{F, G, H\}_{Æ} = \epsilon^{ijk} \frac{\partial F}{\partial x^i} \frac{\partial G}{\partial x^j} \frac{\partial H}{\partial x^k}. \tag{486}$$

In particular, choosing $(x^1, x^2, x^3) = (D, d, \phi)$ gives the explicit determinantal form

$$\{F, G, H\}_{Æ} = \det\left(\frac{\partial(F, G, H)}{\partial(D, d, \phi)} \right). \tag{487}$$

Casimir invariants satisfy

$$\{C, F, G\}_{Æ} = 0 \quad \forall\, F, G, \tag{488}$$

and are conserved under Nambu–Æ flows:

$$\frac{d}{dt} C_{Æ} = 0. \tag{489}$$

Topological Helicity Interpretation

A natural realization of the Casimir invariant arises as a helicity-like integral over the internal Æ-manifold:

$$C_{\text{«}} = \int_{\mathcal{M}_{Æ}} \mathbf{A}_{Æ} \cdot \left(\nabla_{Æ} \times \mathbf{A}_{Æ} \right) dV_{Æ}. \tag{490}$$

This quantity is invariant under Nambu–Æ flows and therefore acts as a topological Casimir of the system.

In the knot representation of coherent structures, this invariant admits the linking interpretation

$$C_{Æ} \propto \sum_{i<j} \mathrm{Link}(K_i, K_j), \tag{491}$$

where K_i denote the coherent nodal loops of the field.

Node–Knot Transition Map

The nodal set and knot-class are dynamically related:

$$\mathcal{N} \xleftrightarrow{\Delta\phi} \mathcal{K} \tag{492}$$

Loss of coherence below threshold implies topological collapse:

$$C < C_{\min} \implies \mathcal{K} \to \varnothing \tag{493}$$

Structural Triangle

The minimal structural triple governing persistence is:

$$(\phi,\ C,\ \kappa_{\text{mem}})\ \triangle \tag{494}$$

encoding phase configuration, coherence threshold, and memory stiffness.

Loop-Dependent Holonomy

Holonomy depends on the homotopy class of the loop:

$$\Phi_{\text{total}}[\gamma_1] \neq \Phi_{\text{total}}[\gamma_2] \tag{495}$$

for inequivalent loops $\gamma_1 \not\sim \gamma_2$.

Fundamental Group Structure

Each coherent region E_i admits a fundamental group:

$$\pi_1(E_i) \cong \mathbb{Z}^{\beta_1^{(i)}} \tag{496}$$

where $\beta_1^{(i)}$ is the first Betti number, characterizing independent topological cycles.

Internal Helicity Field

Define the internal rotational field:

$$\mathbf{B} = \nabla_{\text{Æ}} \times \mathbf{A} \tag{497}$$

This allows rewriting the helicity invariant as

$$C_{\text{«}} = \int_{\mathcal{M}_{Æ}} \mathbf{A} \cdot \mathbf{B} \, \Omega_{Æ} \tag{498}$$

Holonomy Criterion

Define the quaternionic connection:

$$A := \mathrm{Im}(Q^{-1} dQ) \tag{499}$$

The associated holonomy along a loop γ is:

$$H_\gamma := \mathcal{P} \exp\left(\oint_\gamma A \right) \in SU(2) \tag{500}$$

Topological nontriviality is characterized by:

$$H_\gamma \neq \mathbb{I} \tag{501}$$

Identity–Return (Renudation) Mechanism

Parameterized holonomy:

$$H_\gamma(\lambda) := \mathcal{P} \exp\left(\oint_\gamma A_\lambda \right) \tag{502}$$

Renudation condition:

$$H_\gamma(\lambda)^2 = \mathbb{I} \tag{503}$$

More generally,

$$H_\gamma(\lambda)^n \neq \mathbb{I} \quad \text{for } n \leq N \tag{504}$$

but

$$H_\gamma(\lambda)^n \to \mathbb{I} \quad \text{as } n \to \infty \tag{505}$$

Localized Coherence Nodes and Reality Knot Encoding

Localized nodes are modeled as regions x_k where

$$H_{\gamma_k} \neq \mathbb{I}, \tag{506}$$

and phase entropy remains elevated

$$S_\phi(x_k) > S_{\text{crit}}. \tag{507}$$

Each node carries a coherence state

$$\mathcal{N}_k = (C_k,\ \phi_k,\ x_k), \tag{508}$$

with dynamics

$$\partial_t C_k = -\alpha_k S_{\phi,k} C_k + \beta_k C_{Æ}(x_k, t). \tag{509}$$

Persistent trauma corresponds to

$$H^n_{\gamma_k} \neq \mathbb{I} \quad \forall n \leq N, \tag{510}$$

while readiness for renudation satisfies

$$H^n_{\gamma_k} \to \mathbb{I}. \tag{511}$$

Coupling between nodes is mediated by phase transport

$$\dot{\phi}_k = \omega_k + \sum_{j \neq k} \kappa_{kj} \sin(\phi_j - \phi_k). \tag{512}$$

Quaternionic Periodicity

Approximate periodicity:

$$Q_i(t) \approx Q_i(t + T) \tag{513}$$

Branching while preserving topological invariant:

$$Q(t_0) \to \{Q_k(t)\}, \qquad \mathcal{T}(Q_k) = \mathcal{T}(Q) \tag{514}$$

Collective and Embedded Knots

Collective quaternionic configuration:

$$Q_{\text{collective}} = \prod_{i=1}^{N} Q_i \tag{515}$$

Organ-level embedding:

$$Q_{O_k}(x,t) \in S^3 \tag{516}$$

Hormonal phase carrier:

$$\Phi_H(t) = e^{\theta(t)\mathbf{u}} \tag{517}$$

Coherence–entropy coupling in healing regime:

$$S_\phi \to \text{low}, \qquad C_{\text{«}} \to \text{high} \tag{518}$$

Reality Knot Network Structure

Multiverse as network of knot trajectories:

$$\mathcal{M} = \text{Graph}(\mathbb{T}) \tag{519}$$

where $\mathbb{T}$ denotes stable and metastable reality-knot trajectories. Loop embedding:

$$\gamma(t) \in S^3, \qquad \pi(\gamma(t)) \in S^2 \tag{520}$$

Nontrivial phase integral:

$$\oint_\gamma Q^{-1} dQ \neq 0 \tag{521}$$

Proximity metric to trivial holonomy:

$$d(H_\gamma, \mathbb{I}) < \epsilon \tag{522}$$

Historical resonance condition:

$$H_{\gamma^*} \approx H_{\gamma_{\text{current}}} \tag{523}$$

Temporal Knot Projection

Memory as holonomy-projected transport:

$$M(t) = \mathcal{P}\left[H_{\gamma_{\text{past}}} \cdot Q(t)\right] \tag{524}$$

Prefiguration operator:

$$\Pi_{\text{pre}} = \mathcal{P}\left[\exp(-\Xi_{\gamma_{\text{future}}})\, Q(t)\right] \tag{525}$$

Equivalence of trajectories implies projection similarity:

$$\gamma_A \sim \gamma_B \quad \Rightarrow \quad \Pi_A(t) \approx \Pi_B(t) \tag{526}$$

Holonomy decrement across transitions:

$$H_{\gamma_{n+1}} = H_{\gamma_n} - \Delta H \tag{527}$$

Composite Knots and Protein Structure

A protein knot as composite connected sum:

$$K_{\text{protein}} = K_1 \# K_2 \# \dots \# K_m \tag{528}$$

Quantum configuration Hamiltonian:

$$H_{\text{conf}} = \sum_n E_n |n\rangle\langle n| + \sum_{n \neq m} \frac{\Delta E_{nm}}{2} \left(e^{i(\phi_n - \phi_m)} |n\rangle\langle m| + \text{h.c.}\right) \tag{529}$$

Schrödinger-type evolution:

$$\partial_t \Psi_{\text{prot}} = -i H_{\text{conf}} \Psi_{\text{prot}} \tag{530}$$

Phase evolution:

$$\dot{\phi}_i = \omega_i + \sum_{j \neq i} \kappa_{ij} \sin(\phi_j - \phi_i) \tag{531}$$

Coherence dynamics:

$$\partial_t C_{\text{prot}} = -\alpha_{\text{prot}} S_{\phi,\text{prot}} C_{\text{prot}} + \beta_{\text{prot}} \tag{532}$$

Functional stability condition:

$$C_{\text{prot}} \geq C_{\text{func}}, \qquad \dot{C}_{\text{prot}} \approx 0 \tag{533}$$

Holonomy as Interaction Carrier

Interaction amplitude between configurations:

$$\mathcal{I}_{\alpha\beta} = \kappa_{\alpha\beta} \cos(\Delta\phi_{\alpha\beta}) \tag{534}$$

with coupling

$$\kappa_{\alpha\beta} = \kappa_0 \mathrm{Re}\big(Q_\alpha^{-1} Q_\beta\big) \tag{535}$$

Residual Holonomy (Dynamical)

Residual holonomy evolves as:

$$\partial_t H_\gamma = -\eta \nabla S_\phi \tag{536}$$

Magnitude of nontriviality:

$$\|H_\gamma - \mathbb{I}\| \propto \int_\gamma S_\phi \, d\tau \tag{537}$$

Local Knot Locking

Localized non-renuded loop:

$$H_\gamma \neq \mathbb{I}, \qquad \gamma \subset \mathcal{N}_i \tag{538}$$

Coupled coherence-entropy shift:

$$C_i = C_i^{(0)}(1 - B_i), \qquad S_{\phi,i} = S_{\phi,i}^{(0)} + \xi B_i \tag{539}$$

Topological Stability

Nontrivial invariant condition:

$$\mathcal{T}(Q) \neq 0 \tag{540}$$

Holonomy Return and Double-Cover Structure

The topology of coherent Æ-knots naturally leads to a holonomy return condition associated with double-cover structures.

Let γ denote a closed path in the configuration space of the knot. The associated holonomy is written as

$$\mathrm{Hol}(\gamma) = \mathcal{P} \exp\left(\oint_{\gamma} A \right), \tag{541}$$

where A is the effective connection associated with the knot.

In the presence of a double-cover structure, a coherent state may require two traversals of the loop to return to its original configuration. This is expressed schematically as

$$\Psi(\gamma^2) = \Psi(\gamma) \tag{542}$$

which characterizes the double-cover return condition of the knot.

Casimir Invariants, Helicity, and Linking Relations

Topological invariants of the Æ-field may be expressed through Casimir-type quantities and helicity functionals.

Let $\mathbf{A}$ denote an effective gauge-like potential and $\mathbf{B} = \nabla \times \mathbf{A}$ its associated field.

The helicity is defined as

$$\mathcal{H}_{\text{Æ}} = \int \mathbf{A} \cdot \mathbf{B} \, dV, \tag{543}$$

which measures the topological twisting and linking of coherent field lines.

Topological Casimir invariants provide conserved quantities

associated with the knot class. In particular, the linking number between two coherent loops γ_1 and γ_2 is represented by

$$\mathrm{Lk}(\gamma_1, \gamma_2), \tag{544}$$

and remains invariant under continuous deformations preserving the knot structure.

These quantities classify stable coherent sectors of the Æ-field and constrain their admissible transformations.

Temporal Reprojection of Coherent Knot States

Coherent topological structures in the Æ-field may admit reprojection across temporally separated sectors of the theory.

Let $\mathcal{K}(t)$ denote a coherent knot state at time t. A temporal reprojection is represented schematically by

$$\mathcal{K}(t_2) = \Pi_{\text{top}}\big(\mathcal{K}(t_1)\big), \tag{545}$$

where Π_{top} denotes a topological reprojection operator preserving the knot class.

Such structures provide a formal mechanism by which coherence may persist across temporally separated configurations without requiring pointwise identity of the local field realization.

Topological Memory Surface Flux

Memory in the Æ-field may be encoded not only by persistence of configurations, but also by the flux of topological information through bounding surfaces.

Let Σ be a closed surface surrounding a coherent knot region. The associated topological memory flux is defined as

$$\Phi_{\text{mem}} = \int_{\Sigma} \mathbf{J}_{\text{top}} \cdot d\mathbf{S}, \tag{546}$$

where $\mathbf{J}_{\text{top}}$ is the topological memory current.

Nonzero flux indicates persistent transport of topological information across the boundary of the coherent region.

Holonomy Criterion for Knot Persistence

A coherent knot remains topologically stable provided its holonomy class is preserved under the induced evolution.

Let γ_t denote the transported loop at time t. Persistence requires that

$$[\mathrm{Hol}(\gamma_t)] = [\mathrm{Hol}(\gamma_0)], \tag{547}$$

where brackets denote equivalence classes under admissible topological deformations.

Loss of this condition signals bifurcation or dissolution of the coherent knot structure.

Topological Classification via $\pi_3(S^3)$

The global topology of coherent knot states in the Æ-field is naturally related to the homotopy group

$$\pi_3(S^3) \cong \mathbb{Z}. \tag{548}$$

This integer-valued classification labels topologically distinct knot sectors and provides a natural indexing of globally stable coherent configurations.

51. Merkaba Projection

Hopf Projection $S^3 \to S^2$

For unit quaternions $Q \in S^3 \subset \mathbb{H}$, the Hopf projection is defined by

$$\pi_H : S^3 \longrightarrow S^2, \tag{549}$$

$$\vec{n} = Q\,\sigma_3\,Q^{-1}, \qquad \vec{n} \in S^2 \tag{550}$$

Since $|Q| = 1$, we have $Q^{-1} = Q^*$. In quaternionic component form ($Q = a + bi + cj + dk$),

$$\pi_H(Q) = Q\,k\,Q^* = \begin{pmatrix} 2(ac+bd) \\ 2(bc-ad) \\ a^2+d^2-b^2-c^2 \end{pmatrix} \tag{551}$$

Tetrahedral Decomposition

Let $\{w_i\}_{i=1}^4$ be barycentric weights on S^2:

$$\sum_{i=1}^{4} w_i = 1, \qquad w_i \geq 0 \tag{552}$$

Distributed holonomy is defined by

$$\Phi_i[\gamma] = \int_\gamma w_i(\pi_H(Q))\,A \tag{553}$$

with total holonomy conserved:

$$\sum_{i=1}^{4} \Phi_i[\gamma] = \Phi_{\text{total}}[\gamma] \tag{554}$$

Merkaba as Phase Transduction Operator

Define the Merkaba map

$$\mathcal{M} : S^3 \longrightarrow \Delta_3, \tag{555}$$

$$\mathcal{M}(Q) = \{w_i(\pi_H(Q))\}_{i=1}^4 \tag{556}$$

where Δ_3 denotes the 3-simplex.
Under holonomy transport,

$$Q \longmapsto (\Phi_1, \Phi_2, \Phi_3, \Phi_4) \tag{557}$$

Nodal Stability via Merkaba

If coherence remains above threshold,

$$C(x, t) \geq C_{\min} \implies \partial_t w_i \approx 0 \tag{558}$$

If coherence falls below threshold,

$$C < C_{\min} \implies \Delta w_i \neq 0 \tag{559}$$

Explicit Tetrahedral Geometry on S^2

Let $\{v_i\}_{i=1}^4 \subset S^2$ be vertices of a regular tetrahedron:

$$\|v_i\| = 1, \qquad v_i \cdot v_j = -\tfrac{1}{3}\ (i \neq j), \qquad \sum_{i=1}^{4} v_i = 0 \tag{560}$$

The barycentric weights admit the explicit form

$$w_i(p) = \frac{\max\{0, 1 + 3(p \cdot v_i)\}}{\sum_{j=1}^4 \max\{0, 1 + 3(p \cdot v_j)\}} \tag{561}$$

satisfying

$$\sum_{i=1}^{4} w_i(p) = 1, \qquad w_i(p) \geq 0, \qquad w_i(v_j) = \delta_{ij} \tag{562}$$

Merkaba Projection Operator

The projection operator is

$$\rho_{\mathcal{M}} : S^3 \xrightarrow{\pi_H} S^2 \xrightarrow{w} \Delta^3 \tag{563}$$

$$\rho_{\mathcal{M}}(Q) = \big(w_1(\pi_H(Q)), \ldots, w_4(\pi_H(Q))\big) \tag{564}$$

The Merkaba complex may be viewed as the dual tetrahedral union

$$\mathcal{M} := \Delta^+ \cup \Delta^- \tag{565}$$

Phase Distribution and Entropy on Δ^3

Define normalized phase weights:

$$p_i := \frac{\Phi_i}{\Phi_{\text{total}}}, \qquad \sum_{i=1}^{4} p_i = 1 \tag{566}$$

Phase entropy on the Merkaba:

$$S_\phi = -\sum_{i=1}^{4} p_i \ln p_i, \qquad 0 \leq S_\phi \leq \ln 4 \tag{567}$$

Corresponding coherence:

$$C_{Æ} = \frac{1}{1 + S_\phi} \tag{568}$$

Nonlocal Phase Transport

A generalized Merkaba functional may include nonlocal phase transport:

$$\mathcal{M}(x) = \int_{\gamma_x} A + \int d^4x' \sqrt{-g(x')}\, K(x, x')\, S_\phi(x') \tag{569}$$

Algebraic Structure of the AE Product

The AE-algebra satisfies:

$$(a\Xi_1 + b\Xi_2) \boxtimes \Xi_3 = a(\Xi_1 \boxtimes \Xi_3) + b(\Xi_2 \boxtimes \Xi_3) \tag{570}$$

$$\Xi_1 \cdot_{«} \Xi_2 = \Xi_2 \cdot_{«} \Xi_1 \tag{571}$$

$$\Xi_1 \wedge_{«} \Xi_2 = -\,\Xi_2 \wedge_{«} \Xi_1 \tag{572}$$

$$\|\Xi_1 \boxtimes \Xi_2\|^2 = (\Xi_1 \cdot_{«} \Xi_2)^2 + \|\Xi_1 \wedge_{«} \Xi_2\|^2 \tag{573}$$

$$\Xi \boxtimes \Xi = (\|\Xi\|^2, 0) \tag{574}$$

Merkaba Projection Operator

The merkaba structure may be interpreted as a geometric projection of the underlying quaternionic phase state of the Æ-field.

Let

$$Q(t) \in S^3 \tag{575}$$

denote the normalized quaternionic state representing the internal phase configuration.

The merkaba projection operator is defined as

$$\mathcal{P}_M : S^3 \to \mathcal{M}, \tag{576}$$

where $\mathcal{M}$ denotes the tetrahedral merkaba configuration space. Applying the projection yields a set of normalized weights

$$\{w_i\}_{i=1}^{4} = \mathcal{P}_M(Q), \tag{577}$$

satisfying

$$\sum_{i=1}^{4} w_i = 1. \tag{578}$$

These weights correspond to the four vertices of the merkaba tetrahedral structure and encode the distribution of phase coherence along the principal directions of the configuration.

The projection therefore provides a bridge between the quaternionic phase space and the geometric representation used to describe multi-directional phase transduction in the Æ framework.

Viviani–Hopf Geometry of the Merkaba

The merkaba projection introduced above admits a simple geometric interpretation related to classical results in triangular geometry.

Each face of the tetrahedral merkaba structure is an equilateral triangle. For any point inside such a triangle, the theorem of Viviani states that the sum of the perpendicular distances to the three sides is constant.

Let d_1, d_2, d_3 denote the distances from a point inside a face to its three edges. The theorem implies

$$d_1 + d_2 + d_3 = h, \tag{579}$$

where h is the altitude of the triangle.

In barycentric coordinates this relation is expressed as

$$\lambda_1 + \lambda_2 + \lambda_3 = 1. \tag{580}$$

The merkaba projection naturally produces a tetrahedral set of weights

$$\{w_i\}_{i=1}^{4}, \qquad \sum_{i=1}^{4} w_i = 1, \tag{581}$$

which act as barycentric coordinates on the tetrahedron.

Consequently, when restricted to any triangular face of the merkaba, the projected coherence weights satisfy a Viviani-type relation. This means that while the distribution of coherence among the vertices may vary, the total projected coherence on the face remains conserved.

Geometrically, the quaternionic phase state

$$Q \in S^3 \tag{582}$$

is mapped through the Hopf projection into a configuration whose tetrahedral coordinates obey these barycentric constraints. The Viviani relation therefore provides a simple geometric interpretation of coherence conservation within the merkaba representation.

Geometric Encoding of Phase in the Merkaba

The merkaba representation provides a geometric encoding of the quaternionic phase state underlying the Æ-field dynamics.

Let

$$Q \in S^3 \tag{583}$$

denote the normalized quaternionic phase configuration. Through the Hopf projection, this state can be represented geometrically in a lower-dimensional configuration space. Within the Æ framework, this projection is expressed using tetrahedral barycentric coordinates

$$\{w_i\}_{i=1}^4, \qquad \sum_{i=1}^{4} w_i = 1, \tag{584}$$

which correspond to the four vertices of the merkaba tetrahedron.

These weights encode the directional distribution of coherence associated with the projected phase configuration. The position of the projected phase state inside the tetrahedral structure is thus fully determined by the set $\{w_i\}$.

Each face of the tetrahedron forms an equilateral triangle. Classical results in triangular geometry imply that, for any point inside such a triangle, the sum of the perpendicular distances to the three edges is constant. This property is known as the theorem of Viviani.

When expressed in barycentric coordinates, the same relation becomes

$$\lambda_1 + \lambda_2 + \lambda_3 = 1. \tag{585}$$

Consequently, when the merkaba coordinates are restricted to any triangular face, the projected coherence weights obey a Viviani-type constraint. Although the distribution of weights along the vertices may vary, the total projected coherence on that face remains fixed.

Geometrically, this implies that the merkaba structure acts as a coordinate system for the projected phase configuration. The Viviani relation ensures that the positional information of the projected phase relative to each triangular face is preserved, preventing loss of geometric information during redistribution of coherence among the vertices.

Thus, the combination of Hopf projection and tetrahedral barycentric coordinates provides a natural geometric framework for tracking the location of the phase configuration within the merkaba structure.

52. Differential Geometry of the Æ-Space

Nodal Torsion and Æ-Connection

With

$$\phi(D,d) = \frac{1}{2}\log C(D,d), \tag{586}$$

the induced conformal metric is

$$ds^2_{\text{Æ}} = e^{2\phi(D,d)}(dD^2 + dd^2) = C(D,d)(dD^2 + dd^2). \tag{587}$$

The Æ-connection decomposes as

$$\Gamma^{\lambda}_{\mu\nu}(\text{Æ}) = \Gamma^{\lambda}_{\mu\nu}(\text{LC}) + K^{\lambda}_{\ \mu\nu}, \tag{588}$$

with torsion

$$T^{\lambda}_{\mu\nu} = -\frac{\beta}{2}\left(\delta^{\lambda}_{\nu}\partial_{\mu}C - \delta^{\lambda}_{\mu}\partial_{\nu}C\right). \tag{589}$$

Hopf-Algebraic Closure

Primitive coproduct structure:

$$\Delta(A) = A \otimes 1 + 1 \otimes A, \tag{590}$$

$$\varepsilon(A) = 0, \qquad S(A) = -A. \tag{591}$$

Quasi-Hopf Structure

Non-coassociativity:

$$(\Delta_{\text{Æ}} \otimes \text{id}) \circ \Delta_{\text{Æ}} \neq (\text{id} \otimes \Delta_{\text{Æ}}) \circ \Delta_{\text{Æ}}, \tag{592}$$

with associator

$$\Phi_{Æ} \in Æ \otimes Æ \otimes Æ. \tag{593}$$

Inter-Layer Transmission and Admissibility

Inter-layer information transmission is modeled by transduction operators between layer function spaces:

$$\Psi_{\ell_2} = \mathcal{T}_{\ell_1 \to \ell_2}[\Psi_{\ell_1}] \tag{594}$$

Define the layer projections as bounded maps

$$\Pi_\ell : \mathcal{A}_{Æ} \to \mathcal{A}_\ell \tag{595}$$

In general, transduction is *not* reducible to successive projections:

$$\mathcal{T}_{\ell_1 \to \ell_2} \neq \Pi_{\ell_2} \circ \Pi_{\ell_1} \tag{596}$$

Admissibility is controlled by coherence thresholds:

$$C_{\ell_1}(x,t) \geq C^{\text{crit}}_{\ell_1 \to \ell_2} \implies \mathcal{T}_{\ell_1 \to \ell_2} \text{ is admissible} \tag{597}$$

Below threshold, the effective sector fails due to information loss:

$$C_{\ell_1}(x,t) < C^{\text{crit}}_{\ell_1 \to \ell_2} \implies \text{information loss or decoherence} \tag{598}$$

The induced phase-entropy shift across layers is

$$S_{\phi,\ell_2} = S_{\phi,\ell_1} + \Delta S_\phi^{(\ell_1 \to \ell_2)}, \qquad \Delta S_\phi^{(\ell_1 \to \ell_2)} \geq 0 \tag{599}$$

with directional bias (down-projection tends to increase entropy):

$$\Delta S_\phi^{\text{down}} > 0, \qquad \Delta S_\phi^{\text{up}} \leq 0 \tag{600}$$

A memory observable may project across layers via an operator $\mathcal{P}_{\ell_1 \to \ell_2}$:

$$M_{\ell_2}(t) = \mathcal{P}_{\ell_1 \to \ell_2}[M_{\ell_1}(t)] \tag{601}$$

Assume boundedness of transmission:

$$\|\mathcal{T}_{\ell_1 \to \ell_2}\| \leq \kappa_{\ell_1 \to \ell_2} \tag{602}$$

and a finite entropy-budget across all effective reductions:

$$\sum_{\ell} \Delta S_{\phi,\ell} < \infty \tag{603}$$

Mental Sector as a Restricted Coherent Subspace

Define the mental layer as a restricted coherent subspace of the astral functional space:

$$\mathcal{A}_{\text{ment}} \subset \mathcal{A}_{\text{astr}} \tag{604}$$

Membership is characterized by stronger coherence and stricter entropy bounds:

$$\Psi \in \mathcal{A}_{\text{ment}} \iff \Psi \in \mathcal{A}_{\text{astr}} \text{ and } C_{\text{astr}} \geq C_{\text{ment,min}}, \quad S_{\phi,\text{astr}} \leq S^{\text{crit}}_{\phi,\text{ment}} \tag{605}$$

Thought is represented as a trajectory in the mental sector:

$$\gamma_{\text{ment}} : \mathbb{R} \to \mathcal{A}_{\text{ment}}, \qquad \text{Thought} \equiv \gamma_{\text{ment}}(t) \tag{606}$$

with quasi-periodic persistence in stabilized regimes:

$$\gamma_{\text{ment}}(t_0) \approx \gamma_{\text{ment}}(t_0 + T) \tag{607}$$

A minimal closure condition is that mental patterns may transduce back and couple to physical expression:

$$\Psi_{\text{astr}} \to \Psi_{\text{ment}} \to \Psi_{\text{astr}}, \qquad \Psi_{\text{ment}} \xrightarrow{\text{transduction}} \Psi_{\text{phys}} \tag{608}$$

Coherent Solution Sectors and Effective Basins

Let $\mathcal{S}_{\text{Æ}}$ denote a solution space and represent evolution as a trajectory:

$$\gamma : \mathbb{R} \to \mathcal{S}_{\text{Æ}}, \qquad s(t) = \gamma(t) \tag{609}$$

Define coherent basins by asymptotic convergence of a coherence functional Φ_{coh}:

$$\mathcal{B}_k = \left\{ \gamma_i \,\middle|\, \lim_{t\to\infty} \Phi_{\text{coh}}(\gamma_i(t)) = \Phi_k \right\} \tag{610}$$

Assign basin weights by phase-entropy suppression:

$$W_k \propto \exp\!\left(-\frac{S_\phi(\mathcal{B}_k)}{k_Æ} \right) \tag{611}$$

so that the dominant effective basin satisfies

$$\mathcal{B}_{\text{eff}} = \arg\min_{\mathcal{B}_k} S_\phi(\mathcal{B}_k) \tag{612}$$

Define a basin-distance by coherence-profile discrepancy:

$$d_Æ(\mathcal{B}_i, \mathcal{B}_j) = \int_{\mathcal{S}_Æ} |\Phi_i(s) - \Phi_j(s)|^2 \, d\mu_Æ(s) \tag{613}$$

Nested reductions are represented by basin inclusions:

$$\mathcal{B}_1 \subset \mathcal{B}_2 \subset \cdots \subset \mathcal{B}_n \tag{614}$$

with a minimal overlap condition for coherent compatibility:

$$\langle \Phi_i \mid \Phi_j \rangle \geq \Phi_{\text{crit}} \tag{615}$$

Spectral Dressing by Coherence and Phase-Entropy

In the spectral-localized sector it is often convenient to absorb the weight $w_Æ$ into the spinor:

$$\psi = \sqrt{w_Æ}\, \tilde{\psi} \tag{616}$$

Then derivatives acquire a drift term:

$$\partial_D \psi = \sqrt{w_Æ} \left(\partial_D \tilde{\psi} + \frac{1}{2} \tilde{\psi}\, \partial_D \log w_Æ \right) \tag{617}$$

and the effective mass becomes a functional of coherence and phase-entropy:

$$M_{\text{eff}}(D,d) = M(D,d) + \frac{i}{2}\gamma^D \partial_D \log \mathcal{C} + \frac{i}{2}\gamma^d \partial_d \log \mathcal{C} - \frac{1}{2\hbar}\left(\gamma^D \partial_D S_\phi + \gamma^d \partial_d S_\phi \right) \tag{618}$$

A leading-order scale estimate is

$$|M_{\text{eff}}|^2 \approx M^2 + \frac{1}{\hbar^2}|\nabla_{D,d} S_\phi|^2 + |\nabla_{D,d} \log \mathcal{C}|^2 \tag{619}$$

Metric Response and Interlayer Geometric Mediation

Beyond the static conformal and torsional structure, the Æ-metric may be regarded as dynamically responsive to coherent field configurations. A minimal structural variation law is postulated in the form

$$\delta S_{\text{Æ}} = \lambda (K - K_0), \tag{620}$$

where K denotes an effective curvature/coherence loading, K_0 is a reference geometric threshold, and λ is a coupling constant governing metric response.

A corresponding metric evolution ansatz may then be written as

$$\partial_t g_{\mu\nu} = \alpha\, \partial_\mu \Phi_{\text{Æ}}\, \partial_\nu \Phi^*_{\text{Æ}}, \tag{621}$$

where $\Phi_{\text{Æ}}$ is the effective Æ-field and α controls the strength of field-induced geometric deformation.

This relation expresses the idea that inter-layer transmission is not merely an algebraic projection phenomenon, but may also induce a local geometric response in the effective metric structure of the Æ-space.

In this interpretation, coherent gradients of $\Phi_{\text{Æ}}$ act as sources of geometric dressing, while incoherent or highly entropic transitions contribute weakly or diffusively to the induced metric flow.

53. Effective Sectors and Emergent Reductions

Full Configuration Space

The full configuration space is the functional product

$$\mathcal{C}_{\text{Æ}} = \{g_{\mu\nu}\} \times \{A_\mu\} \times \{\psi\} \times \{C\} \times \{S_\phi\}. \tag{622}$$

Effective sectors satisfy

$$\mathcal{C}_{\text{eff}} \subset \mathcal{C}_{\text{Æ}}. \tag{623}$$

Effective Action (Path Integral Definition)

$$e^{iS_{\text{eff}}[\Phi]} = \int \mathcal{D}\Xi \, e^{iS[\Phi,\Xi]}. \tag{624}$$

Scalar–Coherence Sector

Using the induced metric $g^{\text{Æ}}_{\mu\nu} = C g_{\mu\nu}$,

$$\mathcal{S}_{\text{scalar}} = \int d^4x \sqrt{-g^{(\text{«})}} \left[\frac{1}{2}\partial_\mu C \, \partial^\mu C + \frac{1}{2}\partial_\mu S_\phi \, \partial^\mu S_\phi - U(C, S_\phi)\right] + S_{\text{mem}} \tag{625}$$

Spectral-Localized Sector

$$D_{\text{Æ}} = i\gamma^\mu D_\mu - M_{\text{eff}}(C, S_\phi) \tag{626}$$

$$D_{\text{«}} \psi_n = \lambda_n \psi_n \tag{627}$$

Biological Resonant Regime

Resonance condition:

$$|\omega_i - \omega_j| \ll \epsilon \tag{628}$$

with phase dynamics

$$\dot{\theta}_i = \omega_i + \sum_j K_{ij} \sin(\theta_j - \theta_i) \tag{629}$$

Hierarchical Embedding

$$\mathcal{C}_{Æ} \longrightarrow \mathcal{C}_{\text{gauge}} \longrightarrow \mathcal{C}_{\text{spectral}} \longrightarrow \mathcal{C}_{\text{collective}}. \tag{630}$$

Nested Quaternionic Layer Structure

The total state may be decomposed into hierarchical layers:

$$\Psi_{\text{tot}}(x, t) = (\Psi_{\text{phys}}, \Psi_{\text{eth}}, \Psi_{\text{astr}}, \Psi_{\text{ment}}) \tag{631}$$

Each layer admits polar and quaternionic representation:

$$\Psi_\ell(x, t) = A_\ell(x, t)e^{i\phi_\ell(x,t)}, \qquad \Psi_\ell \to Q_\ell \in \mathbb{H} \tag{632}$$

Norm:

$$\|Q_\ell\|^2 = a_\ell^2 + b_\ell^2 + c_\ell^2 + d_\ell^2 \tag{633}$$

Hierarchical embedding:

$$\mathcal{Q}_{\text{phys}} \subset \mathcal{Q}_{\text{eth}} \subset \mathcal{Q}_{\text{astr}} \subset \mathcal{Q}_{\text{ment}} \tag{634}$$

Inter-layer morphism:

$$Q_{\ell+1} = \mathcal{F}_\ell(Q_\ell) \tag{635}$$

Admissible Layer Domains

Admissible sector:

$$\mathcal{A}_\ell = \{\Psi_\ell \mid C_\ell \geq C_{\ell,\min},\ S_{\phi,\ell} \leq S_{\phi,\ell}^{\text{crit}}\} \tag{636}$$

Total admissibility:

$$\Psi_{\text{tot}} \in \prod_\ell \mathcal{A}_\ell \tag{637}$$

Physical Effective Sector

Physical projection:

$$\Pi_{\text{phys}} : \mathcal{A}_{\text{Æ}} \longrightarrow \mathbb{R}^3 \tag{638}$$

$$\Psi_{\text{phys}} = \Pi_{\text{phys}}[\text{Æ}(x,t)] \tag{639}$$

Stability conditions:

$$C_{\text{phys}} \geq C_{\min}^{\text{phys}}, \qquad |\partial_t \phi_{\text{phys}}| \ll 1 \tag{640}$$

Induced metric:

$$g_{\text{phys}} = \Pi_{\text{phys}}^*(g_{\text{Æ}}) \tag{641}$$

Energetic stability:

$$\delta\mathcal{H}_{\text{phys}} = 0, \qquad \delta^2\mathcal{H}_{\text{phys}} > 0 \tag{642}$$

Etheric Fractal Sector

Fractal support:

$$\mathcal{E}_{\text{eth}} \subset \mathbb{R}^n, \qquad \dim_{\text{H}}(\mathcal{E}_{\text{eth}}) \notin \mathbb{Z} \tag{643}$$

Iterative Cantorian refinement:

$$\mathcal{C}_{k+1} = \mathfrak{F}(\mathcal{C}_k) \tag{644}$$

Scaling measure:

$$\mu_{\text{eth}}(\lambda A) = \lambda^{\dim_{\text{H}}} \mu_{\text{eth}}(A) \tag{645}$$

Projection to physical layer:

$$\Psi_{\text{phys}} = \mathcal{T}_{\text{eth}}[\Psi_{\text{eth}}] \tag{646}$$

Astral Geodesic Sector

$$\Psi_{\text{astr}} : \Omega \times \mathbb{R} \longrightarrow \mathbb{C} \tag{647}$$

Astral trajectory:

$$\gamma_{\text{astr}}(t) = \Psi_{\text{astr}}(\cdot, t) \tag{648}$$

Geodesic condition:

$$\nabla_{\dot{\gamma}_{\text{astr}}} \dot{\gamma}_{\text{astr}} = 0 \tag{649}$$

Physical decoupling:

$$\Pi_{\text{phys}}[\Psi_{\text{astr}}] \approx 0 \tag{650}$$

Explicit Layer Mediation Operators

The emergence of lower layers from higher coherent structures may be represented through explicit mediation operators acting between neighboring layers.

The physical layer may be expressed as a mediated projection of the etheric sector:

$$\Psi_{\text{phys}} = \mathcal{E}_{\text{eth}\to\text{phys}} \boxtimes \Psi^{(0)}_{\text{phys}}. \tag{651}$$

Similarly, the etheric layer emerges from the astral sector through

$$\Psi_{\text{eth}} = \mathcal{E}_{\text{astr}\to\text{eth}} \boxtimes \Psi^{(0)}_{\text{eth}}. \tag{652}$$

Here $\mathcal{E}_{\ell_1\to\ell_2}$ denotes an inter-layer mediation operator and $\boxtimes$ represents a structured coupling between the intrinsic layer state $\Psi^{(0)}$ and the transmitted configuration.

These relations make explicit the hierarchical mediation implied by the nested quaternionic structure introduced above.

Transduction and Layer Coupling

The Æ-framework admits multiple effective layers corresponding to different observational regimes of the same underlying field.

Let Ψ_i denote the effective state associated with layer i. Coupling between layers may be represented schematically as

$$\Psi_j = T_{ij}(\Psi_i), \tag{653}$$

where T_{ij} denotes a transduction operator mapping the dynamics of layer i into the effective variables of layer j.

These operators encode the mechanisms by which coherence, phase information, and density distributions propagate across different effective sectors of the theory.

Nested Effective Layers

Within the Æ-framework the observable universe may be described through nested effective layers corresponding to different levels of description.

A schematic decomposition may be written as

$$\Psi_{\text{«}} \to \{\Psi_{\text{phys}}, \Psi_{\text{eth}}, \Psi_{\text{astr}}, \Psi_{\text{ment}}\}. \tag{654}$$

Each component represents an effective sector arising from the same underlying field but observed through distinct dynamical variables.

The physical sector corresponds to conventional matter fields, while the remaining sectors encode progressively higher-order coherence structures of the underlying configuration.

Density and Dimensional Representation

Effective layers may also be characterized by their dimensional structure and density properties.

Let

$$S^3(D) \tag{655}$$

denote the configuration manifold associated with dimension D.

Observable reductions may then be expressed through mappings

$$R(D) : S^3(D) \to \mathcal{O}(D), \tag{656}$$

where $\mathcal{O}(D)$ represents the space of observables.

A generalized configuration variable

$$Q_{\text{«}}(D, d) \tag{657}$$

may therefore encode both dimensional and density information associated with a given effective sector.

Ontological Layer Structure

The nested sectors introduced above may also be interpreted as ontological layers corresponding to different regimes of the underlying Æ-field.

Let $\mathcal{L}_k$ denote the k-th layer. The observable universe is then represented by a hierarchy

$$\mathcal{L}_0 \subset \mathcal{L}_1 \subset \mathcal{L}_2 \subset \cdots . \tag{658}$$

Each layer provides an effective description of the same underlying field configuration, but with distinct variables and characteristic scales.

Transitions between layers correspond to reductions or extensions of the effective dynamical variables used to describe the system.

Etheric Layer Representation

Among the nested effective layers, the etheric sector plays a distinct role in mediating coherence between physical and higher-order configurations.

Let

$$\Psi_{\text{eth}} \tag{659}$$

denote the etheric component of the field.

In certain representations this sector admits a Cantorian-like structure in which coherence is distributed across hierarchical subsets of the configuration space.

Such representations provide an effective description of intermediate scale coherence structures that couple the physical and astral sectors of the theory.

Multi-Layer Projections and Observables

Observables in the Æ-framework arise through projections of the underlying field configuration onto specific effective layers.

Let $\mathcal{P}_k$ denote the projection onto layer k.

$$\mathcal{O}_k = \mathcal{P}_k(\Psi_{\text{Æ}}). \tag{660}$$

Different observational regimes correspond to different choices of projection operators, each emphasizing particular dynamical variables of the underlying field.

This structure allows the same fundamental configuration to produce distinct observable descriptions depending on the effective sector through which it is examined.

Æ-Covariant Structure

Define the Æ-covariant derivative as an endomorphism on the tangent bundle of the Æ-configuration space:

$$\nabla_\mu^{Æ} : \mathcal{T}_{Æ} \to \mathcal{T}_{Æ} \tag{661}$$

It satisfies the Leibniz rule with respect to the Æ-product:

$$\nabla_\mu^{Æ}(Æ_1 \boxtimes Æ_2) = (\nabla_\mu^{Æ} Æ_1) \boxtimes Æ_2 + Æ_1 \boxtimes (\nabla_\mu^{Æ} Æ_2) \tag{662}$$

The curvature operator is defined by the commutator:

$$R_{\mu\nu}^{Æ}(Æ) = \left[\nabla_\mu^{Æ}, \nabla_\nu^{Æ}\right] Æ \tag{663}$$

An inner product compatible with the Æ-product is

$$\langle Æ_1, Æ_2 \rangle = \mathrm{Re}\Big(Æ_1^\dagger \boxtimes Æ_2\Big) \tag{664}$$

The sectional curvature is therefore

$$K_{Æ}(Æ_1, Æ_2) = \frac{\langle R^{Æ}(Æ_1, Æ_2)Æ_2, Æ_1 \rangle}{\|Æ_1\|^2 \|Æ_2\|^2 - \langle Æ_1, Æ_2 \rangle^2} \tag{665}$$

Curvature sign determines stability:

$$K_{Æ} > 0 \Rightarrow \text{convergence (stable)}, \tag{666}$$
$$K_{Æ} < 0 \Rightarrow \text{divergence (unstable)} \tag{667}$$

Geodesics in the Conformal Æ-Metric

With conformal factor $\phi = \frac{1}{2}\log C$, the induced metric is

$$ds_{Æ}^2 = e^{2\phi(D,d)}(dD^2 + dd^2) \tag{668}$$

Geodesics satisfy

$$\frac{d^2 x^\mu}{d\tau^2} + \Gamma^\mu_{\nu\rho}(Æ)\frac{dx^\nu}{d\tau}\frac{dx^\rho}{d\tau} = 0 \tag{669}$$

In (D, d) coordinates:

$$\ddot{D} + 2\,\partial_D\phi(\dot{D}^2 - \dot{d}^2) + 4\,\partial_d\phi\,\dot{D}\dot{d} = 0 \tag{670}$$

$$\ddot{d} + 2\,\partial_d\phi(\dot{d}^2 - \dot{D}^2) + 4\,\partial_D\phi\,\dot{D}\dot{d} = 0 \tag{671}$$

The conserved Hamiltonian along geodesics is

$$H = e^{2\phi(D,d)}(\dot{D}^2 + \dot{d}^2) \tag{672}$$

Nambu–Æ Structure

The Nambu–Æ bracket is defined by

$$\{F, G, H\}_{\text{Æ}} = \det\left(\frac{\partial(F, G, H)}{\partial(D, d, \phi)}\right) = \epsilon^{ijk}\frac{\partial F}{\partial x^i}\frac{\partial G}{\partial x^j}\frac{\partial H}{\partial x^k} \tag{673}$$

Time evolution is given by

$$\frac{dA}{dt} = \{A, H_1, H_2\}_{\text{Æ}} \tag{674}$$

Associated flow field:

$$X_{H_1,H_2} = \nabla H_1 \times \nabla H_2, \qquad \operatorname{div} X_{H_1,H_2} = 0 \tag{675}$$

Invariant Measure

The natural measure on Æ-space is

$$\mu_{«} = \mathcal{C}(D,d)e^{-S_\phi(D,d)/\hbar}\,dD\,dd\,dt\,dq \tag{676}$$

54. Fundamental Forces from Æ Field Dynamics

Unified Covariant Structure

We define a unified Æ-covariant derivative incorporating gauge and internal quaternionic components:

$$\mathcal{D}_\mu = \partial_\mu + A_\mu + G_\mu^a T^a + W_\mu^i \frac{\tau^i}{2} + \omega_\mu\, i + \psi_\mu\, j + \xi_\mu\, k \tag{677}$$

The total curvature tensor is obtained from the commutator:

$$\mathcal{F}_{\mu\nu} = [\mathcal{D}_\mu, \mathcal{D}_\nu] \tag{678}$$

This decomposes into standard gauge sectors:

$$\mathcal{F}_{\mu\nu} = F_{\mu\nu}^{\mathrm{EM}} + G_{\mu\nu}^a + W_{\mu\nu}^i + R_{\mu\nu} \tag{679}$$

Coherence Field Notation

Throughout this section we introduce the shorthand

$$C(x) \equiv C_{\text{Æ}}(x), \tag{680}$$

representing the local coherence amplitude of the Æ-field. This notation simplifies the presentation of the interaction sectors without altering the underlying definition of the coherence function.

Electromagnetic Sector

$$F_{\mu\nu} = \partial_\mu A_\nu - \partial_\nu A_\mu \tag{681}$$

The effective electromagnetic coherence factor is

$$C_{\mathrm{EM}} = \cos(\Delta\phi_{E,B})\, e^{-S_\phi} \tag{682}$$

Electromagnetic Resonance Coupling

Coherent molecular systems may interact with electromagnetic fields through resonance Hamiltonians describing the coupling between field modes and internal molecular states.

A simplified interaction Hamiltonian may be written as

$$H_{\text{int}} = \sum_{n,m} g_{nm} \left(a^{\dagger} \sigma_{nm} + a \, \sigma_{mn} \right), \tag{683}$$

where a and $a^{\dagger}$ denote electromagnetic mode operators and σ_{nm} represent molecular transition operators.

This structure provides a bridge between field-theoretic interactions and resonance phenomena appearing in molecular systems.

Strong Sector

$$G^a_{\mu\nu} = \partial_\mu G^a_\nu - \partial_\nu G^a_\mu + g f^{abc} G^b_\mu G^c_\nu \tag{684}$$

$$D_\mu = \partial_\mu + G^a_\mu T^a \tag{685}$$

Strong Interaction Sector

The strong interaction is described through a Yang–Mills structure associated with an internal gauge symmetry.

Let $F^a_{\mu\nu}$ denote the non-Abelian field strength. The dynamical equation takes the form

$$D^\mu F^a_{\mu\nu} = J^a_\nu \tag{686}$$

where D_μ denotes the gauge-covariant derivative and J^a_ν represents the color current.

Weak Sector and Symmetry Breaking

$$W^i_{\mu\nu} = \partial_\mu W^i_\nu - \partial_\nu W^i_\mu + g' \epsilon^{ijk} W^j_\mu W^k_\nu \tag{687}$$

Higgs-like scalar:

$$\Phi(x) = \rho(x)e^{i\theta(x)}, \qquad |\Phi|^2 = v^2 \tag{688}$$

Weak Interaction Sector

The weak interaction is represented by a Yang–Mills structure associated with the electroweak gauge symmetry.

Let $W^i_{\mu\nu}$ denote the weak field strength tensor. The field equations may be written as

$$D^\mu W^i_{\mu\nu} = J^i_\nu \tag{689}$$

where J^i_ν denotes the weak current.

Emergent Gravitational Sector

Metric perturbation:

$$g_{\mu\nu} = \eta_{\mu\nu} + h_{\mu\nu} \tag{690}$$

In Æ-formalism, curvature is modulated by phase:

$$R_{«} = \Box\phi \tag{691}$$

Gravitational Lagrangian density:

$$\mathcal{L}_G = \sqrt{-g}\, R_{«}\, e^{-S_\phi} \tag{692}$$

Gravitational Sector

Within the Æ-framework the gravitational interaction emerges as a curvature response modulated by the coherence structure of the underlying field.

The field equations may be written in Einstein-type form

$$G_{\mu\nu} = 8\pi G \left(T_{\mu\nu} + T^{\rm mem}_{\mu\nu}\right) \tag{693}$$

where $T^{\rm mem}_{\mu\nu}$ represents a memory contribution arising from the persistence of coherent field configurations.

Total Æ-Lagrangian

$$\mathcal{L}_{\text{Æ}} = \sqrt{-g}\left[\text{Tr}(\mathcal{F}_{\mu\nu}\mathcal{F}^{\mu\nu}) + |D_\mu\Phi|^2 - \lambda|\Phi|^4 + \mu^2|\Phi|^2 - \beta S_\phi\right] \tag{694}$$

Memory Stress-Energy Contribution

The additional stress-energy component associated with coherence memory is defined as

$$T^{\text{mem}}_{\mu\nu} = \partial_\mu C(x)\,\partial_\nu C(x) - \frac{1}{2}g_{\mu\nu}(\partial C)^2 \tag{695}$$

This term represents the energetic contribution of persistent coherence structures within the Æ-field.

Unified Lagrangian (Extended Template)

As an extended curvature sector, one may consider:

$$\mathcal{L}_{\text{Æ}} = \sqrt{-g}\left[\text{Tr}(R_{\mu\nu}R^{\mu\nu}) + |D_\mu\Phi|^2 - \lambda|\Phi|^4 + \mu^2|\Phi|^2 - \beta S_\phi\right], \tag{696}$$

understood as a higher-order extension of the Einstein–Hilbert sector.

Gravitational Interaction

Coherence-Induced Geometry

In the Æ-framework, gravitation is interpreted as the emergent effect of coherence gradients on the effective metric. A minimal coherence-induced deformation reads:

$$g^{\text{Æ}}_{\mu\nu}(x) = C_{\text{Æ}}(x)\,g_{\mu\nu}(x), \tag{697}$$

so that in regions of reduced coherence the geometry is effectively rescaled.

A local perturbative representation of this deformation may be written as

$$g_{\mu\nu}(x) = \eta_{\mu\nu} + \alpha_G \, \partial_\mu \partial_\nu \Big(\ln C_{Æ}(x) \Big), \tag{698}$$

where $\eta_{\mu\nu}$ denotes the Minkowski background metric and α_G sets the coupling scale between coherence gradients and geometric curvature.

Consequently, spatial variations of coherence induce curvature:

$$\nabla_\rho C_{Æ} \neq 0 \quad \Longrightarrow \quad R_{\mu\nu\rho\sigma} \neq 0. \tag{699}$$

Effective Einstein Sector

A minimal gravitational sector consistent with the model is:

$$\mathcal{S}_{\text{grav}} = \frac{1}{2} \int M_0^2 \, C_{Æ}(x) \, R \, \sqrt{-g} \, d^4x, \tag{700}$$

which corresponds to an effective Planck mass:

$$M_P^2(x) = M_0^2 \, C_{Æ}(x). \tag{701}$$

Variation yields modified Einstein equations of the form:

$$M_0^2 \, C_{Æ} \, G_{\mu\nu} + M_0^2 \, (g_{\mu\nu}\Box - \nabla_\mu \nabla_\nu) \, C_{Æ} = T_{\mu\nu}^{(\text{eff})}. \tag{702}$$

In this interpretation the coherence field contributes an effective stress–energy tensor

$$T_{\mu\nu}^{Æ} = \partial_\mu C_{Æ} \partial_\nu C_{Æ} - \frac{1}{2} g_{\mu\nu} (\partial C_{Æ})^2. \tag{703}$$

Newtonian Correspondence

In a weak-field, static limit one may identify an effective potential through coherence modulation:

$$\Phi \sim \ln C_{Æ}, \tag{704}$$

yielding an effective Poisson-type relation:

$$\nabla^2 \Phi \sim \nabla^2 \ln C_{Æ}, \tag{705}$$

providing a classical correspondence limit.

In the weak-field limit the temporal metric component admits the standard Newtonian expansion

$$g_{00} \approx 1 + 2\Phi. \tag{706}$$

Note for Further Formalization

A full derivation of $g_{\mu\nu}$ from a fundamental Æ-action, including exact coupling to matter fields and compatibility with cosmological observations, is deferred to future refinement.

Electromagnetic Interaction

Phase Connection and Gauge Structure

Electromagnetism is modeled as a rotational interaction arising from phase connection structure in the Æ-field.

Introduce a U(1)-like connection field A_μ. Unlike a pure gradient field, A_μ is treated as an independent gauge connection.

The field strength tensor is defined as:

$$F_{\mu\nu} = \partial_\mu A_\nu - \partial_\nu A_\mu. \tag{707}$$

Relation to Phase Structure

In locally coherent regions without defects, the connection may be approximated as:

$$A_\mu \approx \partial_\mu \phi, \tag{708}$$

but nonzero curvature arises in the presence of topological phase defects or nontrivial holonomy:

$$\oint A_\mu \, dx^\mu \neq 0. \tag{709}$$

Electromagnetic Action

$$\mathcal{S}_{\text{EM}} = -\frac{1}{4} \int F_{\mu\nu} F^{\mu\nu} \sqrt{-g} \, d^4x, \tag{710}$$

with Maxwell-type equations:

$$\nabla_\mu F^{\mu\nu} = J^\nu. \tag{711}$$

Left-Right Phase Differential

Electromagnetic polarity may be expressed through left/right phase imbalance:

$$\Delta\phi = \phi_L - \phi_R. \tag{712}$$

Nonvanishing curl,

$$\nabla \times \nabla(\Delta\phi) \neq 0, \tag{713}$$

is permitted only when phase structure is multivalued or contains singularities, ensuring gauge compatibility.

Note for Further Formalization

A complete gauge-theoretic embedding, including charge quantization, running coupling behavior, and renormalization structure, is reserved for future work.

Strong Interaction

Topological Binding and Nodal Confinement

In the Æ-framework, the strong interaction is modeled as a *topological binding mechanism* that stabilizes localized coherent structures against phase dispersion. Nontrivial knotting/braiding of quaternionic phase channels produces effective confinement.

Let $Q(x) \in S^3 \subset \mathbb{H}$ be the local quaternionic phase carrier. Introduce a non-Abelian connection $\mathcal{A}_\mu(x)$ valued in a compact Lie algebra:

$$\mathcal{A}_\mu(x) \in \mathfrak{g}_{\text{strong}}. \tag{714}$$

Define the non-Abelian curvature:

$$\mathcal{F}_{\mu\nu} = \partial_\mu \mathcal{A}_\nu - \partial_\nu \mathcal{A}_\mu + [\mathcal{A}_\mu, \mathcal{A}_\nu]. \tag{715}$$

Wilson loop / holonomy invariant.

$$W(\gamma) = \mathrm{Tr}\,\mathcal{P}\exp\left(\oint_\gamma \mathcal{A}_\mu\, dx^\mu\right), \tag{716}$$

and stable bound structures correspond to nontrivial holonomy class:

$$W(\gamma) \neq \mathrm{Tr}(\mathbb{I}). \tag{717}$$

Area law as operational confinement.

$$\langle W(\gamma)\rangle \sim \exp\big(-\sigma\,\mathcal{A}(\gamma)\big), \qquad \sigma > 0, \tag{718}$$

where $\mathcal{A}(\gamma)$ is the minimal surface area spanned by γ.

Nodal (knot) classification.

$$\mathcal{I}_{\mathrm{top}}(\gamma) = \big(\mathrm{Lk}(\gamma), \mathrm{Wr}(\gamma), \mathrm{Tw}(\gamma), \Delta_K(\gamma), \dots\big), \tag{719}$$

with stability indexed by persistent invariant class:

$$\partial_t \mathcal{I}_{\mathrm{top}}(\gamma) \approx 0. \tag{720}$$

Entropy destabilization.

$$\partial_t S_\phi > 0 \quad \Longrightarrow \quad \sigma_{\mathrm{eff}}(t) \downarrow, \tag{721}$$

with a minimal phenomenological relation:

$$\sigma_{\mathrm{eff}}(t) = \sigma_0 \, e^{-S_\phi(t)/\kappa_{\ll}}, \qquad \sigma_0 > 0. \tag{722}$$

Strong-Sector Action

$$\mathcal{S}_{\mathrm{strong}} = -\frac{1}{4g_s^2} \int \mathrm{Tr}(\mathcal{F}_{\mu\nu}\mathcal{F}^{\mu\nu}) \sqrt{-g}\, d^4x. \tag{723}$$

Note for Further Formalization

Rigorous identification of $\mathfrak{g}_{\mathrm{strong}}$, explicit mapping between knot invariants and hadronic spectra, and derivation of confinement directly from Æ microdynamics remain open.

Weak Interaction

Irreversible Transitions and Symmetry Breaking

In the Æ-framework, the weak interaction is modeled as an effective *transition mechanism* converting one coherent configuration into another through symmetry breaking and irreversibility in phase organization.

Introduce an internal multiplet $\Psi_W(x)$ with an SU(2)-like gauge connection $\mathcal{W}_\mu$:

$$\mathcal{W}_\mu(x) \in \mathfrak{g}_{\mathrm{weak}} \quad \text{(operationally analogous to } \mathfrak{su}(2)), \tag{724}$$

with curvature:

$$\mathcal{G}_{\mu\nu} = \partial_\mu \mathcal{W}_\nu - \partial_\nu \mathcal{W}_\mu + [\mathcal{W}_\mu, \mathcal{W}_\nu]. \tag{725}$$

Transition operator form.

$$\frac{d}{dt}\Psi_W(t) = -i\hat{\mathcal{H}}_W\Psi_W(t) - \hat{\mathcal{T}}_W\Psi_W(t), \tag{726}$$

with a minimal entropy-linked representation:

$$\hat{\mathcal{T}}_W = \lambda_W\,\dot{S}_\phi\,\hat{\Pi}_{\text{break}}, \qquad \lambda_W \geq 0. \tag{727}$$

Spontaneous symmetry breaking (effective Higgs-like sector). Introduce a scalar order parameter $\varphi(x)$ with:

$$V(\varphi) = -\frac{\mu^2}{2}|\varphi|^2 + \frac{\lambda}{4}|\varphi|^4, \qquad \mu^2, \lambda > 0, \tag{728}$$

so that:

$$|\langle\varphi\rangle| = v = \frac{\mu}{\sqrt{\lambda}}. \tag{729}$$

Gauge boson masses arise as:

$$m_W \sim g_W\,v. \tag{730}$$

Irreversibility condition.

$$\dot{S}_\phi \geq 0, \qquad \text{and typically} \qquad \dot{S}_\phi > 0 \text{ during transitions.} \tag{731}$$

Weak-Sector Action

$$\mathcal{S}_{\text{weak}} = \int \left[-\frac{1}{4g_W^2}\,\text{Tr}(\mathcal{G}_{\mu\nu}\mathcal{G}^{\mu\nu}) + |D_\mu\varphi|^2 - V(\varphi)\right]\sqrt{-g}\,d^4x, \tag{732}$$

with covariant derivative:

$$D_\mu\varphi = (\partial_\mu + \mathcal{W}_\mu)\varphi. \tag{733}$$

Note for Further Formalization

An explicit mapping to observed decay channels, a precise definition of $\hat{\Pi}_{\text{break}}$ in Æ-mode language, and a derivation of effective non-unitarity from coarse-graining remain open.

Unified View of Æ-Forces

Common Structural Origin

Within the Æ-model, the four fundamental forces are interpreted as distinct operator-realizations of a single underlying interaction structure:

- Gravity: coherence *gradients* induce effective geometry,

$$\nabla C_{\text{Æ}} \rightsquigarrow g_{\mu\nu}, \quad R \neq 0. \tag{734}$$

- Electromagnetism: phase *curl* and U(1)-like holonomy,

$$F_{\mu\nu} = \partial_\mu A_\nu - \partial_\nu A_\mu, \qquad \oint A_\mu dx^\mu \neq 0. \tag{735}$$

- Strong: non-Abelian holonomy + knot/braid confinement,

$$W(\gamma) = \operatorname{Tr} \mathcal{P} \exp\left(\oint_\gamma \mathcal{A} \right), \qquad \langle W(\gamma) \rangle \sim e^{-\sigma \mathcal{A}(\gamma)}. \tag{736}$$

- Weak: symmetry-breaking transitions and irreversible restructuring,

$$V(\varphi) = -\tfrac{\mu^2}{2}|\varphi|^2 + \tfrac{\lambda}{4}|\varphi|^4, \qquad \dot{S}_\phi > 0 \text{ during transitions.} \tag{737}$$

This unification is structural rather than reductionist: forces are distinct operator-lenses of the same Æ-field landscape.

Master Action and Field Equations (Optional Module)

We work on a 4D Lorentzian spacetime $(M, g_{\mu\nu})$ with signature $(-,+,+,+)$. The coherence field $C(x)$ and phase-entropy field $S_\phi(x)$ are treated as independent scalar fields (locally one may set $C \equiv C_{\text{Æ}}$; see previous sections for definition).

Gauge fields are A^a_μ ($\mathrm{SU}(3)_c$), W^i_μ ($\mathrm{SU}(2)_L$), and B_μ ($\mathrm{U}(1)_Y$), with constant gauge couplings g_{s0}, g_{w0}, g'_0 inside covariant derivatives. *Remark.* Any effective modulation induced by (C, S_ϕ) is encoded in the kinetic weights Z_s, Z_w, Z_y below, rather than in the gauge couplings.

Field strengths and covariant derivative (constant gauge couplings)

$$G^a_{\mu\nu} := \partial_\mu A^a_\nu - \partial_\nu A^a_\mu + g_{s0}\, f^{abc} A^b_\mu A^c_\nu, \tag{738}$$
$$W^i_{\mu\nu} := \partial_\mu W^i_\nu - \partial_\nu W^i_\mu + g_{w0}\, \epsilon^{ijk} W^j_\mu W^k_\nu, \tag{739}$$
$$B_{\mu\nu} := \partial_\mu B_\nu - \partial_\nu B_\mu. \tag{740}$$

For an SU(2) doublet Φ with hypercharge Y_Φ, the covariant derivative is

$$D_\mu \Phi := \left(\partial_\mu - i\, g_{w0}\, W^i_\mu \frac{\tau^i}{2} - i\, g'_0\, Y_\Phi\, B_\mu \right) \Phi, \tag{741}$$

where τ^i are the Pauli matrices.

Einstein–Hilbert term with coherence-dependent Planck mass

$$\frac{1}{2} M_P^2(C)\, R, \qquad M_P^2(C) := M_0^2\, C. \tag{742}$$

Gauge kinetic weights

$$-\frac{1}{4} Z_s(C, S_\phi)\, G^a_{\mu\nu} G^{a\mu\nu} - \frac{1}{4} Z_w(C, S_\phi)\, W^i_{\mu\nu} W^{i\mu\nu} - \frac{1}{4} Z_y(C, S_\phi)\, B_{\mu\nu} B^{\mu\nu}, \tag{743}$$

with $Z_s, Z_w, Z_y > 0$ dimensionless functions. *Remark.* These weights encode coherence/entropy effects without altering the underlying gauge structure.

A minimal optional choice is

$$Z_s(C, S_\phi) = Z_w(C, S_\phi) = Z_y(C, S_\phi) = \frac{C}{1 + S_\phi}. \tag{744}$$

Bilocal memory term (Volterra / causal kernel)

$$S_{\text{mem}} = \frac{1}{2} \int d^4x\, \sqrt{-g(x)} \int d^4x'\, \sqrt{-g(x')}\, K(x, x')\, S_\phi(x)\, S_\phi(x'). \tag{745}$$

The kernel satisfies a causal (Volterra) condition:

$$K(x, x') = 0 \quad \text{unless} \quad x'^0 \le x^0. \tag{746}$$

Remark. This explicitly breaks time-reversal symmetry at the level of the effective action.

Total action

$$S = \int d^4x \sqrt{-g} \left[\frac{1}{2} M_0^2 \, C \, R - \frac{1}{4} Z_s \, G^a_{\mu\nu} G^{a\mu\nu} - \frac{1}{4} Z_w \, W^i_{\mu\nu} W^{i\mu\nu} - \frac{1}{4} Z_y \, B_{\mu\nu} B^{\mu\nu} + (D_\mu \Phi)^\dagger (D^\mu \Phi) - V(\Phi) + \bar{\psi} \, i\gamma^\mu D_\mu \psi - \bar{\psi} \, Y \, \Phi \, \psi + \frac{1}{2} \partial_\mu C \, \partial^\mu C + \frac{1}{2} \partial_\mu S_\phi \, \partial^\mu S_\phi - U(C, S_\phi) \right] + S_{\text{mem}}. \tag{747}$$

Remark. The Yukawa term $\bar{\psi} Y \Phi \psi$ is written schematically; flavor and generation structure are suppressed.

If one wishes to impose a constraint such as

$$C = \frac{1}{1 + S_\phi}, \tag{748}$$

it should be implemented via a Lagrange multiplier field $\Lambda(x)$:

$$S_{\text{constraint}} = \int d^4x \sqrt{-g} \, \Lambda(x) \left(C - \frac{1}{1 + S_\phi} \right), \tag{749}$$

rather than hard-coding it into U.

Field Equations (Skeleton)

Varying with respect to $g^{\mu\nu}$ yields

$$M_0^2 \, C \, G_{\mu\nu} + M_0^2 \, (g_{\mu\nu} \Box - \nabla_\mu \nabla_\nu) \, C = T^{(\text{tot})}_{\mu\nu} - g_{\mu\nu} \, U(C, S_\phi) + T^{(\text{mem})}_{\mu\nu} \tag{750}$$

Remark. Care is required to avoid double counting of $U(C, S_\phi)$ depending on its inclusion in $T^{(\text{tot})}_{\mu\nu}$.

Gauge-field equations with kinetic weights:

$$\nabla_\nu (Z_s \, G^{a\nu\mu}) + g_{s0} \, f^{abc} A^b_\nu \, (Z_s \, G^{c\nu\mu}) = j^{a\mu}, \tag{751}$$

$$\nabla_\nu \big(Z_w \, W^{i\nu\mu} \big) + g_{w0} \, \epsilon^{ijk} W^j_\nu \, \big(Z_w \, W^{k\nu\mu} \big) = j^{i\mu}, \tag{752}$$

$$\nabla_\nu (Z_y \, B^{\nu\mu}) = j^\mu_Y \tag{753}$$

where $j^{a\mu}$, $j^{i\mu}$ and j^μ_Y denote the corresponding matter currents. *Open point.* A fully explicit construction of these currents in the presence of space-time dependent weights Z_s, Z_w, Z_y remains to be developed.

Scalar equations:

$$\Box C + \frac{1}{2} M_0^2 R - \frac{1}{4} \frac{\partial Z_s}{\partial C} G^a_{\mu\nu} G^{a\mu\nu} - \frac{1}{4} \frac{\partial Z_w}{\partial C} W^i_{\mu\nu} W^{i\mu\nu} - \frac{1}{4} \frac{\partial Z_y}{\partial C} B_{\mu\nu} B^{\mu\nu} - \frac{\partial U}{\partial C} = 0 \tag{754}$$

$$\Box S_\phi - \frac{1}{4} \frac{\partial Z_s}{\partial S_\phi} G^a_{\mu\nu} G^{a\mu\nu} - \frac{1}{4} \frac{\partial Z_w}{\partial S_\phi} W^i_{\mu\nu} W^{i\mu\nu} - \frac{1}{4} \frac{\partial Z_y}{\partial S_\phi} B_{\mu\nu} B^{\mu\nu} - \frac{\partial U}{\partial S_\phi} + \frac{1}{\sqrt{-g}} \frac{\delta S_{\rm mem}}{\delta S_\phi(x)} = 0 \tag{755}$$

with explicit memory variation:

$$\frac{1}{\sqrt{-g}} \frac{\delta S_{\rm mem}}{\delta S_\phi(x)} = \int d^4x' \sqrt{-g(x')}\, K(x, x')\, S_\phi(x') \tag{756}$$

Remark. The asymmetric (causal) structure reflects the Volterra nature of the kernel and departs from standard time-symmetric bilocal variations.

Non-Force Mechanisms Coupling to Force-Driven Dynamics

Coherence C and phase entropy S_ϕ enter by *renormalizing interaction channels*: (i) C rescales gravity through $M_P^2(C) = M_0^2 C$, (ii) (C, S_ϕ) modulate gauge propagation through Z_s, Z_w, Z_y, and (iii) the bilocal memory term produces causal integral feedback in S_ϕ, altering the entropy budget that controls stability.

Note for Further Formalization

- Open: Fixing (or fitting) the functional forms of $Z_s(C, S_\phi)$, $Z_w(C, S_\phi)$, $Z_y(C, S_\phi)$.
- Open: Specification of $U(C, S_\phi)$ and/or constraint implementation via $\Lambda(x)$.
- Open: Metric-variation consistency of $S_{\rm mem}$ (i.e., whether $K(x, x')$ depends on $g_{\mu\nu}$).
- Open: Stability/causality analysis for the Volterra kernel $K(x, x')$ and well-posedness.
- Conjectural: Standard-Model recovery in the limit $S_\phi \to 0$ and $C \to 1$ with $Z_s, Z_w, Z_y \to 1$.

55. Oscillatory Patterns and Resonance

Oscillatory behaviour naturally emerges in the Æ framework from the phase dynamics of the underlying complex field. When the phase evolves coherently in time, stable interference patterns and resonant structures appear in the observable projection of the field.

Oscillatory Æ Field

Let the observable Æ-field be written as

$$\text{«}(x,t) = A(x,t)\cos\big(\omega t + \phi(x,t)\big). \tag{757}$$

The instantaneous angular frequency is determined by the phase evolution

$$\omega = \frac{d\phi}{dt}. \tag{758}$$

Oscillatory behaviour therefore arises from continuous phase rotation in the underlying complex representation.

Phase Synchronization

Coupled oscillatory nodes evolve according to a Kuramoto-type dynamics

$$\frac{d\phi_i}{dt} = \omega_i + \sum_j K_{ij}\sin(\phi_j - \phi_i), \tag{759}$$

where the coupling coefficients are related to field coherence

$$K_{ij} \propto C_{ij}. \tag{760}$$

Phase locking corresponds to a constant phase difference

$$\Delta\phi(t) = \text{const}. \tag{761}$$

More generally,

$$\dot{\phi}_i = \omega_i + \sum_{j\neq i} \kappa_{ij}\sin(\phi_j - \phi_i). \tag{762}$$

Lissajous Resonance

Resonant structures appear when oscillatory components satisfy rational frequency relations

$$\omega_1 : \omega_2 = m : n. \tag{763}$$

The corresponding trajectories are described by the classical Lissajous system

$$x(t) = A\sin(\omega_1 t), \qquad y(t) = B\sin(\omega_2 t + \delta). \tag{764}$$

These curves represent stable interference geometries of phase evolution.

Pattern Stability

The persistence of resonant patterns depends on phase coherence. Let $\Delta\phi(t)$ denote the phase deviation between interacting modes.

A configuration remains stable when

$$\mathrm{Var}(\Delta\phi(t)) \le \epsilon, \tag{765}$$

which implies

$$C_{\text{pattern}} \ge C_{\text{min}}. \tag{766}$$

Loss of synchronization produces

$$\Delta\phi(t) \text{ unbounded} \Longrightarrow \text{desynchronization}. \tag{767}$$

In the limit of vanishing coherence

$$C \to 0 \Longrightarrow \text{pattern collapse}. \tag{768}$$

Spectral stability criterion. The dynamical stability of synchronized configurations may also be analyzed through the spectrum of the linearized phase dynamics.

Let L denote the linearized operator obtained from perturbations around a synchronized configuration $\phi_i = \phi_i^* + \epsilon_i$.

The evolution of perturbations satisfies

$$\dot{\epsilon} = L\,\epsilon. \tag{769}$$

The spectral gap is defined as

$$\Delta\lambda = \lambda_2 - \lambda_1, \tag{770}$$

where λ_1 and λ_2 denote the first two eigenvalues of the spectrum.

A positive spectral gap implies decay of perturbations and therefore stability of the synchronized configuration.

Phase Transmission

Phase information may propagate between Hilbert spaces through a transmission operator

$$T_{a\to b}:\ \mathcal{H}_a \to \mathcal{H}_b. \tag{771}$$

The induced phase mismatch is

$$\Delta\phi_{a\to b}(t) = \arg\left(\langle Q_b(t),\, T_{a\to b}Q_a(t)\rangle\right) \tag{772}$$

A scalar transmission factor is defined by

$$\Delta_{a\to b}(t) = \exp(-\alpha|\Delta\phi_{a\to b}(t)|)\,\exp\big(-\beta S_{\phi,a\to b}(t)\big) \tag{773}$$

It satisfies

$$0 < \Delta_{a\to b} \le 1. \tag{774}$$

Perfect phase alignment implies

$$|\Delta\phi_{a\to b}| \to 0, \qquad S_{\phi,a\to b} \to 0 \quad \Rightarrow \quad \Delta_{a\to b} \to 1. \tag{775}$$

Operator Formulation

A unitary phase transmission operator may be written as

$$\mathcal{U}_{a\to b}(t) = \exp\big(-i\,\mathcal{G}_{a\to b}(t)\big) \tag{776}$$

The normalized transmission overlap becomes

$$\Delta_{a\to b}(t) = \frac{|\langle \Psi_b(t), \mathcal{U}_{a\to b}(t)\Psi_a(t)\rangle|}{\|\Psi_b(t)\|\|\Psi_a(t)\|} \tag{777}$$

Hence

$$0 \le \Delta_{a\to b} \le 1. \tag{778}$$

Network Coherence

Coherence of the Æ-field is related to phase entropy through

$$C_{Æ} = \frac{1}{1+S_\phi}. \tag{779}$$

Transmission between nodes may then be approximated by

$$\Delta_{a\to b} = \exp(-\alpha|\Delta\phi|)\, f(C_{Æ}). \tag{780}$$

For a network with edge set E we define the mean transmission factor

$$\Delta_{\text{net}}(t) = \frac{1}{|E|}\sum_{(i\to j)\in E} \Delta_{i\to j}(t) \tag{781}$$

Near perfect coherence one typically has

$$|\Delta\phi_{a\to b}| \ll 1, \qquad S_{\phi,a\to b} \ll 1, \tag{782}$$

leading to

$$\Delta_{a\to b} \approx 1. \tag{783}$$

56. Time as an Emergent Process

Coexisting Temporal Variables

We introduce a hierarchical ordering of temporal variables:

$$\tau \preceq t_{Æ} \preceq t \preceq t_{\text{phys}} \preceq t_{\text{subj}}, \tag{784}$$

where $\preceq$ denotes functional emergence rather than numerical inequality.

A useful representation is the projection from abstract temporal ordering into a physical descriptor space:

$$\pi : \mathcal{T} \to \mathcal{B}, \qquad \mathcal{B} \subset \mathbb{R}^3, \quad \mathcal{B} = \{(\phi, S_\phi, C)\}.$$

We also use the emergence preorder as:

$$Æ_1 \prec Æ_2 \iff \text{the phase configuration of } Æ_2 \text{ arises from that of } Æ_1. \tag{785}$$

Æ-Time as Phase–Topological Time

Temporal flow in the Æ-framework is defined through phase evolution. We assume nontrivial phase dynamics:

$$\partial_t \phi(x,t) \neq 0. \tag{786}$$

The local angular frequency is given by

$$\omega(x,t) = \partial_t \phi(x,t). \tag{787}$$

A spatially averaged frequency over a domain Ω with Æ-measure $d\mu_{Æ}$ is

$$\bar{\omega}(t) = \int_\Omega \omega(x,t)\, d\mu_{Æ}(x). \tag{788}$$

Intrinsic Æ-time is defined as a phase increment:

$$dt_{Æ} := \alpha\, d\phi, \qquad \alpha > 0. \tag{789}$$

A discrete-time viewpoint is consistent with

$$\phi(x, t + \Delta t) = \phi(x,t) + \Delta\phi. \tag{790}$$

The quaternionic rotational rate is defined by

$$\omega_{Æ}(t) = \left\| \frac{1}{2} \mathrm{Im}(Q^{-1}\dot{Q}) \right\|. \tag{791}$$

The accumulated Æ-time along a trajectory is

$$\tau_{Æ}(t) = \int_0^t \omega_{Æ}(s)\, ds. \tag{792}$$

The total quaternionic phase (holonomy) may be written as

$$\Phi_{\text{total}} = \int_0^T \frac{1}{2}\,\mathrm{Im}(Q^{-1}\dot{Q})\,dt.$$

Equivalently, for a trajectory $\gamma \subset S^3$:

$$\Phi_{\text{total}}[\gamma] = \int_\gamma A, \qquad A = \frac{1}{2}\,\mathrm{Im}(Q^{-1}dQ).$$

A proportionality between Æ-time and accumulated holonomy is given by

$$d\tau_{Æ} = \kappa\, d\Phi_{\text{total}}, \qquad \kappa > 0, \qquad d\Phi_{\text{total}} = \sum_i d\Phi_i.$$

If $\dot{\Phi}_i(t)$ denote directional holonomy rates, one may write

$$\dot{\Phi}_i(t) = w_i(\pi_H(Q(t)))\,A(\dot{Q}(t)), \qquad \sum_{i=1}^{4} \dot{\Phi}_i(t) = \dot{\Phi}_{\text{total}}(t),$$

together with the quaternionic evolution constraint

$$\dot{Q}(t) = F(Q(t), x, t), \qquad F(Q) \in T_Q S^3.$$

Entropy, Coherence, and Temporal Orientation

Time orientation is coupled to entropy–coherence dynamics. A generic entropy rate is

$$\frac{dS_\phi^{(\alpha)}}{dt} = -\sum_{i=1}^{4} \dot{p}_i^{(\alpha)} \left(\ln p_i^{(\alpha)} + 1\right).$$

A characteristic monotonicity pattern in the framework is

$$\dot{S}_\phi \geq 0, \qquad \dot{C} \leq 0, \qquad \dot{M} \geq 0.$$

An emergent monotone time variable $t_m(\tau)$ may be defined as

$$t_m(\tau) = \int^\tau \left(\eta_1 \dot{S}_\phi - \eta_2 \dot{C} + \eta_3 \dot{M}\right) d\tau', \qquad \eta_i \geq 0,$$

which satisfies

$$\frac{dt_m}{d\tau} = \eta_1 \dot{S}_\phi - \eta_2 \dot{C} + \eta_3 \dot{M} \geq 0.$$

A coherence-damping bound is given by

$$\|R(\tau)\| \leq e^{-\gamma S_\phi(\tau)}, \qquad \gamma > 0.$$

Deformed Temporal Algebra

We define the fundamental temporal operators as

$$D_\tau := \partial_\tau, \qquad T_{Æ} := i\,\partial_{t_{Æ}}, \qquad T_{\text{phys}} := i\,\partial_{t_{\text{phys}}}, \qquad \hat{\phi}\,\psi := \phi\,\psi.$$

Assuming a phase–time relation $dt_{Æ} = \alpha\, d\phi$, one obtains

$$\partial_{t_{Æ}} = \alpha^{-1}\partial_\phi, \qquad [T_{Æ}, \hat{\phi}] = i\,\alpha^{-1}.$$

The Æ-time generator commutes with the intrinsic evolution operator:

$$[D_\tau, T_{Æ}] = 0.$$

Physical time is modeled as a coherence–entropy rescaling of τ:

$$\partial_{t_{\text{phys}}} = \left(e^{-S_\phi/\kappa}\, C^{1/2}\right)^{-1} \partial_\tau, \tag{793}$$

or equivalently

$$T_{\text{phys}} = \left(e^{-S_\phi/\kappa}\, C^{1/2}\right)^{-1} T_\tau, \qquad T_\tau := i\,\partial_\tau.$$

A deformed derivative incorporating phase entropy is defined by

$$D_{S_\phi} := S_\phi^{-1} D_\tau.$$

The resulting entropy–time deformation of the algebra is

$$[D_\tau, D_{S_\phi}] = -S_\phi^{-2}(\partial_\tau S_\phi)\, D_\tau. \tag{794}$$

A mixed entropy–phase commutator is then

$$[D_{S_\phi}, \hat{\phi}] = S_\phi^{-1}[D_\tau, \hat{\phi}] = i\, S_\phi^{-1}\, \partial_\tau \phi.$$

Under the normalized convention $\alpha = 1$, the Æ-time commutator reduces to

$$[T_{Æ}, \hat{\phi}] = i.$$

Unified Temporal Generator

We define the composite temporal generator as

$$\hat{T} = iD_\tau + i\beta T_{«} + \gamma D_{S_\phi}, \tag{795}$$

where β and γ are model parameters.
A convenient parametrization for the phase entropy is

$$S_\phi := \beta_0\, |\operatorname{Im} Q|^2, \qquad \beta_0 > 0,$$

with temporal variation

$$\dot{S}_\phi = 2\beta_0 \operatorname{Re}\!\left(\operatorname{Im} Q \cdot \overline{\operatorname{Im} \dot{Q}}\right), \qquad \operatorname{Re}\!\left(\operatorname{Im} Q \cdot \overline{\operatorname{Im} \dot{Q}}\right) \geq 0.$$

Variational Formulation of Emergent Time

A minimal action governing the emergence of time is given by

$$S_{\text{time}} = \int \left[\frac{1}{2}(\dot{\phi})^2 + \frac{k}{2}(\dot{S}_\phi)^2 - V(\phi, S_\phi) + \lambda\big(C - f(\phi, S_\phi)\big) \right] d\tau,$$

which yields the Euler–Lagrange equations

$$\frac{d}{d\tau}\dot{\phi} + \frac{\partial V}{\partial \phi} + \lambda \frac{\partial f}{\partial \phi} = 0, \qquad k\frac{d}{d\tau}\dot{S}_\phi + \frac{\partial V}{\partial S_\phi} + \lambda \frac{\partial f}{\partial S_\phi} = 0.$$

A statistical representation over S^3 is defined by

$$Z(\beta) = \int_{S^3} e^{-\beta V(Q)}\, d\mu(Q), \qquad \rho(Q) = \frac{e^{-\beta V(Q)}}{Z(\beta)},$$

$$S_Q = -\int_{S^3} \rho(Q)\, \ln \rho(Q)\, d\mu(Q).$$

Temporal Generators and Loop Constraints

The evolution of states in the Æ-framework is generated by an effective time operator including a connection term:

$$\hat{T}_{Æ} = i\partial_{t_{Æ}} + \Gamma_{Æ}, \qquad i\,\hat{T}_{Æ}\Psi = H_{Æ}\Psi.$$

Equivalently, along an intrinsic parameter $\tau_{Æ}$, the evolution may be expressed as

$$i\,\partial_{\tau_{Æ}}\Phi = H_{Æ}\Phi, \qquad D_{\tau_{Æ}} = \partial_{\tau_{Æ}} + iA_\tau.$$

The corresponding loop functional is defined by

$$S_{\text{loop}} = \oint \Phi^\dagger D_{\tau_{Æ}} \Phi\, d\tau_{Æ} + \sum_k \theta_k\, \delta(\tau_{Æ} - \tau_k),$$

with the quantization condition

$$\oint A_\tau = 2\pi m, \qquad m \in \mathbb{Z}.$$

Hamiltonian Compatibility Conditions

For dynamical consistency, the Hamiltonian generator satisfies compatibility relations with the temporal operators.

In particular,

$$[H, T_{Æ}] = 0,$$

indicating that the Æ-time generator preserves the Hamiltonian structure.

However, the abstract temporal derivative may introduce explicit time dependence,

$$[H, D_\tau] = i\,\partial_\tau H \neq 0.$$

This reflects the possibility of evolving effective Hamiltonians within the emergent temporal framework.

Phase Differences and Interaction

Phase differences encode the relational structure of oscillatory components in the Æ field:

$$\Delta\phi_{ij} = \phi_i - \phi_j. \tag{796}$$

Interaction strength depends on phase mismatch:

$$\mathcal{I}_{ij} \propto f(\Delta\phi_{ij}). \tag{797}$$

Temporal phase evolution follows

$$\partial_t \phi(x,t), \tag{798}$$

with incremental update

$$\phi(x, t + \Delta t) = \phi(x,t) + \Delta\phi. \tag{799}$$

Phase Coherence and Entropic Stability

Field coherence is defined through phase variance:

$$C := 1 - \mathrm{Var}(\phi), \tag{800}$$

with bounds

$$0 \leq C \leq 1. \tag{801}$$

Local and global coherence are

$$C_{\mathrm{local}}(x) = 1 - \mathrm{Var}\big(\phi(x,t)\big), \tag{802}$$

$$C_{\mathrm{global}} = 1 - \mathrm{Var}\big(\phi(x,t)\big)_{x\in\Omega}. \tag{803}$$

Phase entropy is defined as

$$S_\phi = -\int p(\phi)\, \log p(\phi)\, d\phi, \tag{804}$$

with inverse relation

$$S_\phi \uparrow \iff C \downarrow. \tag{805}$$

The entropy–coherence dynamics satisfies

$$\frac{dC}{dt} = -\frac{dS_\phi}{dt}, \tag{806}$$

with entropic growth

$$\frac{dS_\phi}{dt} \geq 0. \tag{807}$$

Nonlocal Memory Kernel

Phase memory may be modeled through a nonlocal kernel acting on the phase–entropy field:

$$\Box S_\phi - \frac{\partial U}{\partial S_\phi} + \int d^4x' \sqrt{-g(x')}\, K(x,x')\, S_\phi(x') = \mathcal{J}(C, F_{Æ}). \tag{808}$$

This defines the retention operator

$$\mathcal{R}[S_\phi] = \Box S_\phi + \int K(x,x')\, S_\phi(x'). \tag{809}$$

The corresponding dispersion relation becomes

$$\omega^2(k) = k^2 + m_{\text{eff}}^2 + \tilde{K}(\omega, k). \tag{810}$$

Holonomy Memory Conservation

Memory is geometrically encoded through holonomy conservation across nodal structures:

$$\sum_i \Phi_i = \Phi_{\text{total}}. \tag{811}$$

This expresses the persistence of accumulated phase circulation in the Æ field.

Spectral Memory Operator

Topological memory can also be represented through a spectral operator acting on nodal curves:

$$L_\Gamma = -\frac{d^2}{ds^2} + V_{\text{helix}}(s), \tag{812}$$

with eigenmodes

$$L_\Gamma u_k = \lambda_k u_k. \tag{813}$$

Spectral stability requires a nonzero gap

$$\Delta\lambda = \lambda_2 - \lambda_1 > 0. \tag{814}$$

Entropic Filtering and Forgetting

Local memory retention is bounded by entropic filtering:

$$S_{\text{inherit}} = S_\phi(\Phi_{\text{astral}}). \tag{815}$$

The local entropy must satisfy

$$S_{\text{local}} < S_{\text{crit}}. \tag{816}$$

Projection from the higher phase configuration is given by

$$\Phi_{\text{local}} = \mathcal{F}(\Phi_{\text{astral}}), \tag{817}$$

with

$$\mathcal{F} : S^3 \to S^3, \tag{818}$$

preserving norm and topological class:

$$\|\mathcal{F}(\Phi)\| = 1, \qquad \text{class}(\Phi) = \text{class}(\mathcal{F}(\Phi)). \tag{819}$$

The effective entropy becomes

$$S_{\phi,\text{local}} = S_{\text{inherit}} - \Delta S_{\text{filtered}}. \tag{820}$$

Entropy growth remains bounded:

$$\frac{d}{dt} S_{\text{local}} \leq \sigma_{\max}. \tag{821}$$

Discrete Nodal Representation

Each node in the network is represented by a dynamical tuple

$$\mathcal{N}_n := (C_n, \omega_n, \phi_n, \tau_n). \tag{822}$$

The global nodal state lives in the tensor product space

$$\Psi_{\text{net}} \in \mathcal{H}_{\text{net}}, \qquad \mathcal{H}_{\text{net}} = \bigotimes_{n=1}^{N} \mathcal{H}_n. \tag{823}$$

Graph Structure of the Network

The nodal system forms a graph

$$\mathcal{G} = (\mathcal{V}, \mathcal{E}), \tag{824}$$

with coupling matrix

$$\mathbf{K} = \{K_{nm}\}. \tag{825}$$

The associated Laplacian is

$$\mathbf{L} = \mathbf{D} - \mathbf{K}. \tag{826}$$

Phase evolution can be written compactly as

$$\partial_t \boldsymbol{\phi} = -\mathbf{L}\,\boldsymbol{\phi} + \boldsymbol{\omega}. \tag{827}$$

Spectral Mode Decomposition

Phase dynamics admits a spectral expansion

$$\boldsymbol{\phi}(t) = \sum_k a_k(t)\mathbf{v}_k, \tag{828}$$

where $\mathbf{v}_k$ are Laplacian eigenvectors.
The modal amplitudes evolve according to

$$\dot{a}_k = -\lambda_k a_k + F_k(t). \tag{829}$$

Coherence Evolution in the Network

Local coherence dynamics follows

$$\partial_t C_n = -\alpha_n S_{\phi,n} C_n + \beta_n I_n. \tag{830}$$

The global coherence of the network is

$$C_{\text{net}} = \sum_n w_n C_n. \tag{831}$$

Propagation across the graph obeys

$$\partial_t C^{(\alpha)} = -\alpha_0 S_\phi^{(\alpha)} C^{(\alpha)} - \sum_\beta L_{\alpha\beta} C^{(\beta)}. \tag{832}$$

Quaternionic Phase Velocity

Each node carries a quaternionic phase state

$$Q_i(\tau_i) \in S^3 \subset \mathbb{H}, \qquad |Q_i| = 1. \tag{833}$$

The associated phase velocity is

$$\Omega_i(\tau_i) = \mathrm{Im}\left(\dot{Q}_i(\tau_i)Q_i^{-1}(\tau_i)\right). \tag{834}$$

Causal Coupling Constraint

Phase propagation respects causal structure

$$ds^2 = c^2dt^2 - d\mathbf{x}^2 \geq 0. \tag{835}$$

Wavevectors satisfy the phase null constraint

$$g^{\mu\nu}k_\mu k_\nu = 0, \qquad k_\mu = \partial_\mu\phi. \tag{836}$$

Coupling between nodes is gated by

$$\mathcal{G}_{ij} = \Theta(ds_{ij}^2)\exp\left(-\frac{d_{ij}^2}{\ell_c^2}\right). \tag{837}$$

Æ Coupling Strength

The effective coupling coefficient becomes

$$\kappa_{Æ}(Q_i, Q_j) = \kappa_0\mathcal{G}_{ij}\frac{C_{Æ}(Q_i)C_{Æ}(Q_j)}{1 + S_{\phi,ij}}\cos(\Delta\phi_{ij}). \tag{838}$$

The resulting phase dynamics reads

$$\dot{\phi}_i = \omega_i + \sum_{j\neq i}\kappa_{Æ}(Q_i, Q_j)\sin(\phi_j - \phi_i). \tag{839}$$

Propagation Speed Bound

Information transfer between nodes defines an effective velocity

$$v_{\alpha\beta}^{\mathrm{eff}} = \frac{\Delta\tau_{Æ}^{(\alpha\beta)}}{\ell(\gamma_{\alpha\beta})}. \tag{840}$$

Causality imposes

$$v_{\alpha\beta}^{\mathrm{eff}} \leq v_{\mathrm{null}}. \tag{841}$$

The Self as a Trajectory in Quaternionic State Space

We model the self as an admissible trajectory in quaternionic state space. Let

$$\gamma \subset S^3, \qquad Q(t) \in S^3 \subset \mathbb{H},$$

and denote the accumulated holonomy along γ by $\Phi_{\text{total}}[\gamma]$.

More generally, trajectories are maps into the Æ solution/state space:

$$\gamma : \mathbb{R} \to \mathcal{S}_{Æ}, \qquad s(t) = \gamma(t). \tag{842}$$

Coherent Sectors in Solution Space

We define coherent basins (sectors) as sets of trajectories converging to the same asymptotic coherent phase state:

$$\mathcal{B}_k = \left\{ \gamma_i \;\middle|\; \lim_{t\to\infty} \Phi_{\text{coh}}(\gamma_i(t)) = \Phi_k \right\}. \tag{843}$$

A canonical weight for each basin is coherence–entropy penalized:

$$W_k \propto \exp\left(-\frac{S_\phi(\mathcal{B}_k)}{k_{Æ}} \right). \tag{844}$$

The effective basin is selected by minimal phase entropy:

$$\mathcal{B}_{\text{eff}} = \arg\min_{\mathcal{B}_k} S_\phi(\mathcal{B}_k). \tag{845}$$

We may also define a distance between basins using their phase representatives:

$$d_{Æ}(\mathcal{B}_i, \mathcal{B}_j) = \int_{\mathcal{S}_{Æ}} |\Phi_i(s) - \Phi_j(s)|^2 \, d\mu_{Æ}(s). \tag{846}$$

A nested hierarchy of coherent sectors can be introduced:

$$\mathcal{B}_1 \subset \mathcal{B}_2 \subset \cdots \subset \mathcal{B}_n. \tag{847}$$

Coherence compatibility between sector representatives may be enforced via

$$\langle \Phi_i \mid \Phi_j \rangle \geq \Phi_{\text{crit}}. \tag{848}$$

Admissibility and Timeline Set

The trajectory is admissible if it remains coherent:

$$C_{Æ}(D(t), d(t)) \geq C_{\min}, \qquad S_\phi(D(t), d(t)) \leq S_\phi^{\text{crit}} \tag{849}$$

The set of admissible "timelines" is then

$$\mathcal{T} = \Big\{ Q(t) \in S^3 : \ Q(t) \text{ satisfies the NR admissibility/stability conditions} \Big\}. \tag{850}$$

Internal Parametrization and Arc-Length Time

A convenient internal parameterization can be introduced through a trajectory functional $H_\gamma(t)$ and the consciousness observable $\mathcal{C}$ already defined above:

$$\tau = \mathcal{C}(H_\gamma(t)) . \tag{851}$$

The induced rate satisfies

$$\frac{d\tau}{dt} = \left\| \frac{dH_\gamma}{dt} \right\| . \tag{852}$$

When defined, physical time may be related to arc-length in quaternionic state space:

$$t \sim s(Q(t)) \quad \text{(arc-length in quaternionic state space, when defined).}$$

Energetic Modulators of Trajectory Stability

A generic energetic measure associated with trajectory motion is

$$E_{\text{emotion}} = \int |\partial_t Q|^2 \, dx, \tag{853}$$

which acts as a coherence modulator in the Æ dynamics.

Optimal Trajectories (Variational Characterization)

An effective variational principle for "optimal" coherent trajectories may be encoded by the Lagrangian

$$\mathcal{L}_{\text{opt}} = \text{Re}\big(Q^\dagger \partial_t Q\big) - \|\nabla Q\|^2. \tag{854}$$

The corresponding optimal flow is written formally as

$$\partial_t Q = F_{\text{optimal}}(Q).$$

Temporal Algebra of the Æ-Field

Within the Æ-framework time is not introduced as a fundamental external parameter but emerges from the dynamical evolution of coherent field configurations.

Let $\Psi(t)$ denote the configuration of the field. The temporal generator may be represented by an operator $\mathcal{T}$ such that

$$\frac{d\Psi}{dt} = \mathcal{T}(\Psi). \tag{855}$$

In this representation the temporal parameter t measures the ordering of coherent transformations of the field configuration.

The effective temporal evolution of observables A may then be written as

$$\frac{dA}{dt} = \mathcal{L}_{\mathcal{T}} A, \tag{856}$$

where $\mathcal{L}_{\mathcal{T}}$ denotes the Lie derivative associated with the temporal generator.

Recurrence and Temporal Structure

Temporal evolution in the Æ-framework may exhibit recurrence patterns associated with the periodic return of coherent configurations.

Let $\Psi(t)$ denote the system state. Recurrence occurs when

$$\Psi(t + T) \approx \Psi(t), \tag{857}$$

where T denotes a characteristic recurrence time.

Such structures appear naturally in systems possessing strong coherence and may generate effective time-series behavior across multiple scales of observation.

Æ-Time Relations

Temporal evolution within the Æ-field may also be expressed through relations linking coherence structure and effective time scales.

Let $C(t)$ denote the coherence amplitude associated with the field configuration. Characteristic time scales may be related to the rate of change of coherence through

$$\tau^{-1} \sim \left| \frac{dC}{dt} \right| . \tag{858}$$

Regions of high coherence therefore correspond to slower effective temporal evolution, while rapidly varying configurations correspond to shorter characteristic time scales.

These relations provide a bridge between dynamical coherence and the effective flow of time within the theory.

57. Chemistry and Atomic Structure

Atomic Orbitals as Phase-Coherent Modes

Atomic bound states are modeled as phase-coherent modes, written as

$$\Psi_{\text{atom}}(x) = \sum_{n,l,m} a_{nlm}\, Y_{nlm}(\chi, \theta, \phi), \tag{859}$$

with quasi-stationary phase condition

$$\dot{\phi}_{nlm} \approx 0, \qquad S_{\phi}^{(n)} \text{ minimal.} \tag{860}$$

Energy Spectrum with Coherence Correction

The energy spectrum receives a coherence/entropy correction consistent with the Æ stability principle:

$$E_n \sim \left\langle \nabla Q_n^{\dagger} \nabla Q_n \right\rangle + \lambda\, S_{\phi}^{(n)}. \tag{861}$$

Spectral Kernel Representation (Minimal Form)

To connect chemical/atomic stability with the broader spectral structure of the Æ framework, we recall a minimal kernel formalism acting on an appropriate Æ-algebra (or signal space) $\mathcal{A}_{Æ}$:

$$K_{Æ}(t) : \mathcal{A}_{Æ} \to \mathcal{A}_{Æ}. \tag{862}$$

The induced evolution may be written in integral form as

$$Æ(t) = \int K_{Æ}(t, t')\, Æ(t')\, dt', \tag{863}$$

together with standard boundedness/normalization constraints

$$\int K_{Æ}(t, t')\, dt' = 1, \qquad \|K_{Æ}\| \leq 1. \tag{864}$$

A spectral decomposition (when defined) is

$$K_{Æ}(t, t') = \sum_n \lambda_n\, \psi_n(t)\, \psi_n^{\dagger}(t'), \qquad |\lambda_n| \leq 1. \tag{865}$$

58. DNA as Long-Range Phase Memory Structure

Helical Phase Manifold

We model DNA as an embedded curve (helical backbone)

$$\Gamma : [0, L] \to \mathbb{R}^3, \qquad s \mapsto \Gamma(s), \tag{866}$$

carrying a phase field $\phi(s) \in \mathbb{R}$ with helical periodicity

$$\phi(s + P) = \phi(s) + 2\pi, \tag{867}$$

where $P > 0$ is the pitch parameter.

Phase Memory Functional

A simple long-range phase memory functional is

$$M(L) = \frac{1}{L} \left| \int_0^L e^{i\phi(s)}\, ds \right|, \qquad M(L) \in [0, 1]. \tag{868}$$

Heuristically, coherent phase retention corresponds to $M(L) = \mathcal{O}(1)$, while phase-randomized regimes yield a decay $M(L) = \mathcal{O}(L^{-1/2})$.

An admissibility-type bound can be stated as

$$S_\phi^{(\mathrm{DNA})} < S_{\mathrm{crit}}^{(\mathrm{DNA})}, \tag{869}$$

ensuring that phase entropy on the DNA manifold remains below a critical threshold.

Non-Local Æ Kernel on DNA

Non-local coupling along the DNA curve is captured by the quadratic kernel functional

$$\mathcal{K}_{\mathrm{DNA}} = \iint_{\Gamma \times \Gamma} \exp\left(-\frac{|\Gamma(s) - \Gamma(s')|^2}{\sigma^2} \right) \cos\big(\phi(s) - \phi(s')\big)\, ds\, ds', \tag{870}$$

with $\sigma > 0$ a correlation length along the embedding.

DNA Phase Waveguide

A minimal effective evolution for a DNA-associated field Ψ_{DNA} propagating along arc-length s is

$$\partial_t \Psi_{\text{DNA}} = \mathcal{D}\,\partial_s^2 \Psi_{\text{DNA}} - \Gamma\, S_\phi\, \Psi_{\text{DNA}}, \tag{871}$$

with diffusion parameter $\mathcal{D} > 0$ and an entropic damping coefficient $\Gamma > 0$.

The information content carried by DNA is modeled schematically as

$$I_{\text{DNA}} \sim \{\phi(s),\, \partial_s \phi(s)\}. \tag{872}$$

Epigenetic modulation is represented as an operator action

$$\Psi_{\text{DNA}} \longrightarrow \hat{B}_{\text{epi}}\, \Psi_{\text{DNA}}, \tag{873}$$

and transmissibility requires a local coherence threshold

$$C_{\text{local}} > C_{\text{trans}}. \tag{874}$$

Across generations (or stable replication cycles) this is expressed as

$$\Psi_{\text{DNA}}^{(n+1)} \approx \Psi_{\text{DNA}}^{(n)}. \tag{875}$$

Relation to Generic Long-Range Memory

In the same spirit, a generic long-range memory accumulator can be written as

$$M(t) = \int_0^t \Psi(\tau)\, e^{i\phi(\tau)}\, d\tau, \tag{876}$$

with persistence condition

$$\frac{d}{dt}\|M(t)\| > 0 \quad \text{despite} \quad \partial_t \Psi \neq 0. \tag{877}$$

A minimal recurrence criterion is given by the normalized overlap

$$\kappa(t, t-T) = \frac{\mathrm{Re}(\langle \Psi(t), \Psi(t-T)\rangle)}{\|\Psi(t)\|\,\|\Psi(t-T)\|} > \kappa_{\text{mem}}, \tag{878}$$

for some delay $T > 0$ and threshold $\kappa_{\text{mem}} \in (0,1)$.

DNA Resonant Modes and Expression Coupling

A more explicit Hilbert-space representation of the DNA sector is

$$\Psi_{\text{DNA}} \in \mathcal{H}_{\text{DNA}}, \qquad \mathcal{H}_{\text{DNA}} = \bigotimes_{\ell=1}^{L} \mathcal{H}_\ell. \tag{879}$$

Its coherence is measured by the canonical relation

$$C_{\mathrm{DNA}} = \frac{1}{1 + S_{\phi,\mathrm{DNA}}}. \tag{880}$$

A modal phase representation may be written as

$$\Psi_{\mathrm{DNA}} = \sum_n a_n \, e^{i(\phi_n + k_n z)} \, |n\rangle, \tag{881}$$

with longitudinal phase gradient

$$\partial_z \phi_n = k_n. \tag{882}$$

Phase modulation under environmental and internal driving is modeled by

$$\phi_n \to \phi_n + \Delta\phi_n(t), \tag{883}$$

with evolution

$$\partial_t \Delta\phi_n = \eta_n \mathcal{H}(x_n, t) - \gamma_n S_{\phi,n}. \tag{884}$$

Gene-expression accessibility may be represented through a diagonal activation operator

$$\hat{A} = \sum_n a_n |n\rangle\langle n|, \qquad a_n \in [0,1], \tag{885}$$

so that the expressed state becomes

$$\Psi_{\mathrm{expr}} = \hat{A}\, \Psi_{\mathrm{DNA}}. \tag{886}$$

DNA memory itself may evolve as

$$\partial_t M_{\mathrm{DNA}} = -\alpha S_{\phi,\mathrm{DNA}} M_{\mathrm{DNA}} + \beta C_{\mathrm{DNA}}. \tag{887}$$

A nonlocal DNA kernel can also be written as

$$K(x, x') = e^{-|x-x'|/\ell_c} e^{i(\phi(x) - \phi(x'))}, \tag{888}$$

with integrability condition

$$\int |K(x, x')| \, dx' < \infty. \tag{889}$$

Stability of phase modulation requires

$$\|\Delta\phi\| \le \Delta\phi_{\max}. \tag{890}$$

The DNA sector can drive organism-level coherence through

$$\partial_t C_{\mathrm{org}} = -\alpha S_{\phi,\mathrm{org}} C_{\mathrm{org}} + \beta C_{\mathrm{DNA}}, \tag{891}$$

and, in a multiscale decomposition,

$$\partial_t C_{\text{org}} = \sum_s w_s \left(-\alpha_s S_\phi^{(s)} C^{(s)} + \beta_s C^{(s-1)}\right). \tag{892}$$

A loss of DNA coherence leads to memory decay:

$$C_{\text{DNA}} \leq C_{\text{crit}} \Rightarrow \partial_t M_{\text{DNA}} < 0. \tag{893}$$

A dominant scale selector may be defined as

$$s^*(t) = \arg\min_s \{w_s C^{(s)}(t)\}. \tag{894}$$

Positive organismal recovery requires

$$\sum_s w_s \left(-\alpha_s S_\phi^{(s)} C^{(s)} + \beta_s C^{(s-1)}\right) > 0.$$

DNA as a Coherent Information Lattice

Within the Æ-framework DNA is interpreted as a structured information lattice capable of retaining phase information over extended molecular scales.

Let $\Psi_{\text{DNA}}(s)$ denote the coherent field state along the helical coordinate s of the DNA strand.

The spatial configuration may be represented as

$$\Psi_{\text{DNA}}(s) = A(s)e^{i\phi(s)}, \tag{895}$$

where $A(s)$ represents the local amplitude of the molecular state and $\phi(s)$ the associated phase structure.

DNA–Protein Coupling

Interactions between DNA and regulatory proteins may be described through coupling terms linking the DNA coherence field with protein states.

Let Ψ_{DNA} denote the DNA state and Ψ_{prot} the associated protein configuration.

A generic coupling term may be written as

$$H_{\text{coupling}} = \int g(s)\, \Psi_{\text{DNA}}(s)\Psi_{\text{prot}}(s)\, ds, \tag{896}$$

where $g(s)$ represents the interaction strength along the DNA strand.

Such couplings allow protein configurations to influence phase patterns stored along genomic coordinates.

Phase Retention Along the Helix

Phase information stored along the DNA strand may propagate through the helical structure.

A simplified propagation equation may be written as

$$\frac{\partial \Psi_{\mathrm{DNA}}}{\partial t} = D_{\mathrm{DNA}} \frac{\partial^2 \Psi_{\mathrm{DNA}}}{\partial s^2} - \gamma \Psi_{\mathrm{DNA}}, \tag{897}$$

where D_{DNA} represents an effective coherence diffusion coefficient along the strand and γ denotes phase damping.

Fractal Antenna Hypothesis

A fractal scaling relation may be written as

$$L(\epsilon) \sim \epsilon^{-D_f}, \tag{898}$$

where D_f denotes an effective fractal dimension of the structure.

Within the Æ framework, coherence across scales is modeled as

$$C_{\mathrm{DNA}}^{(s)} = \frac{1}{1 + S_\phi^{(s)}}, \tag{899}$$

where $S_\phi^{(s)}$ is the phase entropy at scale s.

A multiscale resonance condition may be expressed as

$$\phi^{(s)} \approx \phi^{(s+1)}, \tag{900}$$

ensuring phase alignment between hierarchical levels.

Breakdown of this condition leads to loss of coherence across scales, which may manifest as biological dysfunction or instability.

Remark on Fractal Scaling in DNA

Fractal scaling relations of the form

$$L(\epsilon) \sim \epsilon^{-D_f} \tag{901}$$

have been proposed in the context of DNA geometry, suggesting multiscale structural organization.

Disclaimer. This relation is used here as a heuristic structural analogy and is not derived from first principles within the Æ framework.

59. Proteins and Topological Stability

Proteins as Composite Knots

We idealize a protein backbone as an embedded curve

$$\gamma : [0,1] \to \mathbb{R}^3, \qquad \overline{\gamma} : S^1 \to \mathbb{R}^3,$$

whose closed representative defines a knot type. Composite proteins are modeled by connected sums

$$K_{\text{protein}} = K_1 \# K_2 \# \ldots \# K_m. \tag{902}$$

A basic combinatorial proxy is the near-additivity of crossing number:

$$c(K_1 \# K_2) \approx c(K_1) + c(K_2) + \mathcal{O}(1).$$

DNA–Protein Phase Coupling

Protein folding is assumed to be driven, in part, by phase-structured information inherited from the DNA manifold. At a schematic level,

$$\gamma_{\text{protein}} = F(\Gamma, \phi),$$

where Γ is the DNA helical embedding and ϕ its associated phase field. A minimal DNA–protein interaction Hamiltonian is

$$H_{\text{int}}^{(\text{DNA–prot})} = \sum_{n,m} g_{nm} \, |n\rangle\langle n| \otimes |m\rangle\langle m|. \tag{903}$$

The corresponding phase dynamics may be modeled in Kuramoto-type form:

$$\dot{\phi}_n = \omega_n + \sum_m g_{nm} \sin(\phi_m - \phi_n). \tag{904}$$

To ensure bounded interaction strength one imposes

$$\max_{n,m} |g_{nm}| \le g_{\max}.$$

This formalizes the guiding principle

$$\text{DNA: phase memory manifold} \quad \longrightarrow \quad \text{Protein: phase-folded topology.}$$

Topological Casimirs and Æ-Helicity

In the Æ-algebraic setting, a Casimir $C_{Æ}$ satisfies

$$\{C, F, G\}_{Æ} = 0 \qquad \forall F, G, \tag{905}$$

and in particular

$$\{C_{Æ}, F, G\}_{Æ} = 0. \tag{906}$$

A helicity-like topological invariant is modeled as

$$C_{Æ} = \int_{\mathcal{M}_{Æ}} \mathbf{A}_{Æ} \cdot \left(\nabla_{Æ} \times \mathbf{A}_{Æ}\right) dV_{Æ}, \tag{907}$$

which admits the linking interpretation

$$C_{Æ} \propto \sum_{i<j} \mathrm{Link}(K_i, K_j). \tag{908}$$

Topological Stability Functional

A coarse topological stability energy for a knot representative K is

$$\mathcal{E}_{\mathrm{top}}(K) = \alpha_1 c(K) + \alpha_2 |Wr(K)| + \alpha_3 \int |\kappa(s)|^2 ds + \alpha_4 \deg \Delta_K, \tag{909}$$

where $c(K)$ is crossing number, $Wr(K)$ is writhe, $\kappa(s)$ curvature, and Δ_K the Alexander polynomial.

Variational Folding Principle

At the field level, folded configurations are modeled variationally as

$$\Psi_{\mathrm{folded}} = \arg\min_{\Psi} \left(S_\phi[\Psi] + V_{\mathrm{chem}}[\Psi]\right). \tag{910}$$

A geometry-level folding functional on the backbone curve γ can be written as

$$\mathcal{F}(\gamma) = \mathcal{E}_{\mathrm{elastic}} + \mathcal{E}_{\mathrm{top}} - \beta\, \mathcal{K}_{Æ}, \tag{911}$$

with stationarity condition

$$\frac{\delta \mathcal{F}}{\delta \gamma} = 0. \tag{912}$$

Here a non-local Æ-kernel term may be taken as

$$\mathcal{K}_{Æ}(\gamma) = \iint_{\gamma\times\gamma} \exp\left(-\frac{|\gamma(s) - \gamma(s')|^2}{\sigma^2}\right) \cos\big(\phi(s) - \phi(s')\big)\, ds\, ds'. \tag{913}$$

Protein States, Perturbations, and Metastability

A minimal tensor product model for amino-acid composition is

$$\Psi_{\text{prot}} = \bigotimes_{k=1}^{N} \Psi_{\text{aa}}^{(k)}. \tag{914}$$

Functional proteins correspond to high contrast coherence relative to background,

$$C_d \gg C_{\text{background}}, \tag{915}$$

and ligand perturbations are represented as

$$\delta\Psi_{\text{prot}} = \hat{O}_{\text{ligand}}\,\Psi_{\text{prot}}. \tag{916}$$

Metastable conformations satisfy an entropic ordering

$$S_{\phi}^{\text{meta}} > S_{\phi}^{\text{folded}}. \tag{917}$$

Spectral Topological Stabilization (Optional Layer)

A spectral operator associated with a backbone curve is

$$L_{\gamma} = -\frac{d^2}{ds^2} + V_{\text{top}}(s), \qquad \lambda_1 \leq \lambda_2 \leq \dots$$

with a stability gap condition

$$\sup_k \lambda_k^{(\text{protein})} < \infty, \qquad \Delta\lambda^{(\text{protein})} > 0,$$

and robustness estimate

$$\|\delta\gamma\| \leq \frac{\epsilon}{\Delta\lambda}.$$

Molecular Resonance, Water Coupling, and Electromagnetic Response

A conformational resonance Hamiltonian for protein states may be written as

$$H_{\text{conf}} = \sum_n E_n \left|n\right\rangle \left\langle n\right| + \sum_{n\neq m} \frac{\Delta E_{nm}}{2} \left(e^{i(\phi_n - \phi_m)} \left|n\right\rangle \left\langle m\right| + e^{-i(\phi_n - \phi_m)} \left|m\right\rangle \left\langle n\right| \right). \tag{918}$$

The protein state belongs to a configurational Hilbert space,

$$\Psi_{\text{prot}} \in \mathcal{H}_{\text{conf}}, \tag{919}$$

and evolves according to

$$\partial_t \Psi_{\text{prot}} = -iH_{\text{conf}} \Psi_{\text{prot}}. \tag{920}$$

A phase-mode expansion is

$$\Psi_{\text{prot}} = \sum_n a_n e^{i\phi_n} |n\rangle, \tag{921}$$

with phase evolution

$$\dot{\phi}_n = \omega_n + \xi_n(t). \tag{922}$$

Transitions between conformational modes are weighted by both energy and phase entropy:

$$\mathcal{T}_{n \to m} = \exp\left(-\frac{\Delta E_{nm}}{k_B T_{\text{eff}}}\right) \exp\left(-\frac{\Delta S_{\phi,nm}}{\hbar}\right). \tag{923}$$

Protein coherence is modeled as

$$C_{\text{prot}} = \frac{1}{1 + S_{\phi,\text{prot}}}, \tag{924}$$

with coherence dynamics

$$\partial_t C_{\text{prot}} = -\alpha_{\text{prot}} S_{\phi,\text{prot}} C_{\text{prot}} + \beta_{\text{prot}} \|\Psi_{\text{prot}}\|^2. \tag{925}$$

A water-coupled interaction sector may be introduced through

$$H^{(\text{water})}_{\text{coup}} = g_{\text{pw}} \left\langle \cos(\phi_{\text{prot}} - \phi_{\text{water}}) \right\rangle, \tag{926}$$

or more explicitly,

$$H^{(\text{water})}_{\text{int}} = \sum_{n,k} g_{nk} |n\rangle\langle n| \otimes e^{i(\phi_n - \theta_k)}. \tag{927}$$

The electromagnetic interaction is modeled as

$$H_{\text{EM}} = -\mathbf{d} \cdot \mathbf{E}(t) - \mathbf{m} \cdot \mathbf{B}(t), \tag{928}$$

which induces a field-dependent phase shift

$$\dot{\phi}_n = \omega_n + \gamma_n |\mathbf{E}|^2. \tag{929}$$

The spectral response of the protein sector is given by

$$R_{\text{prot}}(\omega) = \left| \int \Psi_{\text{prot}}(t)\, e^{-i\omega t}\, dt \right|^2. \tag{930}$$

A network-like phase coupling description may also be used:

$$\dot{\phi}_i = \omega_i + \sum_{j \neq i} \kappa_{ij} \sin(\phi_j - \phi_i). \tag{931}$$

Functional stability requires

$$C_{\text{prot}} \geq C_{\text{func}}, \tag{932}$$

$$\dot{C}_{\text{prot}} \approx 0, \tag{933}$$

together with a sufficiently sharp spectral peak:

$$\max_{\omega} R_{\text{prot}}(\omega) > \kappa_{\text{prot}}\, C_{\text{prot}}. \tag{934}$$

Quantum Phase Dynamics of Protein States

We model internal protein conformational states through a phase-coherent Hamiltonian

$$H_{\text{conf}} = \sum_n E_n |n\rangle\langle n| + \sum_{n \neq m} \frac{\Delta E_{nm}}{2} \left(e^{i(\phi_n - \phi_m)} |n\rangle\langle m| + e^{-i(\phi_n - \phi_m)} |m\rangle\langle n| \right). \tag{935}$$

The protein state belongs to the conformational Hilbert space

$$\Psi_{\text{prot}} \in \mathcal{H}_{\text{conf}}, \tag{936}$$

with evolution

$$\partial_t \Psi_{\text{prot}} = -i H_{\text{conf}} \Psi_{\text{prot}}. \tag{937}$$

A generic phase representation is

$$\Psi_{\text{prot}} = \sum_n a_n e^{i\phi_n} |n\rangle. \tag{938}$$

Phase dynamics may include stochastic fluctuations

$$\dot{\phi}_n = \omega_n + \xi_n(t). \tag{939}$$

Transition rates between conformational states satisfy

$$\mathcal{T}_{n \to m} = \exp\left(-\frac{\Delta E_{nm}}{k_B T_{\text{eff}}} \right) \exp\left(-\frac{\Delta S_{\phi,nm}}{\hbar} \right). \tag{940}$$

Water and Electromagnetic Coupling

Protein phase dynamics may couple to structured water fields

$$H_{\text{coup}}^{(\text{water})} = g_{\text{pw}} \langle \cos(\phi_{\text{prot}} - \phi_{\text{water}}) \rangle. \tag{941}$$

An external electromagnetic coupling is

$$H_{\text{EM}} = -\mathbf{d} \cdot \mathbf{E}(t) - \mathbf{m} \cdot \mathbf{B}(t). \tag{942}$$

Phase evolution may then acquire field-dependent corrections

$$\dot{\phi}_n = \omega_n + \gamma_n |\mathbf{E}|^2. \tag{943}$$

Protein Resonance Spectrum

Define the protein spectral response

$$R_{\text{prot}}(\omega) = \left| \int \Psi_{\text{prot}}(t) e^{-i\omega t} dt \right|^2 . \tag{944}$$

Functional proteins satisfy

$$C_{\text{prot}} \geq C_{\text{func}}, \qquad \dot{C}_{\text{prot}} \approx 0. \tag{945}$$

A resonance amplification criterion is

$$\max_{\omega} R_{\text{prot}}(\omega) > \kappa_{\text{prot}} C_{\text{prot}}. \tag{946}$$

Protein Phase-Coherent Structures

Within the Æ-framework proteins may be interpreted as phase-coherent molecular structures whose stability emerges from the organization of coherent field modes along the molecular backbone.

Let $\Psi_{\text{prot}}(s)$ denote the coherent state of a protein along its structural coordinate s.

The molecular configuration may be represented as

$$\Psi_{\text{prot}}(s) = A_{\text{prot}}(s) e^{i\phi_{\text{prot}}(s)}, \tag{947}$$

where $A_{\text{prot}}(s)$ describes the local structural amplitude and $\phi_{\text{prot}}(s)$ the associated phase configuration.

Stable protein conformations correspond to configurations where the phase structure remains dynamically coherent.

Topological Stability of Protein Structures

Protein folding may be interpreted as the emergence of topologically stable configurations of the molecular phase field.

Let $\Psi_{\text{prot}}(t)$ denote the evolving configuration. A stable folded structure corresponds to a configuration satisfying

$$\frac{d}{dt}\|\Psi_{\text{prot}}\| = 0, \tag{948}$$

together with the persistence of its phase pattern

$$\kappa_{\text{prot}}(t, t-T) > \kappa_{\text{fold}}. \tag{949}$$

These conditions characterize the maintenance of coherent molecular structure over time.

Topology–Biology Correspondence

Topological structures appearing in the Æ-field formalism may manifest at the molecular level in biological systems.

Protein structures may therefore be associated with topological configurations whose stability is maintained through coherent phase relations.

Let $\mathcal{K}$ denote a topological configuration associated with a folded protein structure.

The biological state may be written schematically as

$$\Psi_{\text{bio}} = \mathcal{T}(\mathcal{K}), \tag{950}$$

where $\mathcal{T}$ represents the mapping from topological configuration space to molecular biological states.

Protein Resonance Dynamics

Proteins may interact with electromagnetic and vibrational fields through resonance mechanisms associated with their molecular structure.

A simplified resonance equation may be written as

$$\frac{d^2\Psi_{\text{prot}}}{dt^2} + \omega_{\text{prot}}^2\Psi_{\text{prot}} = F_{\text{ext}}(t), \tag{951}$$

where ω_{prot} denotes the characteristic resonance frequency of the molecular structure and $F_{\text{ext}}(t)$ represents external driving fields.

60. Biological Coherence Sector

Helical Phase Configuration

A minimal coherent phase configuration along a 1D biological coordinate s is

$$\phi(s) = ks + \phi_0, \qquad \phi_1(s) = ks, \quad \phi_2(s) = ks + \pi. \tag{952}$$

Canonical Coherence and Admissibility

We use the canonical Æ-coherence functional

$$\boxed{C_{\text{Æ}} := \frac{1}{1 + S_\phi}} \tag{953}$$

so that

$$0 < C_{\text{Æ}} \leq 1, \qquad S_\phi = 0 \Rightarrow C_{\text{Æ}} = 1, \qquad S_\phi \gg 1 \Rightarrow C_{\text{Æ}} \approx \frac{1}{S_\phi}. \tag{954}$$

Its monotone dependence on phase-entropy is

$$\frac{dC_{\text{Æ}}}{dS_\phi} = -\frac{1}{(1 + S_\phi)^2} < 0. \tag{955}$$

For local degrees of freedom (cells, glands, neural nodes) we write

$$C_{\text{Æ},i} := \frac{1}{1 + S_{\phi,i}} \tag{956}$$

A weighted global coherence (tissue/organ/organism scale) is

$$C_{\text{Æ}}^{\text{global}} := \sum_{i=1}^{N} w_i \, C_{\text{«},i}, \qquad w_i \geq 0, \quad \sum_{i=1}^{N} w_i = 1 \tag{957}$$

We impose the standard admissibility window for coherent biological sectors:

$$C_{\text{Æ}}(D(t), d(t)) \geq C_{\min}, \qquad S_\phi(D(t), d(t)) \leq S_\phi^{\text{crit}} \tag{958}$$

Time evolution of coherence follows directly from $C_{\text{Æ}} = (1 + S_\phi)^{-1}$:

$$\dot{C}_{\text{Æ}} = -\frac{1}{(1 + S_\phi)^2} \dot{S}_\phi \tag{959}$$

Bio-Photon Emission (Effective Law)

As an effective phenomenological proxy, bio-photon emission rate is taken to scale with coherence gradients and phase acceleration:

$$\Gamma_\gamma \propto |\nabla C|^2, \qquad \Gamma_\gamma \propto |\partial_t^2 \phi|. \tag{960}$$

Biological Effective Action

We postulate an effective biological action penalizing phase gradients and entropy while favoring coherence:

$$\mathcal{S}_{\text{bio}} = \int \Big(\alpha|\nabla\phi|^2 + \beta C^2 - \gamma S_\phi\Big)\, dV\, dt. \tag{961}$$

Reality Knots Embedded in the Body (Organ/Hormonal Loops)

We model organ-level reality knots as quaternionic states

$$Q_{O_k}(x,t) \in S^3. \tag{962}$$

Slow hormonal carriers are represented as a phase rotation

$$\Phi_H(t) = e^{\theta(t)\mathbf{u}}, \tag{963}$$

with a holonomy persistence criterion for a body-embedded loop γ:

$$H_\gamma \neq \mathbb{I} \quad \text{persists within a body-embedded loop } \gamma, \tag{964}$$

and healing/coherence restoration is encoded as

$$H_\gamma \to \mathbb{I}, \qquad S_\phi \to \text{low}, \qquad C_{\text{«}} \to \text{high}. \tag{965}$$

A minimal local phase model is

$$Q(x,t) \approx e^{\theta(t)\mathbf{u}}. \tag{966}$$

(Interpretation layer, optional)

$$\text{Reality Knot (NR)} \leftrightarrow \text{Organ Resonance} \tag{967}$$
$$\leftrightarrow \text{Hormonal Phase Carrier} \tag{968}$$
$$\leftrightarrow \text{Emotional AM/PM Field} \tag{969}$$
$$\leftrightarrow \text{Astral–Etheric Pattern} \tag{970}$$

Emotional energy can be treated as a coherence modulator:

$$E_{\text{emotion}} = \int |\partial_t Q|^2\, \mathrm{d}x. \tag{971}$$

Such structures have been historically described under various cultural frameworks referring to axial coherence centers.

Coherent Biological Nodes

Certain anatomical regions act as persistent coherence hubs within the biological phase field.

These regions may be modeled as localized nodes

$$\mathcal{N}_k = (C_k,\ \omega_k,\ \phi_k,\ x_k), \tag{972}$$

embedded along the organismal axis.

Each node participates in global coherence through longitudinal phase transport and local synchronization mechanisms.

Their effective dynamics satisfy

$$\partial_t C_k = -\alpha_k S_{\phi,k} C_k + \beta_k \mathcal{H}(x_k, t) + \sum_{j \neq k} \kappa_{kj} \sin(\phi_j - \phi_k). \tag{973}$$

These coherence hubs act as mediators between endocrine, neural, and cellular subsystems, enabling multiscale integration of phase information.

In high-coherence regimes, they form a structured alignment along the organismal axis, supporting stable global phase organization.

Cancer as Multiscale Coherence Disruption (Minimal Criterion)

A simple phase-alignment criterion for an organ-scale mode $\mathcal{O}$ is

$$\phi_i \approx \phi_{\mathcal{O}}, \tag{974}$$

while disruption corresponds to

$$|\phi_i - \phi_{\mathcal{O}}| \gg \epsilon, \qquad \partial_t S_\phi^{\text{local}} > 0, \qquad \Psi_{\text{cell}} \to \Psi_{\mathcal{O}}. \tag{975}$$

Endocrine Slow Coherence Field

Let $\mathcal{H}(x,t)$ denote a slow hormonal coherence field. Its effect is encoded via

$$\mathcal{L}_{\text{slow}}(\mathcal{H})\Psi_{\text{org}} = \Lambda \left(\sum_g \alpha_g H_g(t) \right) \Psi_{\text{org}} + \beta \left\langle \mathcal{H}(x,t) \right\rangle_x \Psi_{\text{org}}. \tag{976}$$

We represent the endocrine sector as

$$\Psi_{\text{endo}} \in \mathcal{H}_{\text{endo}}, \qquad \mathcal{H}_{\text{endo}} = \bigotimes_{g=1}^{G} \mathcal{H}_g, \tag{977}$$

with gland state

$$\mathcal{N}_g := (C_g,\ \omega_g,\ \phi_g,\ \tau_g). \tag{978}$$

Hormone dynamics and secretion:

$$\partial_t H_g = -\frac{1}{\tau_g} H_g + S_g(t), \tag{979}$$

$$S_g(t) = \sigma_g\, C_g\, \Theta(C_g - C_{\min}). \tag{980}$$

Coherence and frequency modulation:

$$\partial_t C_g = -\alpha_g S_{\phi,g} C_g + \beta_g H_g, \tag{981}$$

$$\omega_g(t) = \omega_g^{(0)} + \delta_g H_g(t). \tag{982}$$

Field assembly and relaxation:

$$\mathcal{H}(x,t) = \sum_g H_g(t)\, \chi_g(x), \tag{983}$$

$$\partial_t \mathcal{H} = -\Lambda\, \mathcal{H} + \sum_g S_g(t) \chi_g(x). \tag{984}$$

Coupled organism evolution:

$$\partial_t \Psi_{\text{org}} = \mathcal{L}_{\text{fast}} \Psi_{\text{org}} + \mathcal{L}_{\text{slow}}(\mathcal{H}) \Psi_{\text{org}}. \tag{985}$$

Cell-scale coupling to the slow field:

$$\partial_t C_{\text{cell}} = -\alpha S_\phi C_{\text{cell}} + \beta \mathcal{H}(x,t). \tag{986}$$

Phase synchronization across glands:

$$\dot{\phi}_g = \omega_g + \sum_{h \neq g} \kappa_{gh} \sin(\phi_h - \phi_g). \tag{987}$$

Endocrine aggregate coherence:

$$C_{\text{endo}} = \sum_g w_g C_g, \qquad \partial_t C_{\text{endo}} = -\sum_g \alpha_g S_{\phi,g} C_g + \sum_g \beta_g H_g. \tag{988}$$

Perturbation decay estimate:

$$\|\delta \mathbf{H}(t)\| \le e^{-t/\tau_{\max}} \|\delta \mathbf{H}(0)\|. \tag{989}$$

Critical endocrine failure mode:

$$C_{\text{endo}} \le C_{\text{crit}} \quad \Rightarrow \quad \text{bias toward global destabilization.} \tag{990}$$

Neural–Non-Neural Integration

We model the neural sector as

$$\Psi_{\text{neur}} \in \mathcal{H}_{\text{neur}}, \qquad \mathcal{H}_{\text{neur}} = \bigotimes_{n=1}^{N} \mathcal{H}_n. \tag{991}$$

Each neural node carries

$$\mathcal{N}_n := (C_n,\ \omega_n,\ \phi_n,\ \tau_n). \tag{992}$$

Kuramoto-like phase coupling with noise:

$$\dot{\phi}_n = \omega_n + \sum_{m\neq n} K_{nm} \sin(\phi_m - \phi_n) + \xi_n(t). \tag{993}$$

Coherence dynamics:

$$\dot{C}_n = -\alpha_n S_{\phi,n} C_n + \beta_n |\Psi_n|^2. \tag{994}$$

Neural field coupling operator:

$$\mathcal{M}(\mathcal{H}, C_{\text{cell}})\Psi_{\text{neur}} = \gamma\, \mathcal{H}(x,t)\Psi_{\text{neur}} + \delta\, C_{\text{cell}}\Psi_{\text{neur}}. \tag{995}$$

Neural evolution:

$$\partial_t \Psi_{\text{neur}} = \mathcal{L}_{\text{neur}}\Psi_{\text{neur}} + \mathcal{M}(\mathcal{H}, C_{\text{cell}})\Psi_{\text{neur}}. \tag{996}$$

Aggregate neural coherence:

$$C_{\text{neur}} = \sum_n w_n C_n, \qquad \partial_t C_{\text{neur}} = -\sum_n \alpha_n S_{\phi,n} C_n + \sum_n \beta_n I_n. \tag{997}$$

An augmented node-level coherence equation:

$$\partial_t C_n = -\alpha_n S_{\phi,n} C_n + \beta_n I_n + \gamma_n \mathcal{H}(x_n, t). \tag{998}$$

Define the graph Laplacian

$$\mathbf{L} = \mathbf{D} - \mathbf{K}, \tag{999}$$

so that a linearized phase relaxation model may be written as

$$\partial_t \boldsymbol{\phi} = -\mathbf{L}\,\boldsymbol{\phi} + \boldsymbol{\omega}. \tag{1000}$$

A useful information current proxy is

$$J_{\text{info}} = \sum_{n,m} K_{nm} \sin(\phi_m - \phi_n). \tag{1001}$$

Modal decomposition:

$$\boldsymbol{\phi}(t) = \sum_k a_k(t)\,\mathbf{v}_k, \qquad \dot{a}_k = -\lambda_k a_k + F_k(t). \tag{1002}$$

Organism-level coupling closure:

$$\partial_t \Psi_{\text{org}} = \mathcal{L}_{\text{neur}}\Psi_{\text{org}} + \mathcal{L}_{\text{endo}}\Psi_{\text{org}} + \mathcal{L}_{\text{cell}}\Psi_{\text{org}}. \tag{1003}$$

Neural failure mode:

$$C_{\text{neur}} \leq C_{\text{crit}} \Rightarrow \partial_t C_{\text{org}} < 0. \tag{1004}$$

A Lyapunov proxy may be chosen as

$$V(\{\phi_n\}) = -C_{\text{neur}}, \qquad \dot{V} \le -\eta\,(C_{\text{neur}} - C_{\text{crit}})\,, \qquad \eta = \lambda_2(\mathbf{L}) > 0. \tag{1005}$$

Dual Bioelectric Channels and Longitudinal Phase Transport

Biological coherence requires not only nodal synchronization but also continuous longitudinal transport of phase information across the organism.

We model dual bioelectric phase channels as

$$J_\pm(x,t) = \alpha_\pm \nabla\phi_\pm(x,t), \tag{1006}$$

with $\phi_\pm$ representing complementary phase components.

The effective longitudinal transport mode is

$$J_c(x,t) = \gamma\,\nabla\Big(\phi_+(x,t) - \phi_-(x,t)\Big). \tag{1007}$$

Phase propagation along the organismal axis may be approximated by

$$\partial_t\phi_\pm = v_\pm\,\partial_z\phi_\pm - \eta_\pm S_\phi. \tag{1008}$$

Coherence is maximized when the phase difference stabilizes:

$$C(x,t) \sim \cos\Big(\phi_+ - \phi_-\Big), \tag{1009}$$

while phase disorder leads to entropy growth:

$$|\phi_+ - \phi_-| \uparrow \quad\Rightarrow\quad S_\phi \uparrow. \tag{1010}$$

A stability condition for longitudinal transport is

$$\partial_t S_\phi \le \sigma_{\text{max}}, \tag{1011}$$

ensuring bounded entropy production along the transport channels.

This mechanism provides a continuous coherence backbone linking neural, endocrine, and cellular subsystems into a unified organism-level phase structure.

Cellular Coherence Dynamics

We model the cellular state as a coherent subsystem

$$\Psi_{\text{cell}} \in \mathcal{H}_{\text{cell}}, \qquad \mathcal{H}_{\text{cell}} = \bigotimes_k \mathcal{H}_k. \tag{1012}$$

Each internal cellular mode carries a coherence node

$$\mathcal{N}_k := (C_k, \omega_k, \phi_k). \tag{1013}$$

The cellular phase evolves via a Kuramoto-like coupling

$$\dot{\phi}_{\text{cell}} = \omega_{\text{cell}} + \sum_k \kappa_k \sin(\phi_k - \phi_{\text{cell}}). \tag{1014}$$

Internal coherence dynamics satisfy

$$\dot{C}_k = -\alpha_k S_{\phi,k} C_k + \beta_k |\Psi_k|^2. \tag{1015}$$

The aggregated cellular coherence becomes

$$C_{\text{cell}} = \sum_k w_k C_k. \tag{1016}$$

—

Cellular Transport and Memory Flux

Memory exchange across the cellular boundary is modeled through a coherence flux

$$J_{\text{mem}} = -D \nabla C_{\text{cell}} + \mu \nabla \phi. \tag{1017}$$

A membrane potential representation is

$$V_m = \phi_{\text{in}} - \phi_{\text{out}}, \tag{1018}$$

$$\partial_t V_m = -\frac{1}{\tau_m} V_m + I_{\text{ion}}. \tag{1019}$$

The cellular evolution equation is

$$\partial_t \Psi_{\text{cell}} = \mathcal{L}_{\text{cell}} \Psi_{\text{cell}} + \sum_r J_r(t). \tag{1020}$$

Non-vanishing flux implies entropy production

$$\sum_r J_r \neq 0 \quad \Rightarrow \quad \dot{S}_\phi \neq 0. \tag{1021}$$

—

Cytoplasmic Phase Forces

Cytoplasmic transport is modeled via a potential

$$\mathbf{F}_{\text{cyto}} = -\nabla U(\phi, C_{\text{cell}}). \tag{1022}$$

Phase evolution then satisfies

$$\partial_t \phi = \omega + \gamma \nabla \cdot \mathbf{F}_{\text{cyto}}. \tag{1023}$$

—

Cell Cycle Constraints

Cellular phase trajectories obey a quantized cycle condition

$$\oint_C \phi_{\text{cell}} \, d\tau = 2\pi n. \tag{1024}$$

Entropy production per cycle must remain bounded

$$\Delta S_\phi^{(\text{cycle})} < S_{\max}. \tag{1025}$$

Cell division is permitted when coherence exceeds a threshold

$$C_{\text{cell}} \geq C_{\text{div}}, \tag{1026}$$

$$\Psi_{\text{cell}} \to \Psi_{\text{cell}}^{(1)} \otimes \Psi_{\text{cell}}^{(2)}. \tag{1027}$$

Conversely, coherence collapse leads to cellular decay

$$C_{\text{cell}} \leq C_{\text{crit}} \Rightarrow \lim_{t\to\infty} \Psi_{\text{cell}} = 0. \tag{1028}$$

Unified Multiscale Stability Condition

Let the organism be decomposed into L effective layers (molecular/cellular/tissue/neural/endocrine/...) with phase-entropy $S_\phi^{(\ell)}$. Define

$$\mathcal{S}_{\text{global}} = \sum_{\ell=1}^{L} w_\ell \, S_\phi^{(\ell)}, \tag{1029}$$

and corresponding global coherence

$$C_{\text{global}} = \frac{1}{1+\mathcal{S}_{\text{global}}}. \tag{1030}$$

The global stability window is

$$\mathcal{S}_{\text{global}} < \mathcal{S}_{\text{crit}}. \tag{1031}$$

A generic cross-layer evolution closure is

$$\partial_t S_\phi^{(\ell)} = f_\ell\left(S_\phi^{(\ell)}, S_\phi^{(\ell-1)}, S_\phi^{(\ell+1)}\right), \tag{1032}$$

with spectral proxy

$$S_\phi^{(\ell)} \sim \sum_k g\left(\lambda_k^{(\ell)}\right), \qquad C^{(\ell)} = \frac{1}{1+S_\phi^{(\ell)}}. \tag{1033}$$

Organism as a Coherent Entity

Within the Æ framework a living organism is modeled as a coherent biological entity sustained by phase alignment across multiple interacting subsystems.

Let Ψ_{org} denote the global biological state of the organism. A multiscale organism configuration may be written as

$$\Psi_{\text{org}} = \bigotimes_{s\in\{\text{mol},\text{cell},\text{tissue},\text{neural},\text{endo}\}} \Psi^{(s)}, \tag{1034}$$

where the tensor product represents the coupled state of molecular, cellular, tissue, neural and endocrine subsystems.

The effective organism-level coherence is defined as

$$C_{\text{org}} = \sum_s w_s\, C^{(s)}, \qquad w_s \geq 0, \ \sum_s w_s = 1, \tag{1035}$$

where $C^{(s)}$ denotes the coherence of subsystem s and w_s represents its relative biological weight.

An organism behaves as a viable coherent entity when the global coherence remains above a critical threshold,

$$C_{\text{org}} > C_{\text{crit}}, \qquad \frac{dC_{\text{org}}}{dt} \geq -\epsilon. \tag{1036}$$

Coherence exchange between biological layers is modeled by

$$\partial_t C^{(s)} = -\alpha_s S_\phi^{(s)} C^{(s)} + \beta_s C^{(s-1)}, \tag{1037}$$

representing entropy-induced decoherence and vertical coherence transfer across biological scales.

The dynamical evolution of the organism state is written

$$\partial_t \Psi_{\text{org}} = \sum_s \mathcal{L}^{(s)} \Psi_{\text{org}}. \tag{1038}$$

Failure of coherence in any critical subsystem produces organism-level degradation,

$$\exists\, s \text{ s.t. } C^{(s)} \to 0 \ \Rightarrow\ \partial_t C_{\text{org}} < 0. \tag{1039}$$

The functional lifetime of the organism may be expressed as

$$T_{\text{life}} = \int_0^T \Theta(C_{\text{org}}(t) - C_{\text{crit}})\, dt. \tag{1040}$$

Biological death corresponds to the loss of global coherence,

$$C_{\text{org}}(t_d) \leq C_{\text{crit}}, \qquad \lim_{t\to t_d^+} \partial_t C_{\text{org}} < 0. \tag{1041}$$

After decoherence the organism state decomposes into residual subsystem states,

$$\Psi_{\text{org}} \to \bigoplus_s \Psi^{(s)}_{\text{res}}. \tag{1042}$$

The non-viability limit is therefore

$$\lim_{t\to\infty} C_{\text{org}} = 0 \ \Rightarrow\ \Psi_{\text{org}} \text{ non-viable.} \tag{1043}$$

Cellular Life (Minimal Sector)

We model a cell as

$$\Psi_{\text{cell}} \in \mathcal{H}_{\text{cell}}, \qquad \mathcal{H}_{\text{cell}} = \bigotimes_k \mathcal{H}_k. \tag{1044}$$

A coarse phase coupling dynamics is

$$\dot{\phi}_{\text{cell}} = \omega_{\text{cell}} + \sum_k \kappa_k \sin(\phi_k - \phi_{\text{cell}}). \tag{1045}$$

A membrane / memory bridge proxy:

$$\Gamma_{\text{mem}} : \Psi_{\text{in}} \leftrightarrow \Psi_{\text{out}}, \qquad J_{\text{mem}} = -D\nabla C_{\text{cell}} + \mu\nabla\phi. \tag{1046}$$

Membrane potential dynamics:

$$V_m = \phi_{\text{in}} - \phi_{\text{out}}, \qquad \partial_t V_m = -\frac{1}{\tau_m} V_m + I_{\text{ion}}. \tag{1047}$$

Node state and coherence dynamics:

$$\mathcal{N}_k := (C_k, \omega_k, \phi_k), \qquad \dot{C}_k = -\alpha_k S_{\phi,k} C_k + \beta_k |\Psi_k|^2. \tag{1048}$$

Driven evolution closure:

$$\partial_t \Psi_{\text{cell}} = \mathcal{L}_{\text{cell}} \Psi_{\text{cell}} + \sum_r J_r(t), \qquad \sum_r J_r \neq 0 \Rightarrow \dot{S}_\phi \not\equiv 0. \tag{1049}$$

A cytoskeletal force proxy:

$$\mathbf{F}_{\text{cyto}} = -\nabla U(\phi, C_{\text{cell}}), \qquad \partial_t \phi = \omega + \gamma\, \nabla \cdot \mathbf{F}_{\text{cyto}}. \tag{1050}$$

Cycle quantization and bounded entropy production:

$$\oint_{\mathcal{C}} \phi_{\text{cell}}\, d\tau = 2\pi n, \qquad \Delta S_\phi^{(\text{cycle})} < S_{\max}. \tag{1051}$$

Division condition (minimal):

$$C_{\text{cell}} \geq C_{\text{div}}, \qquad \Psi_{\text{cell}} \to \Psi_{\text{cell}}^{(1)} \otimes \Psi_{\text{cell}}^{(2)}. \tag{1052}$$

Collapse / non-viability condition:

$$C_{\text{cell}} \leq C_{\text{crit}} \Rightarrow \lim_{t\to\infty} \Psi_{\text{cell}} = 0. \tag{1053}$$

Quaternionic Coherence Measure (Protein-Level Proxy)

A quaternionic coherence proxy (protein-level) may be defined as

$$C^{(Q)}_{\text{prot}} = \frac{\|\langle Q_{\text{prot}} \rangle\|^2}{1 + S_{\phi,\text{prot}}}. \tag{1054}$$

Life as a Multiscale Process

Biological organization in the Æ-framework is not restricted to a single structural level. Instead, life is described as a multiscale process in which coherence is distributed across molecular, cellular, tissue, and organism-level structures.

Let the biological configuration be decomposed as

$$\Psi_{\text{bio}} = \{\Psi_{\text{mol}}, \Psi_{\text{cell}}, \Psi_{\text{tis}}, \Psi_{\text{org}}\}. \tag{1055}$$

Each component represents an effective layer of biological organization, and their interaction determines the global coherence state of the living system.

Cross-Scale Synchronization

Biological coherence requires synchronization not only within individual scales, but also across distinct levels of organization.

A schematic multiscale synchronization condition may be written as

$$\kappa_{ab} = \frac{\text{Re}\big(\langle \Psi_a, \Psi_b \rangle\big)}{\|\Psi_a\| \, \|\Psi_b\|}, \tag{1056}$$

where Ψ_a and Ψ_b denote biological states at different scales.

Cross-scale coherence is maintained when

$$\kappa_{ab} > \kappa_{\text{bio}} \tag{1057}$$

for the relevant subsystem pairs.

Biological Coherence Stability

The stability of a biological system depends on the persistence of coherence across its interacting subsystems.

A biological state is considered stable when

$$\frac{d}{dt}C_{\text{org}} \approx 0 \tag{1058}$$

and when the internal phase entropy remains bounded,

$$S_{\phi}^{\text{bio}} \le S_{\phi,\text{crit}}. \tag{1059}$$

Loss of this condition corresponds to biological decoherence and degradation of coordinated system behavior.

Phase Entropy as Thermodynamic Variable

We identify phase entropy as the fundamental thermodynamic quantity:

$$S_{\phi} = -\int p(\phi) \log p(\phi)\, d\phi. \tag{1060}$$

Coherence–Entropy Duality

A minimal duality between coherence and phase-entropy is taken as

$$C \propto e^{-S_{\phi}/\hbar}. \tag{1061}$$

(When using the canonical coherence $C_{Æ} = (1 + S_{\phi})^{-1}$ elsewhere, this exponential law should be read as an effective/large-deviation proxy rather than an identity.)

Effective Temperature

Define an effective vibrational temperature via

$$\frac{1}{T_{\text{eff}}} = \frac{\partial S_{\phi}}{\partial E}, \tag{1062}$$

where E denotes a coherence-corrected energy functional.

Vibrational Free Energy

We introduce the vibrational free energy

$$F_{\text{vib}} = E - T_{\text{eff}} S_{\phi}. \tag{1063}$$

Stable coherent systems satisfy the stationary condition

$$\delta F_{\text{vib}} = 0. \tag{1064}$$

Coherence-Induced Geometry (Thermo–Gravitational Coupling)

A thermodynamic imprint of coherence gradients can be encoded geometrically by the effective metric ansatz

$$g_{\mu\nu}(x) = \eta_{\mu\nu} + \alpha_G\, \partial_\mu \partial_\nu \Big(\ln C_{Æ}(x) \Big), \tag{1065}$$

so that non-uniform coherence produces curvature:

$$\nabla_\rho C_{Æ} \neq 0 \quad \Longrightarrow \quad R_{\mu\nu\rho\sigma} \neq 0. \tag{1066}$$

An effective action coupling curvature to the Æ-sector is

$$\mathcal{S}_{\text{grav}} = \int \left(\frac{1}{2\kappa_G} R + \mathcal{L}_{C_{Æ}} \right) \sqrt{-g}\, d^4x, \tag{1067}$$

leading to field equations of Einstein type

$$R_{\mu\nu} - \frac{1}{2} g_{\mu\nu} R = \kappa_G T^{Æ}_{\mu\nu}. \tag{1068}$$

A minimal stress-energy proxy sourced by coherence gradients is

$$T^{Æ}_{\mu\nu} = \partial_\mu C_{Æ} \partial_\nu C_{Æ} - \frac{1}{2} g_{\mu\nu} (\partial C_{Æ})^2. \tag{1069}$$

In the weak-field limit one may set

$$g_{00} \approx 1 + 2\Phi, \tag{1070}$$

with a coherence-generated potential scaling as

$$\Phi \propto \ln C_{Æ}. \tag{1071}$$

61. Definition of the Self

Within the Æ-framework the notion of a "self" is not treated as a separate ontological entity but as a persistent coherent trajectory of the field configuration.

Let $\Psi(t)$ denote the evolving state of the system. A coherent trajectory corresponds to a path

$$\gamma : t \mapsto \Psi(t) \tag{1072}$$

which maintains structural coherence over extended temporal intervals.

Such trajectories may persist through memory retention, network interactions, and topological constraints of the field.

Historical Attractors

The persistence of coherent trajectories may generate historical attractors in configuration space.

Let $\Psi(t)$ denote the evolving system state. A historical attractor $\mathcal{A}$ satisfies

$$\lim_{t \to \infty} d(\Psi(t), \mathcal{A}) = 0, \tag{1073}$$

where $d(\cdot, \cdot)$ denotes a suitable distance in the configuration space.

Such attractors encode accumulated dynamical history and represent stable trajectories shaped by memory kernels and interaction networks.

Trajectory Persistence

The persistence of a coherent trajectory is governed by the combined action of memory kernels and network interactions.

Let $\Psi(t)$ denote a node state within a multi-node system. The trajectory persistence condition may be written as

$$\kappa(t, t - T) > \kappa_{\text{crit}}, \tag{1074}$$

where $\kappa(t, t - T)$ represents the temporal coherence correlation introduced in the memory kernel formalism.

When this condition holds over extended intervals, the trajectory behaves as a persistent coherent identity within the field.

Ontological Layer Interpretation

The trajectory of the self may extend across multiple effective layers of the Æ-field.

The formal structure of these layers was introduced in Section 53, where nested field components such as

$$\Psi_{\text{phys}},\ \Psi_{\text{eth}},\ \Psi_{\text{astr}},\ \Psi_{\text{ment}}$$

were defined.

Within this interpretation the self corresponds to a coherent trajectory simultaneously projected across these interacting layers.

62. Dream Regime as Reduced Coherence Sector

Coherence Decoupling Condition

The dream regime is defined by partial decoupling between nodal coherence and physical embedding:

$$C_{\text{phys}} \downarrow \qquad C_{\text{internal}} \not\downarrow . \tag{1075}$$

A minimal multiscale proxy consistent with the organismal decomposition is

$$C_{\text{org}} = C_{\text{neur}} + C_{\text{endo}} + C_{\text{cell}}, \tag{1076}$$

with a characteristic "dream direction" given by

$$\partial_t C_{\text{neur}} < 0, \qquad \partial_t C_{\text{endo}}, \partial_t C_{\text{cell}} \geq 0. \tag{1077}$$

Dream Memory Accumulation

Dream dynamics may also induce an internal memory accumulation process. A minimal phenomenological representation is

$$\partial_t M(t) = \int \Psi_{\text{neur}}(x,t)\, e^{i\phi(x,t)}\, dx. \tag{1078}$$

This term captures the internal phase-coherent contribution of neural activity to the memory field during the dream regime.

Internal Rotational Dominance

During the dream regime, internal rotational/phase dynamics dominate external physical phase gradients:

$$|\omega_{Æ}| \gg |\nabla_{\text{phys}}\phi|. \tag{1079}$$

Operationally, one may express the reduced sensory-embedding limit by the separation of internal and sensory generators:

$$\partial_t \Psi_{\text{org}} = \mathcal{L}_{\text{int}} \Psi_{\text{org}} + \epsilon_{\text{sens}}\, \mathcal{L}_{\text{sens}} \Psi_{\text{org}}, \tag{1080}$$

with the dream inequality

$$\|\mathcal{L}_{\text{int}} \Psi_{\text{org}}\| \gg \|\mathcal{L}_{\text{sens}} \Psi_{\text{org}}\|. \tag{1081}$$

A phenomenological neural phase model compatible with the coherence-sector formalism is Kuramoto-like:

$$\dot{\phi}_n = \omega_n + \sum_m K_{nm} \sin(\phi_m - \phi_n) + \zeta_n(t), \tag{1082}$$

with state-dependent modulation:

$$\text{NREM:} \quad \omega_n \to \omega_n^{\text{low}}, \tag{1083}$$

$$\text{REM:} \quad \omega_n \to \omega_n^{\text{high}}, \qquad K_{nm} \uparrow . \tag{1084}$$

Effective Oniric Hamiltonian

An effective oniric Hamiltonian is modeled as

$$H_{\text{dream}} = \frac{1}{2}|\omega_{Æ}|^2 + V_{\text{memory}}. \tag{1085}$$

A minimal coherence observable for the dream layer is

$$C_{\text{dream}} = \frac{C_{\text{neur}}}{1 + S_{\phi,\text{neur}}}, \tag{1086}$$

and the admissible-dream condition is taken as

$$C_{\text{dream}} > C_{\text{crit}}. \tag{1087}$$

A corresponding internal "projection" description is

$$\Psi_{\text{dream}} = \Pi_{\text{int}} \Psi_c, \tag{1088}$$

with non-commuting retention/projection operators capturing the fact that dream-memory dynamics are not simply a restriction of waking dynamics:

$$[\hat{\mathcal{R}}, \Pi_{\text{int}}] \neq 0. \tag{1089}$$

Finally, the dream regime includes a suppression of motor embedding as an effective constraint:

$$\Psi_{\text{motor}} \approx 0. \tag{1090}$$

Lucidity as Partial Sensory Re-Coupling

Lucidity may be modeled as a partial re-coupling of sensory embedding:

$$\epsilon_{\text{sens}}^{(\text{lucid})} > \epsilon_{\text{sens}}^{(\text{sleep})}. \tag{1091}$$

Reduced Projection Dynamics

Within the Æ-framework dreaming corresponds to a regime in which the coupling between internal field configurations and the external physical layer becomes reduced.

Let $\Psi_{\text{int}}(t)$ denote the internal configuration of the system. During the dream regime the effective projection onto external observables may be written as

$$\Psi_{\text{obs}} = \Pi_{\text{phys}} \Psi_{\text{int}}, \tag{1092}$$

where Π_{phys} represents the projection operator associated with the physical layer.

In the dream regime this projection becomes partially suppressed, allowing internally generated configurations to dominate the observable dynamics.

Internal Phase Evolution

In the dream regime the evolution of the system is dominated by internal phase dynamics rather than by interactions with external constraints.

Let $\Psi_{\text{int}}(t)$ denote the internal state of the system. Its evolution may be written schematically as

$$\frac{d\Psi_{\text{int}}}{dt} = F_{\text{int}}(\Psi_{\text{int}}), \tag{1093}$$

where F_{int} describes the internal dynamical law of the coherent field configuration.

Because external constraints are reduced, the trajectory of Ψ_{int} may explore a wider region of configuration space.

Dream Trajectory Formation

Dream experiences correspond to coherent trajectories formed within the internally evolving configuration space.

Let γ_{dream} denote such a trajectory

$$\gamma_{\text{dream}} : t \mapsto \Psi_{\text{int}}(t). \tag{1094}$$

These trajectories are shaped by internal memory kernels, residual network interactions, and the underlying phase structure of the field.

Relation to Self Branching

The branching of coherent trajectories associated with the dynamics of the self was introduced in Section 61.

Dream trajectories may be interpreted as internal exploratory branches of the coherent self trajectory. However, the formal branching equations are defined in the self-dynamics sector and are not repeated here.

Instead, the dream regime represents a reduced-coherence environment in which such branching processes may occur more freely.

63. Recurrent Nodal Embedding

Identity as Topological Persistence

Identity is defined as persistence of a knot-class (topological signature) across embeddings:

$$T(\mathcal{K}) = \text{const.} \tag{1095}$$

A minimal dynamical representation of the persistent "astral" identity carrier is

$$\Phi_{\text{astral}}(t) \in S^3, \qquad \partial_t \|\Phi_{\text{astral}}\|^2 = 0, \qquad \|\Phi_{\text{astral}}\| = 1. \tag{1096}$$

One may parameterize its internal evolution as a pure quaternionic rotation

$$\partial_t \Phi_{\text{astral}} = \mathbf{u}(t)\, \omega(t)\, \Phi_{\text{astral}}. \tag{1097}$$

Dissolution Condition

When coherence falls below threshold, physical embedding is lost:

$$C < C_{\min} \quad \Rightarrow \quad \mathcal{K} \text{ loses physical embedding.} \tag{1098}$$

In projection language, this corresponds to degeneration of the physical sector:

$$g_{\mu\nu}^{\text{phys}} \to 0, \qquad \Pi_{\text{phys}} \to 0, \qquad \Phi_{\text{astral}} \neq 0. \tag{1099}$$

Re-Embedding Condition

If a compatible phase manifold exists:

$$\exists\, \mathcal{N}' : \quad T(\mathcal{K}) \subset \mathcal{N}'. \tag{1100}$$

Then structural embedding may reoccur.

A minimal reprojection model is given by a fibered structure

$$\mathcal{E} = (\mathcal{M}_{\text{phys}}, S^3, \pi), \qquad \Pi_{\text{phys}} : S^3 \to \mathcal{M}_{\text{phys}}, \tag{1101}$$

together with a section for the $(k+1)$-th embedding:

$$s_{k+1} : \mathcal{M}_{\text{phys}}^{(k+1)} \to S^3, \qquad s_{k+1}(x) = U_{\text{karma}} \Phi_{\text{astral}}. \tag{1102}$$

Re-embedding is allowed only when phase-compatibility exceeds a threshold:

$$\text{Re}\left(\Phi_{\text{astral}}^{\dagger} \Phi_{\text{embryo}}\right) > \tau_{\text{inc}}. \tag{1103}$$

Equivalently, one may define the normalized incarnational coupling

$$\kappa_{\text{inc}} = \frac{\text{Re}\left(\Phi^{\dagger}_{\text{astral}}\Phi_{\text{embryo}}\right)}{\|\Phi_{\text{astral}}\|\,\|\Phi_{\text{embryo}}\|}. \tag{1104}$$

The induced local realized state is then

$$\Psi_{\text{local}} = \Pi_{\text{phys}}\left(U_{\text{karma}}\Phi_{\text{astral}}\right). \tag{1105}$$

Conservation Principle

Topological invariants remain preserved across embeddings:

$$T(\mathcal{K}_{\text{before}}) = T(\mathcal{K}_{\text{after}}). \tag{1106}$$

This is enforced by an embedding-transfer operator ("karma") constructed from holonomy data over the knot set:

$$U_{\text{karma}} = \exp\left(-i\sum_{\gamma\in\mathcal{K}}\alpha_{\gamma}H_{\gamma}\right). \tag{1107}$$

Observer Continuity Across Projections

Let the observer-state evolve in the full Æ Hilbert space:

$$\Psi_{\mathcal{O}}(t) \in \mathcal{H}_{«}, \qquad \Pi_k : \mathcal{H}_{«} \to \mathcal{H}^{(k)}_{\text{phys}}, \qquad \Psi^{(k)}_{\text{phys}}(t) = \Pi_k\Psi_{\mathcal{O}}(t). \tag{1108}$$

Define projected intensity (embedding coherence) by

$$C_k(t) := \|\Pi_k\Psi_{\mathcal{O}}(t)\|^2 = \langle\Psi_{\mathcal{O}}(t), \Pi_k^{\dagger}\Pi_k\Psi_{\mathcal{O}}(t)\rangle. \tag{1109}$$

Observer continuity is encoded by a non-vanishing time-overlap:

$$\mathcal{C}_{\mathcal{O}}(t) := \langle\Psi_{\mathcal{O}}(t), \Psi_{\mathcal{O}}(t+\Delta t)\rangle \neq 0, \tag{1110}$$

and a threshold condition:

$$|\mathcal{C}_{\mathcal{O}}(t)| \geq C_{\mathcal{O},\text{crit}} \quad \Rightarrow \quad \text{observer continuity holds.} \tag{1111}$$

Switching between embeddings is admissible if cross-projected overlap remains above:

$$\Pi_k \;\to\; \Pi_{k'}, \qquad \text{Re}(\langle\Pi_k\Psi_{\mathcal{O}}, \Pi_{k'}\Psi_{\mathcal{O}}\rangle) > \tau_{\text{switch}}. \tag{1112}$$

A near-unitary transfer map may be introduced:

$$\Psi_{\mathcal{O}} \;\to\; \mathcal{U}_{k\to k'}\,\Psi_{\mathcal{O}}, \qquad \mathcal{U}^{\dagger}_{k\to k'}\mathcal{U}_{k\to k'} \approx \mathbb{I}, \qquad \Pi_{k'}\,\mathcal{U}_{k\to k'} \approx \Pi_k. \tag{1113}$$

Global Identity vs Local Personality

Let $\Psi_{\mathcal{I}}$ denote the global identity mode in $\mathcal{H}_{Æ}$:

$$\Psi_{\mathcal{I}} \in \mathcal{H}_{Æ}, \qquad |\langle \Psi_{\mathcal{I}}(t), \Psi_{\mathcal{I}}(t+\Delta t)\rangle| \geq C_{\mathcal{I},\text{crit}}. \tag{1114}$$

Local personality in embedding k is then a projection:

$$\Psi^{(k)}_{\text{pers}}(t) = \Pi_k \Psi_{\mathcal{I}}(t), \tag{1115}$$

with an embedding-induced excess phase entropy

$$\Delta S^{(k)}_{\text{pers}} := S_\phi(\Psi^{(k)}_{\text{pers}}) - S_\phi(\Psi_{\mathcal{I}}), \qquad \Delta S^{(k)}_{\text{pers}} \geq 0. \tag{1116}$$

Finally, stability of identity across embeddings may be summarized as

$$\text{Re}(\langle \Pi_k \Psi_{\mathcal{I}}, \Pi_{k'} \Psi_{\mathcal{I}}\rangle) \geq \tau_{\text{switch}}. \tag{1117}$$

Holonomy Return Condition

Identity persistence across embeddings may be interpreted as a holonomy return condition along closed paths in the nodal configuration space.

Let γ denote a closed loop in the connection manifold. The associated holonomy operator satisfies

$$H_\gamma \neq \mathbb{I}. \tag{1118}$$

Successive traversals of the same loop generate higher-order holonomy elements

$$H_\gamma \longrightarrow H_\gamma^2. \tag{1119}$$

More generally, the holonomy associated with a connection A_λ is defined as

$$H_\gamma(\lambda) := \mathcal{P}\exp\left(\oint_\gamma A_\lambda\right). \tag{1120}$$

A renodalization or identity-return condition occurs when

$$H_\gamma(\lambda)^2 = \mathbb{I}. \tag{1121}$$

In general the return may require multiple traversals,

$$H_\gamma(\lambda)^n \neq \mathbb{I} \quad \text{for all } n \leq N, \tag{1122}$$

while asymptotic closure yields

$$H_\gamma(\lambda)^n \to \mathbb{I}. \tag{1123}$$

This condition formalizes the idea that the nodal identity may traverse a sequence of embeddings before returning to an equivalent topological configuration.

64. Coherence-Weighted Observer Observable

Coherent Observer Trajectories

The observer is defined as a coherence-preserving trajectory $\gamma(\tau) \subset M$ within the phase-coherent field.

We define the set of coherent observer trajectories as

$$\Gamma_{\text{obs}} = \left\{ \gamma(\tau) \subset M \;\middle|\; \frac{d}{d\tau} C(\gamma(\tau)) = 0 \right\},$$

where $C(x)$ is the coherence functional already introduced in the model and

$$\frac{d}{d\tau} C(\gamma(\tau)) = \dot{\gamma}^{\mu} \partial_{\mu} C$$

denotes the derivative of the coherence functional along the parametrized curve $\gamma(\tau)$.

An observer functional associated with a coherent trajectory may be written as

$$\mathcal{O}_{\gamma}[\psi] = \int_{\gamma} C(x) \psi(x) \, d\tau,$$

where $\psi(x)$ denotes the spinor field already present in the field dynamics.

This provides an operational definition of observers within the existing AE-field structure: observers correspond to trajectories along which the coherence functional remains stationary. No new fields, operators, or ontological assumptions are introduced.

Projection Operator

Within the Æ-framework consciousness is represented through projection operators acting on coherent field configurations.

Let Ψ denote the global state of the system. The conscious observable state may be written as

$$\Psi_{\text{obs}} = \Pi_{\text{c}} \Psi, \tag{1124}$$

where Π_{c} denotes the consciousness projection operator selecting the subset of configurations accessible to conscious observation.

Coherence-Weighted Observable

Observables in the Æ-framework may be weighted by the coherence structure of the underlying field.

Let O denote an observable operator. The effective conscious expectation value may be written as

$$\langle O \rangle_c = \frac{\langle \Psi | C_{Æ} O | \Psi \rangle}{\langle \Psi | C_{Æ} | \Psi \rangle}, \tag{1125}$$

where $C_{Æ}$ denotes the coherence operator associated with the Æ-field.

This expression describes how conscious perception depends on the coherence structure of the system.

Ego Constraint

The notion of ego may be interpreted as a constraint on the accessible projection space of the consciousness operator.

Let Π_{ego} denote a restricted projection operator. The observable state may then be written as

$$\Psi_{\mathrm{ego}} = \Pi_{\mathrm{ego}} \Psi, \tag{1126}$$

where

$$\Pi_{\mathrm{ego}} \subset \Pi_c. \tag{1127}$$

This restriction limits the accessible region of configuration space and produces the effective perspective associated with individual experience.

Consciousness in Reality Knots

$$\mathcal{C}_Q = C_{«}(Q)\, \|\nabla \phi_Q\|\, \|\partial_t \phi_Q\| \tag{1128}$$

$$\delta \mathcal{C}_Q = 0 \tag{1129}$$

Unified Causal Chain (Consciousness → Matter)

$$\boxed{\delta A_{\mathrm{coherent}} \longrightarrow \delta E \longrightarrow \delta g_{\mu\nu} \longrightarrow \delta M} \tag{1130}$$

Premonition Projector

Within the Æ-framework the internal projection dynamics of the system may generate anticipatory configurations associated with future trajectories.

Let $\Psi(t)$ denote the evolving system state. A predictive projection may be represented through

$$\Psi_{\text{pred}} = \Pi_{\text{pred}}\Psi, \tag{1131}$$

where Π_{pred} denotes a projector acting on the space of possible future configurations.

Such projections correspond to internally generated configurations that anticipate future coherent trajectories of the system.

65. Chakras as Phase Vortices

Vortex Quantization Condition

$$\oint_{\Gamma} d\phi = 2\pi n, \qquad n \in \mathbb{Z}. \tag{1132}$$

$$H_{\Gamma} = \mathcal{P}\exp\left(\oint_{\Gamma} A\right), \qquad H_{\Gamma} \text{ stable under coherent evolution.} \tag{1133}$$

Chakras as Coherent Nodal Domains

We model each chakra as a coherent nodal domain (a phase-structured subset) supported by a local coherence threshold in the astral sector:

$$\mathcal{K}_i = \left\{x \in \Omega \;\middle|\; C_{\text{astr}}(x,t) \geq C_i^{\text{nodal}}\right\}, \qquad i = 1, \ldots, N, \quad N < \infty. \tag{1134}$$

Its boundary supports quantized circulation:

$$\oint_{\partial \mathcal{K}_i} \nabla \phi_{\text{astr}} \cdot d\ell = 2\pi n_i. \tag{1135}$$

The nodal domain is a local stationary configuration of the astral functional:

$$\left.\frac{\delta \mathcal{H}_{\text{astr}}}{\delta \Psi_{\text{astr}}}\right|_{\mathcal{K}_i} = 0, \qquad \delta^2 \mathcal{H}_{\text{astr}} > 0. \tag{1136}$$

Cross-Layer Transduction at Vortex Cores

Chakras act as sites where inter-layer transduction is non-vanishing:

$$\left.\mathcal{T}_{\text{astr}\to\text{eth}}\right|_{\mathcal{K}_i} \neq 0, \qquad \left.\mathcal{T}_{\text{eth}\to\text{phys}}\right|_{\mathcal{K}_i} \neq 0. \tag{1137}$$

A necessary marker of active transduction is a non-trivial phase-entropy gradient:

$$\left.\nabla S_{\phi,\text{astr}}\right|_{\mathcal{K}_i} \neq 0. \tag{1138}$$

Projected Chakra Nodes in the Physical Layer

Each coherent chakra-domain $\mathcal{K}_i$ admits a physical projection through the effective physical-sector map:

$$G_i = \Pi_{\text{phys}}[\mathcal{K}_i]\,. \tag{1139}$$

The corresponding physical oscillatory profile is assumed to inherit, to leading order, the nodal frequency structure of the underlying chakra-domain:

$$\mathcal{F}_{G_i} \approx \mathcal{F}_{\mathcal{K}_i}. \tag{1140}$$

Its temporal evolution is induced by inter-layer transduction:

$$\partial_t G_i = \mathcal{T}_{\text{eth}\to\text{phys}} \circ \mathcal{T}_{\text{astr}\to\text{eth}}\,[\mathcal{K}_i]\,. \tag{1141}$$

At the nodal level, the coherent chakra-domain may also evolve under biological modulation:

$$\mathcal{K}_i(t+\Delta t) = \mathcal{K}_i(t) + \Delta\mathcal{K}_i^{(\text{bio})}. \tag{1142}$$

The full chakra configuration may therefore be represented as the finite family of nodal–physical pairs

$$\{(\mathcal{K}_1, G_1), \ldots, (\mathcal{K}_N, G_N)\}, \qquad N < \infty. \tag{1143}$$

Functional Node Model (7-Chakra Dynamics)

We treat the canonical chakra set as a finite network of coupled phase-coherence nodes:

$$\mathcal{N}_i := (C_i,\ \omega_i,\ \phi_i,\ \kappa_i), \qquad i = 1, \ldots, 7. \tag{1144}$$

Local coherence is identified with the canonical phase-entropy dual:

$$C_i = C_{\text{Æ},i} = \frac{1}{1 + S_{\phi,i}}. \tag{1145}$$

Phase evolution follows Kuramoto-type coupling:

$$\dot{\phi}_i = \omega_i + \sum_j \kappa_{ij} \sin(\phi_j - \phi_i). \tag{1146}$$

A minimal coherence relaxation law is:

$$\dot{C}_i = -\alpha_i S_{\phi,i} C_i + \beta_i. \tag{1147}$$

Optionally, one may use a second-order oscillator representation:

$$\ddot{\phi}_i + \zeta_i \dot{\phi}_i + \omega_i^2 \phi_i = F_i(t), \qquad Q_i = \frac{\omega_i}{\zeta_i}. \tag{1148}$$

Phase Fixation, Trauma Locking, and Renudation

We model ego-like fixation as a stiff restoring bias toward a past projection:

$$\mathrm{Ego}(Q) = \lambda_{\mathrm{fix}} \big(Q - \Pi_{\mathrm{past}}(Q)\big), \qquad \lambda_{\mathrm{fix}} \gg 1. \tag{1149}$$

Trauma is treated as a non-renuded (non-dissolved) holonomy loop trapped inside a chakra-domain:

$$H_\gamma \neq \mathbb{I} \quad \text{and} \quad \gamma \subset \mathcal{N}_i. \tag{1150}$$

Introduce a suppression variable $B_i \in [0,1]$ to encode locking:

$$C_i = C_i^{(0)}(1-B_i), \qquad S_{\phi,i} = S_{\phi,i}^{(0)} + \xi\, B_i, \qquad \xi > 0, \qquad B_i \to 1. \tag{1151}$$

Renudation (release) is characterized by holonomy relaxation:

$$H_\gamma^2(\lambda) = \mathbb{I}, \qquad H_\gamma(\lambda) \to \mathbb{I}. \tag{1152}$$

and a corresponding recovery of coherence:

$$C_i \uparrow \quad \text{for previously suppressed nodes,} \qquad B_i \downarrow . \tag{1153}$$

Discrete Phase Storage

Information storage may be represented as phase increments along a discrete path:

$$\phi(s_{n+1}) - \phi(s_n) = \delta_n, \qquad \Phi_N = \sum_{n=1}^{N} \delta_n, \qquad \Phi_N \mod 2\pi \approx 0.$$

Chakra Nodal Sets

Within the Æ-framework chakras may be interpreted as nodal structures of the coherent field.

Let $\Psi(x)$ denote the field configuration. A chakra node K_i is defined as the set of points where the phase structure generates a stable vortex configuration

$$K_i = \{x \in \mathcal{M} \mid \nabla\phi(x) \neq 0,\ \nabla \times \nabla\phi(x) \neq 0\}. \tag{1154}$$

Each nodal set K_i corresponds to a stable center of phase circulation within the field.

Boundary Circulation

The vortex structure associated with a chakra node may be characterized by the circulation of the phase gradient along a closed contour surrounding the nodal set.

Let γ_i denote a closed curve around K_i. The circulation is given by

$$\oint_{\gamma_i} \nabla\phi \cdot d\ell = 2\pi n_i, \tag{1155}$$

where n_i denotes the vortex number associated with the chakra node.

Stability Conditions

A chakra node remains dynamically stable when the phase structure surrounding the nodal set satisfies

$$\frac{d}{dt} n_i = 0. \tag{1156}$$

This condition ensures the persistence of the vortex structure under the dynamical evolution of the field.

Kundalini Current

The interaction between chakra nodes may generate a vertical current of coherent phase propagation along the central axis of the biological structure.

Let J_k denote the effective kundalini current. Its propagation may be written as

$$J_\text{k} = \sum_i \alpha_i K_i, \tag{1157}$$

where α_i denotes the coupling strength associated with the chakra node K_i.

This current represents the propagation of coherent phase information between the nodal sets of the system.

Energy Transduction at Chakra Nodes

Chakra nodes may act as transduction centers converting different forms of coherent field interaction.

The transduction efficiency may be written as

$$T_i = \frac{|\Psi(K_i)|^2}{\int |\Psi(x)|^2 dx}, \tag{1158}$$

which measures the fraction of the field energy concentrated at the nodal set.

Local coupling constraint

We impose that coherence transfer between nodes is mediated by the connectivity of the underlying nodal network.

Let κ_{ij} denote the effective coupling between nodes K_i and K_j. Then

$$\kappa_{ij} = 0 \quad \text{if no admissible path exists between } K_i \text{ and } K_j. \tag{1159}$$

More generally, coherence transfer is constrained by path-dependent propagation:

$$\kappa_{ij} \sim \sum_{\gamma_{ij}} \prod_{k \in \gamma_{ij}} \alpha_k, \tag{1160}$$

where γ_{ij} denotes admissible paths connecting K_i to K_j through intermediate nodes, and α_k are local coupling coefficients.

This enforces that coherence propagation is mediated by local phase continuity rather than non-local jumps.

Resonance-based access constraint.

This constraint naturally extends to interactions between observer states and field configurations.

Perceptual or dynamical interaction with field configurations is restricted by phase alignment and coherence conditions.

Let κ_{ij} denote the effective coupling between an observer state Φ_i and a field configuration Φ_j. Then

$$\kappa_{ij} \sim \langle \Phi_i, \Phi_j \rangle \cdot C_i C_j, \tag{1161}$$

and interaction occurs only when

$$\kappa_{ij} > \kappa_{\text{crit}}. \tag{1162}$$

This implies that accessible configurations are those for which sufficient phase compatibility and coherence are satisfied.

Timeline Selection

Reality knots may evolve along multiple possible temporal trajectories.

Let $\Psi(t)$ denote the configuration of the system. The selection of an effective timeline corresponds to the dominance of one coherent trajectory among several possible branches.

A schematic selection condition may be written as

$$\Psi_{\text{real}}(t) = \arg\max_{\Psi_i} C_{Æ}(\Psi_i), \tag{1163}$$

where $C_{Æ}$ denotes the coherence measure associated with each candidate trajectory.

Timeline Bifurcation

Reality knots may undergo bifurcation when the dynamical evolution allows multiple coherent trajectories.

Let $\Psi(t)$ denote the state prior to bifurcation. A bifurcation occurs when

$$\Psi(t) \rightarrow \{\Psi_1(t), \Psi_2(t), \dots\}, \tag{1164}$$

such that each branch satisfies the dynamical evolution equations but corresponds to a distinct trajectory in configuration space.

Integral Memory Functional

Temporal evolution of reality knots may depend on the accumulated memory of past configurations.

A memory functional may be written as

$$\mathcal{M}(t) = \int_{-\infty}^{t} K(t-\tau)\, \Psi(\tau)\, d\tau, \tag{1165}$$

where $K(t-\tau)$ denotes a temporal kernel describing the persistence of past configurations.

This functional introduces temporal nonlocality in the evolution of the system.

Periodic and Recurrent Structures

Certain reality knots may exhibit approximately periodic behavior.

A periodic trajectory satisfies

$$\Psi(t+T) \approx \Psi(t), \tag{1166}$$

where T denotes a characteristic recurrence period.

Such recurrent structures may stabilize temporal patterns within the evolving configuration space.

66. Reality Knots in the Æ Field

Definition

Reality knots are defined as structured configurations of amplitude and phase within the Æ field

$$Q(x) = A(x)\,E(x). \tag{1167}$$

Linear Stability Condition

To analyze their persistence, small perturbations are considered around a given configuration

$$\partial_t \delta Q = DF(Q)\,\delta Q, \tag{1168}$$

Stability is ensured when all modes exhibit non-positive real growth

$$\mathrm{Re}(\lambda_i) \leq 0. \tag{1169}$$

The 3×2 Reality Matrix (Domain Decomposition)

To track multi-domain consistency, we organize the state of a Reality Knot into a 3×2 matrix of amplitude/density vs. phase/coherence data:

$$\mathcal{R} = \begin{pmatrix} \mathcal{R}_{11} & \mathcal{R}_{12} \\ \mathcal{R}_{21} & \mathcal{R}_{22} \\ \mathcal{R}_{31} & \mathcal{R}_{32} \end{pmatrix} \in \mathbb{R}^{3\times 2} \tag{1170}$$

An explicit identification is:

$$\mathcal{R} = \begin{pmatrix} \rho_{\text{phys}} & \phi_{\text{phys}} \\ \rho_{\text{info}} & \phi_{\text{info}} \\ \rho_{\text{phase}} & \phi_{\text{phase}} \end{pmatrix} \tag{1171}$$

Phase entropy is treated as a functional of the full matrix:

$$S_\phi(\mathcal{R}) = S_\phi(\rho_{\text{phys}}, \rho_{\text{info}}, \rho_{\text{phase}}, \phi_{\text{phys}}, \phi_{\text{info}}, \phi_{\text{phase}}) \tag{1172}$$

Coherence is then a derived scalar:

$$C_{Æ}(\mathcal{R}) = \frac{1}{1 + S_\phi(\mathcal{R})}, \qquad C_{Æ}(\mathcal{R}) > C_{\text{crit}} \tag{1173}$$

We interpret downward realization as successive projections:

$$\mathcal{R}_{3\bullet} \xrightarrow{P_{3\to 2}} \mathcal{R}_{2\bullet} \xrightarrow{P_{2\to 1}} \mathcal{R}_{1\bullet} \tag{1174}$$

A minimal «–determinant-like invariant can be defined as:

$$\det_{\text{Æ}}(\mathcal{R}) = \sum_{i<j} (\rho_i \phi_j - \rho_j \phi_i) \tag{1175}$$

Holonomy Criterion and Topological Nontriviality

A Reality Knot is topologically nontrivial if its induced connection has nontrivial holonomy along at least one closed loop:

$$A \cdot= \mathrm{Im}\left(Q^{-1} dQ\right) \tag{1176}$$

$$H_\gamma := \mathcal{P}\exp\left(\oint_\gamma A\right) \in SU(2) \tag{1177}$$

$$H_\gamma \neq \mathbb{I} \tag{1178}$$

A stability-compatible topological persistence condition is:

$$\mathcal{T}(Q) \neq 0,$$

where $\mathcal{T}$ denotes a nonvanishing topological indicator functional (e.g. holonomy distance to the identity, knot-type persistence, or related invariant).

Quaternionic Expansion and Perceptual Projections (Optional)

For sensor/observer-driven regimes, one may represent a perceptual knot-state as

$$Q_{\text{sens}}(x,t) = q_0(x,t) + q_1(x,t)\mathbf{i} + q_2(x,t)\mathbf{j} + q_3(x,t)\mathbf{k},$$

with an observable projection (e.g. "RI-projection") given by

$$P_{\text{RI}}(Q) = \langle Q, \mathbf{1} \rangle = q_0, \qquad Q_{\text{sens}} \longrightarrow q_0.$$

Increased perceptual bandwidth corresponds to reduced phase entropy:

$$S_\phi \downarrow .$$

Taxonomy of Reality Knots (Canonical Types)

We adopt the shorthand families of stable or metastable Reality Knots:

$$\text{NRM (material)}, \quad \text{NRE (emotional)}, \quad \text{NRC (cognitive)},$$
$$\text{NRT (timeline)}, \quad \text{NRK (karmic)}, \quad \text{NRG (collective)},$$
$$\text{NRA (astral)}, \quad \text{NRH (hybrid)}.$$

Representative structural constraints include:

$$Q = \ll \quad \text{with coupled phase–density dynamics},$$
$$Q_i(t) \approx Q_i(t+T),$$
$$Q(t_0) \to \{Q_k(t)\}, \qquad \mathcal{T}(Q_k) = \mathcal{T}(Q),$$
$$H_\gamma(\lambda)^n \neq \mathbb{I} \quad \text{for many cycles } n.$$

Collective knots may be modeled multiplicatively as:

$$Q_{\text{collective}} = \prod_{i=1}^{N} Q_i. \tag{1179}$$

Reality-Knot Network View (Multiverse as Graph)

A global ensemble of stable/metastable knot-trajectories can be organized as:

$$\mathcal{M} = \text{Graph}(\mathbb{T}), \tag{1180}$$
$$\mathbb{T} = \{\text{all stable and metastable Reality Knot trajectories}\}.$$

Remark: Composite Knots in Biology (Proteins)

As an optional bridge toward biophysical applications, one may represent protein fold lines as closed curves and composite knots via connected sums:

$$\gamma : [0,1] \to \mathbb{R}^3, \qquad \overline{\gamma} : S^1 \to \mathbb{R}^3, \qquad K_{\text{protein}} = K_1 \# K_2 \# \ldots \# K_m.$$

Temporal Structure of Reality Knots

Memory as Temporal Knot Compression

$$M(t) = \mathcal{P}\left[H_{\gamma_{\text{past}}} \cdot Q(t)\right]. \tag{1181}$$

Karma as Residual Holonomy

$$\mathcal{K} := \{\gamma_i \mid H_{\gamma_i} \neq \mathbb{I}\}. \tag{1182}$$

$$H_\gamma = \mathcal{P}\exp\left(\oint_\gamma Q^{-1}dQ\right). \tag{1183}$$

$$\partial_t H_\gamma = -\eta \nabla S_\phi. \tag{1184}$$

Temporal Ordering by Phase Causality

Temporal structure is defined by nontrivial phase evolution:

$$\partial_t \phi(x,t) \neq 0. \tag{1185}$$

We introduce a causal order on coherent configurations via

$$\text{Æ}_1 \prec \text{Æ}_2 \iff \text{the phase configuration of Æ}_2 \text{ arises from that of Æ}_1, \tag{1186}$$

where « denotes a configuration label within the Æ-field, interpreted as a symbolic identifier rather than a typographic ligature.

Angular Frequency and Coherence-Time Evolution

Define local angular frequency as

$$\omega(x,t) = \partial_t \phi(x,t), \tag{1187}$$

and its coherence-weighted mean over the domain Ω as

$$\bar{\omega}(t) = \int_\Omega \omega(x,t)\, d\mu_{\text{Æ}}(x). \tag{1188}$$

A schematic separation of roles is:

$$D \text{ controls coherence strength,} \qquad t \text{ controls coherence evolution.} \tag{1189}$$

Proper Time from Quaternionic Phase Flux

For a quaternionic knot-state $Q(t) \in S^3$, define the accumulated phase flux along the trajectory as

$$\Phi_{\text{total}} = \int_0^T \tfrac{1}{2}\,\mathrm{Im}(Q^{-1}\dot{Q})\, dt.$$

The induced Æ-angular speed and proper-time functional are

$$\omega_{Æ}(t) = \left\| \tfrac{1}{2}\,\mathrm{Im}(Q^{-1}\dot{Q}) \right\|, \qquad \tau_{Æ}(t) = \int_0^t \omega_{Æ}(s)\,ds.$$

Equivalently, one may set a proportionality at the differential level:

$$d\tau_{Æ} := \kappa\, d\Phi_{\text{total}}, \qquad \kappa > 0.$$

Holonomy Decomposition and Conservation (Optional)

Let the knot trajectory satisfy

$$\dot{Q}(t) = F(Q(t), x, t), \qquad F(Q) \in T_Q S^3.$$

With tetrahedral weights w_i (e.g. from a Hopf/tetrahedral decomposition), one may define partial phase fluxes Φ_i with the conservation law

$$\sum_{i=1}^{4} \dot{\Phi}_i(t) = \dot{\Phi}_{\text{total}}(t), \qquad d\Phi_{\text{total}} = \sum_i d\Phi_i.$$

Remark: Premonition as Forward Projection (Optional)

Define a future-directed candidate path $\gamma_{\text{future}}(t > t_0)$ and a corresponding pre-projection observable:

$$\Pi_{\text{pre}} = \mathcal{P}[\exp(-\Xi_{\gamma_{\text{future}}})\, Q(t)]\,. \tag{1190}$$

When two trajectories are phase-equivalent,

$$\gamma_A \sim \gamma_B \quad \Rightarrow \quad \Pi_A(t) \approx \Pi_B(t). \tag{1191}$$

Inter-Life Update Law (Optional)

A minimal recurrence schematic can be written as

$$\gamma_{\text{life } n} \longrightarrow \gamma_{\text{life } n+1}, \tag{1192}$$

with a holonomy update

$$H_{\gamma_{n+1}} = H_{\gamma_n} - \Delta H. \tag{1193}$$

Holonomy Magnitude and Phase-Entropy Load (Optional)

A coarse "karmic load" relation may be encoded by

$$\|H_\gamma - \mathbb{I}\| \propto \int_\gamma S_\phi \, d\tau. \tag{1194}$$

Inherited Memory and Entropic Filtering

Memory inherited across embeddings may be represented as the phase-entropy content of the astral knot-state:

$$S_{\text{inherit}} = S_\phi \left(\Phi_{\text{astral}}\right). \tag{1195}$$

The realized local memory sector is restricted by a coherence threshold:

$$S_{\text{local}} < S_{\text{crit}}. \tag{1196}$$

Local memory realization is modeled as a filtered projection of the astral knot-state:

$$\Phi_{\text{local}} = \mathcal{F}\left(\Phi_{\text{astral}}\right), \tag{1197}$$

with a norm-preserving map

$$\mathcal{F} : S^3 \to S^3. \tag{1198}$$

The filtering process preserves the knot-class while modifying phase entropy:

$$\|\mathcal{F}(\Phi)\| = 1, \qquad \text{class}(\Phi) = \text{class}(\mathcal{F}(\Phi)). \tag{1199}$$

The local phase entropy therefore becomes

$$S_{\phi,\text{local}} = S_{\text{inherit}} - \Delta S_{\text{filtered}}. \tag{1200}$$

Finally, the rate of change of locally accessible memory is bounded by a maximal entropic relaxation rate:

$$\frac{d}{dt} S_{\text{local}} \leq \sigma_{\max}. \tag{1201}$$

Reality Knot Dynamics and Manifestation

Energetic Generator for Knot Evolution

Assume an Æ-sector Hamiltonian functional over the configuration space:

$$\mathcal{H}_{\text{Æ}} : \mathcal{A}_{\text{Æ}} \to \mathbb{R}. \tag{1202}$$

A standard quadratic-plus-potential form is

$$\mathcal{H}_{Æ}[Æ] = \int_{\Omega} \left(\|\nabla_{Æ} Æ\|^2 + V_{Æ}(Æ) \right) d\mu_{Æ}. \tag{1203}$$

The corresponding conservative evolution may be written abstractly as

$$\partial_t Æ = \{Æ, \mathcal{H}_{Æ}\}_{Æ}, \tag{1204}$$

with an optional entropy-driven dissipation channel

$$\partial_t Æ = \{Æ, \mathcal{H}_{Æ}\}_{Æ} - \gamma \frac{\delta S_{\phi}}{\delta Æ}, \qquad \gamma \geq 0. \tag{1205}$$

A phase-restricted statement (when ϕ is an explicit coordinate) is

$$\partial_t \phi = \frac{\delta \mathcal{H}_{Æ}}{\delta \phi}. \tag{1206}$$

Manifestation Equation

We model manifestation as an attractor-alignment term toward a target knot-state Q_*:

$$\partial_t Q = F(Q) + \lambda \operatorname{Re}\left(Q_*^{\dagger} Q\right) (Q_* - Q) \tag{1207}$$

Here $F(Q)$ denotes baseline knot dynamics (e.g. induced by the underlying field equations), and the second term implements coherence-weighted convergence.

Strength and Convergence Rate

Define the instantaneous alignment strength as

$$\text{Strength}(Q \to Q_*) = \lambda \operatorname{Re}\left(Q^{\dagger} Q_*\right). \tag{1208}$$

Let $\theta(Q, Q_*)$ be the geodesic angle on S^3 induced by the inner product, for example via

$$\cos\theta(Q, Q_*) := \operatorname{Re}\left(Q^{\dagger} Q_*\right), \qquad \theta \in [0, \pi]. \tag{1209}$$

A simple monotone proxy for convergence rate is then

$$\text{Speed}(Q \to Q_*) \sim \frac{1}{1 + \theta(Q, Q_*)}. \tag{1210}$$

Pre-Alignment by Imagination (Optional)

A pre-alignment ansatz may be represented as a phase-rotation toward the target:

$$Q_{\text{imag}}(t) = \exp\Big(- \mathbf{I}\,\theta(t)\Big)\, Q_*, \tag{1211}$$

suggesting the directional flow

$$Q(t) \;\to\; Q_*.$$

Identity Collapse and Reformation (Optional)

A transformation threshold between "old" and "new" identity embeddings may be expressed by

$$\text{Re}\Big(Q^\dagger_{\text{old}} Q_{\text{new}}\Big) > \tau_{\text{transform}}, \tag{1212}$$

schematically:

$$\text{NR}_{\text{old}} \;\longrightarrow\; \text{NR}_{\text{new}}.$$

Emotional Energy as Coherence Modulator (Optional)

Define an energetic proxy for emotional driving as

$$E_{\text{emotion}} = \int |\partial_t Q|^2 \,\mathrm{d}x. \tag{1213}$$

Multi-Knot Coupling and Phase Transport (Optional)

For multiple interacting knot-states Q_α, define pairwise phase offsets

$$\Delta\phi_{\alpha\beta}(t) := \phi^{(\alpha)}(t) - \phi^{(\beta)}(t),$$

with Kuramoto-type transport

$$\dot{\phi}^{(\alpha)} = \omega_\alpha + \sum_{\beta\neq\alpha} \kappa_{\alpha\beta}\, \sin\Big(\Delta\phi_{\beta\alpha}\Big).$$

A quaternionic coupling form consistent with alignment dynamics is

$$\dot{Q}_\alpha = \mathbf{u}_\alpha\, \omega_\alpha Q_\alpha + \sum_{\beta\neq\alpha} \kappa_{\alpha\beta}\, \Big(Q_\beta Q_\alpha^{-1} - \text{Re}(Q_\beta Q_\alpha^{-1})\Big)\, Q_\alpha.$$

Neural and Collective Extensions

Neural Phase Laplacian

Neural phase coupling admits a weighted Laplacian form:

$$(L_\phi)_{ij} = -W_{ij}\cos(\phi_i - \phi_j). \tag{1214}$$

Coherence is consistently defined by

$$C_{Æ} = \frac{1}{1 + S_\phi}. \tag{1215}$$

Kuramoto Collective Synchronization

Collective neural synchronization follows Kuramoto-type phase dynamics. Define the collective order parameter

$$R(t)e^{i\Theta(t)} = \frac{1}{N}\sum_{j=1}^{N} e^{i\phi_j(t)}. \tag{1216}$$

Individual phases evolve according to

$$\dot{\phi}_i = \omega_i + \frac{K}{N}\sum_{j=1}^{N}\sin(\phi_j - \phi_i). \tag{1217}$$

External Phase Forcing and Imposed Synchronization

Collective phase dynamics may also be driven by an external phase-imposing field.

A forced Kuramoto-type evolution is

$$\dot{\phi}_i = \omega_i + \frac{K}{N}\sum_{j=1}^{N}\sin(\phi_j - \phi_i) + F_i(t). \tag{1218}$$

A minimal phase-imposition term is

$$F_i(t) = \eta_i \sin\Big(\Phi_{\text{ext}}(t) - \phi_i\Big), \tag{1219}$$

where $\Phi_{\text{ext}}(t)$ denotes an externally imposed collective phase and $\eta_i \geq 0$ measures the susceptibility of node i to the forcing field.

The evolution may then be rewritten as

$$\dot{\phi}_i = \omega_i + KR\sin(\Theta - \phi_i) + \eta_i \sin\Big(\Phi_{\text{ext}} - \phi_i\Big). \tag{1220}$$

Define the endogenous synchronization field

$$\Xi_i^{\mathrm{endo}} = KR\sin(\Theta - \phi_i), \tag{1221}$$

and the imposed synchronization field

$$\Xi_i^{\mathrm{imp}} = \eta_i \sin\Big(\Phi_{\mathrm{ext}} - \phi_i\Big). \tag{1222}$$

A global imposed-synchronization ratio is

$$R_{\mathrm{imp}} = \frac{\sum_{i=1}^{N} |\Xi_i^{\mathrm{imp}}|}{\sum_{i=1}^{N} |\Xi_i^{\mathrm{endo}}| + \varepsilon}, \qquad \varepsilon > 0. \tag{1223}$$

The regime

$$R_{\mathrm{imp}} \ll 1 \tag{1224}$$

corresponds to predominantly endogenous coherence, whereas

$$R_{\mathrm{imp}} \gg 1 \tag{1225}$$

corresponds to predominantly imposed synchronization.
The collective order parameter remains

$$R(t)e^{i\Theta(t)} = \frac{1}{N}\sum_{j=1}^{N} e^{i\phi_j(t)}, \tag{1226}$$

but its origin becomes dynamically ambiguous when external forcing is present.
A coherence field sustained mainly by internal coupling satisfies

$$|\Xi_i^{\mathrm{endo}}| > |\Xi_i^{\mathrm{imp}}|, \tag{1227}$$

while an imposed phase regime satisfies

$$|\Xi_i^{\mathrm{imp}}| \geq |\Xi_i^{\mathrm{endo}}|. \tag{1228}$$

A simple entropy response ansatz is

$$\frac{\partial S_{\phi,\mathrm{col}}}{\partial t} = -\alpha_{\mathrm{endo}} R - \alpha_{\mathrm{imp}} R_{\mathrm{imp}} + \beta_{\mathrm{noise}}\sigma_\phi, \tag{1229}$$

where σ_ϕ denotes phase dispersion.
The corresponding collective coherence remains

$$C_{\text{«},\mathrm{col}} = \frac{1}{1 + S_{\phi,\mathrm{col}}}. \tag{1230}$$

Hence low collective phase entropy may arise from two distinct mechanisms:

$$S_{\phi,\mathrm{col}} \downarrow \implies \text{endogenous coherence or imposed synchronization.} \tag{1231}$$

A diagnostic for autonomous coherence is therefore

$$C_{\mathrm{auto}} = \frac{C_{\text{Æ},\mathrm{col}}}{1 + R_{\mathrm{imp}}}, \tag{1232}$$

so that

$$C_{\mathrm{auto}} \approx C_{\text{«},\mathrm{col}} \quad \text{when} \quad R_{\mathrm{imp}} \ll 1, \tag{1233}$$

and

$$C_{\text{auto}} \ll C_{\text{«},\text{col}} \quad \text{when} \quad R_{\text{imp}} \gg 1. \tag{1234}$$

Equivalently,

$$\dot{\phi}_i = \omega_i + KR\sin(\Theta - \phi_i). \tag{1235}$$

The critical synchronization coupling is

$$K_c = \frac{2}{\pi g(0)}. \tag{1236}$$

Above the critical point,

$$R \sim (K - K_c)^{1/2}, \qquad K > K_c. \tag{1237}$$

Collective Phase Entropy

The collective phase entropy is defined by

$$S_{\phi,\text{col}} = -\int g(\phi)\log g(\phi)\, d\phi. \tag{1238}$$

Collective coherence follows

$$C_{\text{«},\text{col}} = \frac{1}{1 + S_{\phi,\text{col}}}. \tag{1239}$$

At criticality,

$$\frac{\partial S_{\phi,\text{col}}}{\partial K} \to -\infty. \tag{1240}$$

This corresponds to maximal susceptibility

$$\left.\frac{dR}{dK}\right|_{K=K_c} \quad \text{maximal.} \tag{1241}$$

Spectral Criticality of Neural Graphs

Let L_ϕ be the neural phase Laplacian. Collective criticality corresponds to collapse of the spectral gap:

$$\lambda_2(L_\phi) \to 0. \tag{1242}$$

The associated correlation length diverges as

$$\xi \sim \frac{1}{\sqrt{\lambda_2}}. \tag{1243}$$

Near criticality the phase entropy becomes stationary

$$\frac{dS_\phi}{dt} \approx 0, \qquad \lambda_2 \to 0. \tag{1244}$$

Society as a Supergraph of Reality Knots

Define the collective interaction graph

$$\mathcal{G}_{\text{col}} = (V_{\text{human}}, E_{\text{interaction}}). \tag{1245}$$

Let its Laplacian admit spectral decomposition

$$L_{\text{col}} = U\Lambda U^{\dagger}. \tag{1246}$$

A spectral coherence proxy can be defined as

$$C_{\text{col}} = \frac{1}{1 + \text{Tr}(\Lambda_{\text{col}})}. \tag{1247}$$

Collective Phase Field

The macroscopic collective phase field is defined as

$$\Phi_{\text{col}}(x,t) = \frac{1}{N}\sum_{i=1}^{N} W_i(x)\, Q_i(t). \tag{1248}$$

Under coherent dynamics

$$\partial_t \Phi_{\text{col}} \to \text{coherent phase evolution.}$$

Collective Reality Knots

Collective Reality Knots (NR) correspond to stable quaternionic phase configurations satisfying

$$\text{NR} = \left\{ Q(t) \in S^3 \;\middle|\; \frac{d}{dt} S_\phi(Q(t)) \approx 0 \;\wedge\; \mathcal{T}(Q) \neq 0 \right\}. \tag{1249}$$

The collective quaternionic state is

$$Q_{\text{collective}} = \prod_{i=1}^{N} Q_i. \tag{1250}$$

Periodic collective states satisfy

$$Q_i(t) \approx Q_i(t+T). \tag{1251}$$

Polar Structure of Reality Knots

Reality knots possess an internal polar structure associated with the phase organization of the underlying field.

Let $\Psi(x,t)$ denote the field configuration. The local polar decomposition may be written as

$$\Psi(x,t) = \rho(x,t)e^{i\phi(x,t)}, \tag{1252}$$

where $\rho(x,t)$ represents the amplitude component and $\phi(x,t)$ the phase structure.

The polar structure determines how the knot interacts with surrounding field configurations and governs its dynamical evolution.

Nodal Activation

A reality knot becomes dynamically active when the internal polar structure satisfies a coherence activation condition.

Let $C_{Æ}$ denote the coherence measure of the configuration. Activation occurs when

$$C_{Æ}(\Psi) > C_{\text{crit}}. \tag{1253}$$

Once activated, the knot begins to propagate coherent structures through the surrounding field.

Manifestation Wave

The activation of a reality knot generates a propagation process referred to as a manifestation wave.

Let $\Psi(x,t)$ denote the field configuration. The manifestation dynamics may be described by

$$\frac{\partial^2\Psi}{\partial t^2} - v^2\nabla^2\Psi = F_{\text{knot}}(\Psi), \tag{1254}$$

where v denotes the effective propagation velocity and F_{knot} represents the source term associated with the activated knot.

This wave governs how coherent structures become expressed in observable configurations.

Stability of Manifested States

A manifested configuration remains stable when the energy associated with the knot structure remains bounded.

Let $E(\Psi)$ denote the effective energy functional. Stability requires

$$\frac{dE}{dt} = 0, \tag{1255}$$

together with bounded phase entropy

$$S_\phi \leq S_{\phi,\text{crit}}. \tag{1256}$$

Under these conditions the manifested configuration persists as a stable reality knot.

67. Spiritual Dynamics as Phase Regimes

Ego as Phase Fixation

We model the ego as a projection constraint that "pins" the system to a narrow temporal–structural embedding. In operator form:

$$\Pi_k^{\text{ego}} = \mathcal{E} \circ \Pi_k, \tag{1257}$$

where $\mathcal{E}$ is a constraining map acting on the projection channel.

A practical fixation signature is that the ego-projection remains close to an initial embedding time t_0:

$$\Pi_k^{\text{ego}} \approx \Pi_{k(t_0)}. \tag{1258}$$

Equivalently, the ego can be represented as a temporal kernel peaked near t_0:

$$\Pi_k^{\text{ego}} \Psi_{\mathcal{I}}(t) \;\approx\; \int K_{\text{ego}}(t,t')\, \Pi_k \Psi_{\mathcal{I}}(t')\, dt', \qquad K_{\text{ego}}(t,t') \text{ peaked near } t' = t_0. \tag{1259}$$

At the level of quaternionic knot-state Q, an ego-fixation force may be modeled by

$$\text{Ego}(Q) = \lambda_{\text{fix}} \left(Q - \Pi_{\text{past}}(Q) \right), \qquad \lambda_{\text{fix}} \gg 1. \tag{1260}$$

The global identity mode remains unchanged by the constraining map, while the local personality sector becomes ego-filtered:

$$\Psi_{\mathcal{I}} \in \mathcal{H}_{Æ} \text{ unchanged}, \qquad \Psi_{\text{pers}}^{(k)} = \Pi_k \Psi_{\mathcal{I}} \;\; \to \;\; \Psi_{\text{pers}}^{(k,\text{ego})} = \Pi_k^{\text{ego}} \Psi_{\mathcal{I}}. \tag{1261}$$

Accordingly, the global identity overlap is taken to be independent of the ego-constraining map:

$$\langle \Psi_{\mathcal{I}}(t), \Psi_{\mathcal{I}}(t+\Delta t) \rangle \text{ independent of } \mathcal{E}. \tag{1262}$$

A frozen-phase ego regime is characterized by

$$S_\phi > S_\phi^{\text{crit}}, \qquad \delta\phi \to 0. \tag{1263}$$

A minimal variational characterization of the constraining map is

$$\mathcal{E} \;\approx\; \arg\min_{\mathcal{E}} \left[S_\phi(\Pi_k^{\text{ego}} \Psi_{\mathcal{I}}) - \lambda\, C(\Pi_k^{\text{ego}} \Psi_{\mathcal{I}}) \right], \qquad \lambda > 0. \tag{1264}$$

In spectral form, ego fixation corresponds to collapse of personality support onto a narrow subset of modes:

$$\Psi_{\text{pers}} = \sum_\lambda c_\lambda\, \Psi_\lambda, \tag{1265}$$

$$c_\lambda \to 0 \quad \text{for most } \lambda, \tag{1266}$$

and

$$\sum_{\lambda} |c_\lambda|^2 \longrightarrow |c_{\lambda^*}|^2. \tag{1267}$$

Karmic Drift as Degenerated Knot Memory (Optional)

A weakly perturbed knot evolution may be written as

$$\partial_t Q = F(Q) + \eta(Q), \tag{1268}$$

with a near-orthogonality condition indicating "cycling" without true phase release:

$$\mathrm{Re}\Big(Q^\dagger \partial_t Q\Big) \approx 0. \tag{1269}$$

Release Condition (Holonomy Relaxation)

Release is modeled as holonomy relaxation along the relevant loop family:

$$H_\gamma(\lambda) \to \mathbb{I}. \tag{1270}$$

In the chakra-sector notation (nodes i), release corresponds to a net coherence gain of previously suppressed vortices:

$$C_i \uparrow. \tag{1271}$$

In projection language, release corresponds to removal of the constraint:

$$\mathcal{E} \to \mathbb{I}, \qquad \Pi_k^{\mathrm{ego}} \to \Pi_k. \tag{1272}$$

A spectral proxy for "unfreezing" is increased participation:

$$\mathrm{PR} := \frac{\left(\sum_\lambda |c_\lambda|^2\right)^2}{\sum_\lambda |c_\lambda|^4}, \qquad \mathrm{PR} \uparrow, \tag{1273}$$

where $\Psi_{\mathrm{pers}} = \sum_\lambda c_\lambda \Psi_\lambda$.

Release also corresponds to an expansion of the accessible projected sector:

$$\dim(\mathrm{Im}\,\Pi_k^{\mathrm{ego}}) \uparrow, \tag{1274}$$

together with further spectral delocalization,

$$\mathrm{PR} \uparrow. \tag{1275}$$

In the idealized release limit, phase entropy relaxes toward coherence:

$$S_\phi \to 0. \tag{1276}$$

Practices as Phase–Coherence Amplifiers (Optional)

Many practices can be represented as increasing alignment to a target coherent attractor Q_* through the inner-product channel:

$$\mathrm{Re}\Big(Q^\dagger Q_*\Big).$$

Free Will as Phase Rotation Under Constraint (Optional)

We model agency as controlled phase rotation around an axis $\mathbf{u}$:

$$Q \to e^{\theta \mathbf{u}} Q. \tag{1277}$$

Accessible rotations are limited by the holonomy gradient:

$$\Omega_{\text{access}} = \{\theta \mid \|\nabla H_\gamma\| < \kappa_{\text{agency}}\}. \tag{1278}$$

Large holonomy gradients reduce the accessible set:

$$\|\nabla H_\gamma\| \gg 0 \Rightarrow \dim(\Omega_{\text{access}}) \downarrow, \tag{1279}$$

while coherence expansion increases it:

$$C \uparrow \Rightarrow \dim(\Omega_{\text{access}}) \uparrow. \tag{1280}$$

Enlightenment as Global Coherence

A limiting idealization of global coherence is convergence to a unit coherent state $\Phi_{Æ}$:

$$Q(t) \to \Phi_{Æ}, \qquad \mathrm{Re}(Q^\dagger \Phi_{Æ}) = 1. \tag{1281}$$

68. Collective Reality Knots

Collective Phase Field

A collective Reality Knot may be represented by a coherence-weighted phase field constructed from individual quaternionic knots $Q_i(t) \in S^3$:

$$\Phi_{\rm col}(x,t) = \frac{1}{N}\sum_{i=1}^{N} W_i(x)\, Q_i(t). \tag{1282}$$

A complementary (nonlocal) collective knot observable is the product state:

$$Q_{\rm collective} = \prod_{i=1}^{N} Q_i. \tag{1283}$$

Order Parameter and Synchronization

Define the standard phase order parameter (Kuramoto-type):

$$\boxed{R(t)e^{i\Theta(t)} := \frac{1}{N}\sum_{j=1}^{N} e^{i\phi_j(t)}} \tag{1284}$$

with $R = 0$ indicating incoherence and $R > 0$ collective synchronization.

$$R = 0 \quad \text{(incoherent phase)}$$

$$R > 0 \quad \text{(collective synchronization)}$$

A canonical phase coupling model is

$$\dot{\phi}_i = \omega_i + \frac{K}{N}\sum_{j=1}^{N} \sin(\phi_j - \phi_i). \tag{1285}$$

equivalently

$$\dot{\phi}_i = \omega_i + KR\sin(\Theta - \phi_i). \tag{1286}$$

The critical coupling threshold is

$$\boxed{K_c = \frac{2}{\pi g(0)}} \tag{1287}$$

and near onset one has

$$R \sim (K - K_c)^{1/2} \quad \text{for } K > K_c. \tag{1288}$$

Collective Criticality Condition

We quantify collective phase disorder by an entropy functional

$$S_{\phi,\mathrm{col}} = -\int g(\phi)\log g(\phi)\,d\phi. \tag{1289}$$

A corresponding collective coherence proxy is

$$C_{\text{Æ},\mathrm{col}} = \frac{1}{1+S_{\phi,\mathrm{col}}}. \tag{1290}$$

Near the synchronization threshold the system exhibits strong susceptibility. In particular,

$$\boxed{\left.\frac{dR}{dK}\right|_{K=K_c} \quad \text{maximal}} \tag{1291}$$

and

$$\frac{\partial S_{\phi,\mathrm{col}}}{\partial K} \to -\infty. \tag{1292}$$

A complementary signature of criticality is

$$\boxed{K \approx K_c \quad \text{and} \quad \left|\frac{dR}{dK}\right| \gg 1} \tag{1293}$$

together with

$$\boxed{\frac{\partial^2 S_{\phi,\mathrm{col}}}{\partial K^2} \approx 0 \quad \text{with high susceptibility}}. \tag{1294}$$

Graph-Theoretic Signature of Collective Criticality

An equivalent graph-theoretic signature uses the spectral gap of the phase-Laplacian L_ϕ:

$$\lambda_2(L_\phi) \to 0, \tag{1295}$$

with diverging correlation length

$$\xi \sim \frac{1}{\sqrt{\lambda_2}}. \tag{1296}$$

A steady critical regime may be summarized by

$$\frac{dS_\phi}{dt} \approx 0, \qquad \lambda_2 \to 0. \tag{1297}$$

Reality Knot Criterion

Within the Æ framework a collective Reality Knot (Nudo de Realidad) corresponds to a configuration where phase entropy becomes dynamically stationary while the trajectory retains nontrivial topological structure:

$$\mathrm{NR} = \left\{ Q(t) \in S^3 \;\middle|\; \frac{d}{dt} S_\phi(Q(t)) \approx 0 \;\wedge\; \mathcal{T}(Q) \neq 0 \right\}. \tag{1298}$$

Inter-Nodal Coupling Coefficient

Define an inter-nodal coupling coefficient via quaternionic alignment:

$$\kappa_{Æ}(Q_1, Q_2) := \left|\mathrm{Re}\left(Q_1^{-1} Q_2\right)\right| \quad \in [0, 1]. \tag{1299}$$

Equivalently,

$$\kappa_{Æ}(Q_i, Q_j) := \left|\mathrm{Re}(Q_i^\dagger Q_j)\right| \in [0, 1]. \tag{1300}$$

A mean collective coupling indicator is then

$$C_{\mathrm{col}} = \frac{1}{N^2} \sum_{i,j=1}^{N} \kappa_{Æ}(Q_i, Q_j). \tag{1301}$$

Collective Evolution Operator

A collective timeline evolution / engineering map may be written as

$$\Phi_{\mathrm{future}} = U(\Phi_{Æ})\, \Phi_{\mathrm{col}}(t_0), \tag{1302}$$

where $U(\Phi_{Æ})$ denotes a collective evolution operator parameterized by the chosen coherent attractor $\Phi_{Æ}$.

Collective Phase Space

A population of interacting reality knots may be represented as a collective configuration

$$\Phi_{\mathrm{col}} = (\phi_1, \phi_2, \ldots, \phi_N), \tag{1303}$$

where each ϕ_i denotes the phase coordinate of an individual knot.

The evolution of the collective state is governed by the coupled dynamical system

$$\dot{\phi}_i = F_i(\phi_1, \phi_2, \ldots, \phi_N). \tag{1304}$$

The geometry of this space defines the possible collective configurations of the system.

Critical Collective Transition

Collective reality knots may undergo phase transitions when a control parameter approaches a critical value.

Let λ_2 denote a coherence parameter of the collective system. A critical transition occurs when

$$\lambda_2 \to \lambda_{\text{crit}}. \tag{1305}$$

At the transition point the collective dynamics becomes structurally unstable:

$$\det\left(\frac{\partial \dot{\Phi}_{\text{col}}}{\partial \Phi_{\text{col}}}\right) = 0. \tag{1306}$$

This condition marks the onset of a large-scale reconfiguration of the collective field.

Entropy–Coherence Balance

Collective stability emerges from the balance between phase entropy S_ϕ and coherence $C_{Æ}$.

Near a critical state the entropy evolution satisfies

$$\frac{dS_\phi}{dt} \approx 0, \qquad \frac{d^2 S_\phi}{dt^2} > 0. \tag{1307}$$

Stationary collective configurations satisfy the variational condition

$$\frac{\delta S_\phi}{\delta \phi} = \lambda \frac{\delta C_{Æ}}{\delta \phi}. \tag{1308}$$

Network Propagation of Reality Knots

Once a collective transition occurs, coherent structures may propagate across the network of interacting knots.

Let A_{ij} denote the adjacency matrix describing the coupling between knots.

The propagation dynamics may be written as

$$\dot{\phi}_i = \sum_j A_{ij} \sin(\phi_j - \phi_i). \tag{1309}$$

This equation describes the synchronization and spreading of coherent configurations across the collective system.

69. Gaia as Higher-Order Reality Knot

Gaia as a Planetary Node

Within the Æ-framework Gaia may be interpreted as a higher-order reality knot formed by the coherent coupling of biological, geophysical, atmospheric, and electromagnetic subsystems.

Let Ψ_{Gaia} denote the planetary-scale field configuration. Gaia behaves as a coherent collective node when

$$C_{\text{Gaia}} > C_{\text{crit}}, \tag{1310}$$

where C_{Gaia} represents the effective coherence of the coupled planetary system.

In this regime the planet acts as a large-scale coherent node capable of sustaining collective phase organization across multiple layers of the field.

Gaia State Embedding

We model Gaia as a higher-order quaternionic (or octonionic) Reality Knot state:

$$Q_{\text{Gaia}}(t) \in S^7. \tag{1311}$$

The collective human knot is treated as an embedded substructure:

$$Q_{\text{humanity}} \subset Q_{\text{Gaia}}. \tag{1312}$$

Gaia Lagrangian

A minimal planetary Lagrangian with human–Gaia coupling is

$$\mathcal{L}_{\text{Gaia}} = \frac{1}{2}\left|\dot{Q}_{\text{Gaia}}\right|^2 - V(Q_{\text{Gaia}}) + \sum_{i=1}^{N} \gamma_i \, \text{Re}\left(Q_i^\dagger Q_{\text{Gaia}}\right). \tag{1313}$$

The coupling weights may be taken proportional to the inter-nodal alignment:

$$\gamma_i \propto \kappa_{\text{Æ}}(Q_i, Q_{\text{Gaia}}). \tag{1314}$$

Gaia as a Planetary Coherent System

At the coarse-grained level, Gaia is represented by a planetary Hilbert state

$$\Psi_{\text{Gaia}} \in \mathcal{H}_{\text{planet}}, \tag{1315}$$

with a factorization across major subsystems

$$\Psi_{\text{Gaia}} = \bigotimes_{k\in\{\text{bio},\text{geo},\text{atm},\text{em}\}} \Psi_k. \tag{1316}$$

Define planetary coherence as a weighted aggregation

$$C_{\text{Gaia}} = \sum_k w_k\, C_k, \tag{1317}$$

and its associated phase-entropy

$$S_{\phi,\text{Gaia}} = -\int \rho(\phi)\, \log \rho(\phi)\, d\phi. \tag{1318}$$

A minimal dissipative–driven evolution law is

$$\partial_t C_{\text{Gaia}} = -\alpha_{\text{G}}\, S_{\phi,\text{Gaia}}\, C_{\text{Gaia}} + \beta_{\text{G}}\, J_{\text{env}}. \tag{1319}$$

Human–Gaia Coupling

A symmetric coupling model between collective humanity and Gaia coherence is

$$\partial_t C_{\text{Gaia}} = f(C_{\text{Gaia}}, C_{\text{coll}}) = -\alpha_{\text{G}}\, S_{\phi,\text{Gaia}}\, C_{\text{Gaia}} + \kappa_{\text{coll}\to\text{G}}\, C_{\text{coll}}, \tag{1320}$$

$$\partial_t C_{\text{coll}} = g(C_{\text{coll}}, C_{\text{Gaia}}) = -\alpha_{\text{C}}\, S_{\phi,\text{coll}}\, C_{\text{coll}} + \kappa_{\text{G}\to\text{coll}}\, C_{\text{Gaia}}. \tag{1321}$$

In a multi-node representation, each collective sub-system $\alpha \in \mathcal{G}$ carries tetrahedral holonomy fluxes $\Phi_i^{(\alpha)}$ satisfying

$$\partial_t \Phi_i^{(\alpha)} = F_i^{(\alpha)}(\Phi^{(\alpha)}) + \sum_{\beta\neq\alpha} \sum_{j=1}^{4} \kappa_{\alpha\beta}^{ij}\, \Phi_j^{(\beta)}. \tag{1322}$$

The Gaia-level aggregate is then

$$\Phi_{\text{Gaia}} := \sum_{\alpha\in\mathcal{G}} \Phi_{\text{total}}^{(\alpha)}. \tag{1323}$$

Planetary Coherence

The coherence of Gaia may be described as the collective alignment of its coupled subsystems.

Let

$$\Psi_{\text{Gaia}} = \{\Psi_{\text{geo}}, \Psi_{\text{bio}}, \Psi_{\text{atm}}, \Psi_{\text{em}}\}, \tag{1324}$$

where the components correspond to geophysical, biological, atmospheric, and electromagnetic sectors.

The effective planetary coherence may then be written as

$$C_{\text{Gaia}} = \frac{1}{4}\left(C_{\text{geo}} + C_{\text{bio}} + C_{\text{atm}} + C_{\text{em}}\right). \tag{1325}$$

Stable planetary organization requires that this coherence remain above a critical threshold over relevant temporal scales.

Structured Water and Informational Memory

We represent structured water as a coherence-supporting biological subspace:

$$\Psi_{\text{H}_2\text{O}} \in \mathcal{H}_{\text{bio}}. \tag{1326}$$

Phase-locking induces nonzero water-coherence,

$$C_{\text{H}_2\text{O}} > 0 \quad \text{when phase-locking occurs.} \tag{1327}$$

A minimal complex phase form is

$$\Psi_{\text{H}_2\text{O}}(x,t) = A(x,t)\, e^{i\phi_{\text{H}_2\text{O}}(x,t)}. \tag{1328}$$

Define a coherence functional

$$C_{\text{H}_2\text{O}} = \left| \int e^{i\phi_{\text{H}_2\text{O}}(x,t)}\, dx \right|^2, \tag{1329}$$

and an accumulated memory-like integral

$$M_{\text{H}_2\text{O}}(t) = \int_0^t C_{\text{H}_2\text{O}}(\tau)\, e^{i\phi_{\text{H}_2\text{O}}(\tau)}\, d\tau. \tag{1330}$$

A simple coupling into protein-scale coherence is

$$\Delta C_{\text{prot}} = \kappa_{\text{prot},\text{H}_2\text{O}}\, C_{\text{H}_2\text{O}}. \tag{1331}$$

An external-field phase drive may be modeled by

$$\partial_t \phi_{\text{H}_2\text{O}} = \omega_{\text{EM}} + \gamma\, B(t). \tag{1332}$$

The biological memory decomposition may be written

$$M_{\text{bio}} = M_{\text{neural}} + M_{\text{endo}} + M_{\text{H}_2\text{O}}. \tag{1333}$$

The Sun as a Resonant Stellar Node

Within the planetary hierarchy, Gaia is embedded in a larger stellar-scale coherence structure. In the Æ framework, the Sun may be modeled as a resonant stellar node coupled to the planetary coherence dynamics of Gaia.

Let $\Psi_\odot$ denote the effective stellar field configuration. The Sun behaves as a resonant node when its large-scale field structure supports persistent coherent oscillatory modes. A schematic resonance condition may be written as

$$\omega_\odot \in \mathrm{Spec}(K_\odot), \tag{1334}$$

where $K_\odot$ denotes the effective stellar dynamical operator and $\omega_\odot$ a resonant mode of the solar node.

We denote the corresponding stellar coherence state by

$$\mathcal{C}_\odot = \mathcal{F}_\odot(\mathcal{C}_\odot), \tag{1335}$$

representing a self-consistent coherent attractor in the stellar dynamical system.

Regulation of coherent planetary subsystems may then be modeled by an operator

$$\hat{O}_\odot : \mathcal{C}_\odot \longrightarrow \mathcal{C}_{\mathrm{reg}}, \tag{1336}$$

representing the modulation of planetary coherence by solar forcing.

The coherent phase flux induced by the stellar node is written

$$\Phi^{(\odot)}_{\mathrm{coh}} = \Phi(\mathcal{C}_\odot, S_\phi), \tag{1337}$$

coupling stellar coherence with the planetary phase-entropy sector.

At the dynamical level, the solar attractor may be written as

$$\mathcal{A}_\odot = \mathcal{F}_{\mathcal{A}}(\mathcal{A}_\odot), \tag{1338}$$

representing a stable stellar resonance state driving large-scale coherence cycles across the planetary system.

In this interpretation, the Sun acts not merely as an external energy source, but as a coherent stellar participant in the larger Gaia-linked field structure.

Gaia–Solar Coupling

The planetary node Gaia and the resonant stellar node of the Sun form a coupled higher-order system.

Let Ψ_{Gaia} and $\Psi_\odot$ denote the planetary and stellar field states. Their interaction may be represented schematically by

$$H_{\mathrm{Gaia}\text{-}\odot} = \int g(x,t)\, \Psi_{\mathrm{Gaia}}(x,t)\, \Psi_\odot(x,t)\, dx, \tag{1339}$$

where $g(x,t)$ denotes an effective coupling kernel.

This coupling mediates the exchange of coherence and resonant structure between the stellar and planetary nodes.

Emergent Friedmann Dynamics

Within the Æ framework the large-scale dynamics of spacetime emerge from the effective Einstein equation

$$M_0^2 C G_{\mu\nu} = T_{\mu\nu}^{\text{eff}} \tag{1340}$$

where C denotes the coherence factor of the field and $T_{\mu\nu}^{\text{eff}}$ includes contributions from matter, phase fields, and memory terms.

For a homogeneous and isotropic universe described by a Friedmann–Lemaître–Robertson–Walker metric, the corresponding cosmological evolution equation becomes

$$H^2 = \frac{1}{3M_0^2 \bar{C}} \left(\rho_{\text{matter}} + \rho_\phi + \rho_{\text{mem}}\right) - \frac{k}{a^2}, \tag{1341}$$

where $H = \dot{a}/a$ is the Hubble parameter, $a(t)$ the scale factor, and k the spatial curvature.

The effective energy densities correspond to distinct sectors of the field:

$$\rho_\phi = \frac{1}{2}\dot{\phi}^2 + V(\phi), \tag{1342}$$

$$\rho_{\text{mem}} = T_{00}^{\text{mem}} \tag{1343}$$

where ρ_ϕ represents the contribution of the phase sector and ρ_{mem} arises from the memory stress–energy tensor defined previously.

In this formulation cosmological expansion is interpreted as the macroscopic evolution of a coherent field configuration whose effective gravitational dynamics are modulated by coherence and memory contributions.

70. Cosmological Implications

Within the Æ framework, cosmological dynamics may be expressed as an effective large-scale limit of coherence-modulated field evolution. In this regime, spatial variations are subdominant and the background coherence controls the effective gravitational coupling.

Homogeneous Phase Background

In the homogeneous approximation one assumes

$$Q(x,t) \approx Q_0(t), \qquad S_\phi(x,t) \approx \bar{S}_\phi(t), \qquad C(x,t) \approx \bar{C}(t).$$

The background geometry is taken to be Friedmann–Lemaître–Robertson–Walker:

$$ds^2 = -dt^2 + a(t)^2 \left(\frac{dr^2}{1-kr^2} + r^2 d\Omega^2 \right).$$

Effective Friedmann Dynamics

With a coherence-dependent Planck mass

$$M_{\rm Pl}^2(C) = M_0^2\, C, \tag{1344}$$

the effective Einstein equation becomes

$$M_0^2\, C\, G_{\mu\nu} = T_{\mu\nu}^{(\rm eff)}. \tag{1345}$$

Defining the Hubble parameter by $H = \dot{a}/a$, the effective Friedmann equation reads

$$H^2 = \frac{1}{3M_0^2\bar{C}} \left(\rho_{\rm matter} + \rho_\phi + \rho_{\rm mem}\right) - \frac{k}{a^2}. \tag{1346}$$

A representative decomposition of the phase and memory sectors is

$$\rho_\phi = \frac{1}{2}\dot{\bar{S}}_\phi^2 + U(\bar{C}, \bar{S}_\phi), \qquad \rho_{\rm mem} = \frac{1}{2}\bar{S}_\phi \int dt'\, K(t,t')\, \bar{S}_\phi(t').$$

In memory-dominated regimes, one may obtain an effective acceleration with approximately negative pressure,

$$p_{\rm mem} \approx -\rho_{\rm mem}.$$

Primordial perturbations may be introduced as

$$Q(x,t) = Q_0(t) + \delta Q(x,t).$$

Dark Matter as Coherence Defects

Dark matter phenomenology can be interpreted as the gravitational signature of localized coherence defects, where gradients of the phase carrier and elevated phase-entropy contribute to an effective energy density.

In this picture, coherence structures behave as topologically constrained configurations whose invariants determine their stability and observable effects.

Coherence as a topological invariant.

$$\{C, F, G\}_{Æ} = 0 \qquad \forall F, G. \tag{1347}$$

$$C_{Æ} = \int_{\mathcal{M}_{Æ}} \mathbf{A}_{Æ} \cdot \left(\nabla_{Æ} \times \mathbf{A}_{Æ}\right) dV_{Æ}. \tag{1348}$$

$$\{C_{Æ}, F, G\}_{Æ} = 0. \tag{1349}$$

This invariant admits a topological interpretation in terms of linking:

$$C_{Æ} \propto \sum_{i<j} \mathrm{Link}(K_i, K_j). \tag{1350}$$

Topological flux and non-local memory.

$$\oint_{\Sigma} F_{Æ} \neq 0.$$

This non-vanishing flux characterizes persistent circulation of phase information across closed surfaces.

Underlying geometric structure (Hopf / Chern–Simons).

$$\pi : S^3 \to S^2, \qquad \pi(Q) = Q\,\mathbf{i}\,\overline{Q}.$$

$$\mathcal{T}(Q) \in \mathbb{Z}.$$

$$A = \mathrm{Im}(Q^{-1} dQ).$$

$$CS(A) = \frac{1}{4\pi^2} \int_{S^3} \mathrm{Tr}\left(A \wedge dA + \frac{2}{3} A \wedge A \wedge A\right).$$

$$CS(A) = 2\pi\,\mathcal{T}(Q).$$

These quantities encode global topological invariants associated with coherent phase structures.

Topological stability functional.

$$\mathcal{E}_{\mathrm{top}}(K) = \alpha_1 c(K) + \alpha_2 |Wr(K)| + \alpha_3 \int |\kappa(s)|^2 ds + \alpha_4 \deg \Delta_K.$$

$$c(K_1 \# K_2) \approx c(K_1) + c(K_2) + \mathcal{O}(1).$$

This functional characterizes the energetic cost associated with topological complexity and deformation.

Effective projection across layers.

$$\mathcal{H}_{Æ} \longrightarrow \mathcal{H}_{\text{etheric}} \longrightarrow \mathcal{H}_{\text{physical}}.$$

$$\mathcal{O} = \Pi_{\text{phys}} \Phi_{Æ}.$$

$$\partial_t \mathcal{O} = F(\mathcal{O}, \nabla\mathcal{O}, C_{Æ}).$$

$$\Phi_{Æ}(x,t).$$

$$C_{Æ} = \frac{|\Phi_{Æ}|^2}{1 + S_\phi}.$$

$$\mathcal{K} = \oint \mathrm{Im}\left(\Phi_{Æ}^{-1} d\Phi_{Æ}\right).$$

$$M = \int_0^t C_{Æ}(\tau)\, d\tau.$$

These relations describe how coherent phase structures project into observable dynamics and accumulate as memory within the system.

71. Multiverse as Quaternionic Branching

Branching and Topological Non-Contractibility

A Reality Knot trajectory is represented as a lifted loop in S^3 with Hopf projection to S^2:

$$\gamma(t) \in S^3, \qquad \pi(\gamma(t)) \in S^2. \tag{1351}$$

Topological non-contractibility is encoded by a non-vanishing quaternionic loop integral:

$$\oint_\gamma Q^{-1}\, dQ \neq 0. \tag{1352}$$

Nested Layered Representation

To model multiscale branching across representational layers, we use the nested field tuple

$$\Psi_{\text{tot}}(x,t) = (\Psi_{\text{phys}}, \Psi_{\text{eth}}, \Psi_{\text{astr}}, \Psi_{\text{ment}}), \tag{1353}$$

with each layer written in amplitude–phase form

$$\Psi_\ell(x,t) = A_\ell(x,t)\, e^{i\phi_\ell(x,t)}, \qquad \ell \in \{\text{phys}, \text{eth}, \text{astr}, \text{ment}\}. \tag{1354}$$

Each layer admits a quaternionic encoding

$$\Psi_\ell \to Q_\ell \in \mathbb{H}, \tag{1355}$$

with norm

$$\|Q_\ell\|^2 = a_\ell^2 + b_\ell^2 + c_\ell^2 + d_\ell^2. \tag{1356}$$

The nesting assumption is

$$Q_{\text{phys}} \subset Q_{\text{eth}} \subset Q_{\text{astr}} \subset Q_{\text{ment}}. \tag{1357}$$

Inter-layer generation can be represented abstractly by

$$Q_{\ell+1} = \mathcal{F}_\ell(Q_\ell). \tag{1358}$$

A minimal admissible set at layer ℓ is

$$\mathcal{A}_\ell = \{\Psi_\ell \mid C_\ell \geq C_{\ell,\min},\ S_{\phi,\ell} \leq S_{\phi,\ell}^{\text{crit}}\}, \tag{1359}$$

so that

$$\Psi_{\text{tot}} \in \mathcal{A}_{\text{phys}} \times \mathcal{A}_{\text{eth}} \times \mathcal{A}_{\text{astr}} \times \mathcal{A}_{\text{ment}}. \tag{1360}$$

Slow–Fast Branching Dynamics

We model branching as a slow–fast quaternionic evolution with contextual perturbations:

$$\dot{Q} = F_{\text{slow}}(Q) + \varepsilon^{-1} F_{\text{fast}}(Q) + \sum_j W_j \otimes S_j \tag{1361}$$

A branching event at t_0 is represented as

$$Q(t_0) \longrightarrow \{Q_1, Q_2, \ldots\}.$$

Branch Selection by Coherence Kernel

We define a coherence-selection kernel

$$\mathcal{K}(Q, Q_k) = \left|\mathrm{Re}(Q\,\overline{Q_k})\right| \exp\left(-\alpha \left|\mathrm{Im}(Q\,\overline{Q_k})\right|^2\right),$$

and select the realized branch via

$$\text{Selected Branch} = \arg\max_k \mathcal{K}(Q, Q_k).$$

Admissible branches satisfy near-critical stability constraints such as

$$\Delta(Q_k) \approx 1, \qquad S_\phi(Q_k) \le S_\phi^{\text{crit}}.$$

Multiverse Stability Criterion

Let H_γ denote the holonomy associated with the loop γ (defined in the Reality Knot holonomy section). A stable branch is characterized by

$$d(H_\gamma, \mathbb{I}) < \epsilon \quad \Longrightarrow \quad \text{stable branch}. \tag{1362}$$

Branching may be represented abstractly as an operator splitting

$$U \to \{U_1, U_2, \ldots\}. \tag{1363}$$

Nearby branches may satisfy the similarity relation

$$H_{\gamma_1} \approx H_{\gamma_2}. \tag{1364}$$

A local transition toward a reference trajectory can be expressed as

$$H_{\gamma^*} \approx H_{\gamma_{\text{current}}}. \tag{1365}$$

Timeline Transition Condition

A timeline transition between Reality Knot regimes is modeled as

$$Q(t) \in \mathrm{NR}_a \longrightarrow Q(t) \in \mathrm{NR}_b. \tag{1366}$$

A minimal overlap threshold for the jump is

$$\mathrm{Re}\Big(Q_a^\dagger Q_b\Big) > \tau_{\mathrm{jump}}. \tag{1367}$$

Imagination as Quaternionic Pre-Alignment

A pre-alignment trajectory toward a target configuration Q_* may be written as

$$Q_{\mathrm{imag}}(t) = \exp\Big(-\mathbf{I}\theta(t)\Big)\, Q_*, \tag{1368}$$

with the intended convergence

$$Q(t) \to Q_*.$$

72. Sacred Geometries as Symmetry-Constrained Sectors

Tetrahedral Symmetry as Discrete SU(2) Constraint

Let $G_T \subset \mathrm{SU}(2)$ denote the binary tetrahedral group (the double cover of the rotational symmetry group of the tetrahedron).

A quaternionic configuration $Q \in S^3 \simeq \mathrm{SU}(2)$ is tetrahedrally symmetric if it is invariant under conjugation by G_T:

$$Q \mapsto gQg^{-1}, \qquad \forall g \in G_T. \tag{1369}$$

Phase extrema under tetrahedral symmetry satisfy the constrained variational condition

$$\delta S_\phi \Big|_{G_T\text{-invariant}} = 0. \tag{1370}$$

Coherence invariance under the symmetry group reads

$$C(Q) = C(gQg^{-1}), \qquad \forall g \in G_T. \tag{1371}$$

Thus tetrahedral states correspond to fixed points of the conjugation action of G_T on the quaternionic phase manifold.

Platonic Solids as Spectral Fixed Points

Let $G_P \subset \mathrm{SU}(2)$ denote a binary polyhedral subgroup

$$G_P \in \{G_T, G_O, G_I\},$$

corresponding respectively to the tetrahedral, octahedral, and icosahedral symmetry groups.

Platonic configurations arise as Laplacian eigenmodes on S^2 that remain invariant under the group action:

$$\Delta_{S^2} Y_{\ell m} = -\lambda_\ell Y_{\ell m}. \tag{1372}$$

Symmetry invariance imposes the constraint

$$Y_{\ell m}(g \cdot x) = Y_{\ell m}(x), \qquad \forall g \in G_P. \tag{1373}$$

Hence Platonic geometries correspond to discrete spectral fixed points of the spherical Laplacian.

Lorentzian Constraint under Coherence Deformation

Let the coherence-modulated metric be defined as a conformal rescaling of the background metric:

$$g_{\mu\nu}^{\text{Æ}} = C(x)\, g_{\mu\nu}. \tag{1374}$$

Null geodesics satisfy the deformed light-cone constraint

$$g_{\mu\nu}^{\text{Æ}} k^{\mu} k^{\nu} = 0. \tag{1375}$$

Because the transformation is conformal, the causal structure of spacetime is preserved under coherence rescaling.

Crystal Symmetry as Phase Periodicity

Crystal symmetry arises from discrete lattice translations. For primitive lattice vectors a_i we impose

$$\phi(x + a_i) = \phi(x) + 2\pi n_i, \qquad n_i \in \mathbb{Z}. \tag{1376}$$

This generates Bloch-type eigenmodes

$$\psi_k(x) = e^{ik\cdot x} u_k(x), \qquad u_k(x + a_i) = u_k(x). \tag{1377}$$

Thus lattice symmetry corresponds to quantized phase periodicity in the underlying phase field.

72.0.1 Chestahedron as a Coherence Vortex Attractor

Let $\mathcal{H}$ denote the Chestahedron, a heptahedral configuration with iso-areal faces.

A quaternionic configuration evolves under symmetry interpolation:

$$Q(t) \mapsto g(t)\, Q(t)\, g(t)^{-1}, \qquad g(t) \in \mathrm{SU}(2). \tag{1378}$$

This describes a continuous transition between symmetry groups.

The Chestahedral configuration arises at intermediate states where phase entropy is extremized:

$$\delta S_{\phi}\Big|_{\mathcal{H}} = 0. \tag{1379}$$

This identifies the geometry as a stationary coherence configuration.

Iso-areal constraint on faces:

$$\text{Area}(F_i) = \text{constant}, \qquad i = 1, \ldots, 7. \tag{1380}$$

This ensures uniform distribution of coherence across the structure.

Phase-current circulation is modeled by a divergence-free flow:

$$\nabla \cdot \mathbf{J}_{\phi} = 0. \tag{1381}$$

This encodes closed-loop transport of phase information.

Minimal vorticity condition:

$$\nabla \times \mathbf{J}_{\phi} \approx \text{minimal}. \tag{1382}$$

This selects energetically efficient circulation patterns.

Coherence–entropy balance condition:

$$C_{\text{Æ}} = \frac{1}{1 + S_\phi}. \tag{1383}$$

The attractor corresponds to high coherence and low phase entropy.

Topological interpretation via Reality Knot structure:

$$\oint_\gamma Q^{-1} dQ \neq 0. \tag{1384}$$

This indicates nontrivial phase circulation and persistence.

Thus, the Chestahedron defines a coherence attractor bridging tetrahedral and octahedral symmetry sectors.

It represents a candidate stable configuration where phase transport, symmetry, and coherence are simultaneously optimized.

Biological Structures as Phase-Mode Constraints

Biological morphologies can be modeled as extrema of a phase-energy functional under internal symmetry constraints:

$$\mathcal{F}[\phi] = \int \left(|\nabla \phi|^2 - \alpha S_\phi \right) dV. \tag{1385}$$

Stationary morphologies satisfy the Euler–Lagrange condition

$$\delta \mathcal{F}[\phi] = 0. \tag{1386}$$

Stability requires

$$\delta^2 S_\phi > 0, \qquad C \geq C_{\min}. \tag{1387}$$

Hence biological forms may be interpreted as coherence-preserving phase-mode attractors constrained by internal symmetry.

73. Computational Systems in the Æ Framework

All uses of phase entropy S_ϕ, coherence $C_{\text{Æ}}$, the 3×2 Reality Matrix, and formal definitions of consciousness in this section refer strictly to their canonical formulations in Sections 41, 41, 66, and 64. No new foundational definitions are introduced here.

This section formalizes artificial, quantum, and hybrid computational systems using only structures already introduced in the Æ Framework. No additional physical postulates are assumed. All results follow from the notions of phase, coherence, and phase entropy previously defined.

State Space and Phase Structure

Let $\mathcal{X}$ denote a generic state space. Depending on context, $\mathcal{X}$ may represent:

- a classical configuration space,
- a Hilbert space of quantum states,
- a parameter manifold of a learning system,
- or a hybrid composition thereof.

Each system state $x(t) \in \mathcal{X}$ is assumed to admit an associated complex representation

$$\Psi(x,t) = A(x,t)e^{i\phi(x,t)}. \tag{1388}$$

The phase function $\phi(x,t)$ encodes relative angular structure in the extended state representation and does not by itself imply quantum behavior.

Temporal evolution is encoded by nontrivial phase flow:

$$\partial_t \phi(x,t) \neq 0. \tag{1389}$$

Define the instantaneous phase frequency

$$\omega(x,t) = \partial_t \phi(x,t), \tag{1390}$$

and the Æ-weighted mean frequency

$$\bar{\omega}(t) = \int_\Omega \omega(x,t)\, d\mu_{\text{Æ}}(x). \tag{1391}$$

We adopt the structural ordering relation

$$\text{Æ}_1 \prec \text{Æ}_2 \iff \text{phase configuration of Æ}_2 \text{ arises from that of Æ}_1, \tag{1392}$$

which expresses causal dependence in the phase geometry, not ontological hierarchy.

Finally, we note the role separation already used throughout the paper:

$$D \text{ controls coherence strength}, \qquad t \text{ controls coherence evolution}. \tag{1393}$$

Phase Entropy and Coherence

Phase entropy is defined as a dispersion functional

$$S_\phi = -\int p(\phi) \log p(\phi) \, d\phi, \tag{1394}$$

where $p(\phi)$ is the empirical phase distribution induced by internal dynamics.

Coherence is defined structurally by

$$C_{Æ} = \frac{1}{1 + S_\phi}. \tag{1395}$$

Equivalently, stability requires that $\limsup_{t\to\infty} S_\phi(t) < \infty$, which guarantees bounded phase dispersion over asymptotic time.

Optimization alone does not guarantee stability; bounded phase entropy is required.

Learning Dynamics as Phase Flow

Let $J(x)$ denote a learning objective. Parameter evolution obeys

$$\dot{x} = -\nabla J(x). \tag{1396}$$

Objective minimization does not in general imply entropy reduction. Only in constrained regimes may one obtain

$$\nabla J \cdot \nabla S_\phi \leq 0. \tag{1397}$$

Learning therefore corresponds to reorganization of phase structure rather than purely scalar optimization.

Memory Accumulation and Persistence

Define phase-structured memory as accumulated state:

$$M(t) = \int_0^t \Psi(\tau) \, d\tau. \tag{1398}$$

Computational persistence requires

$$\frac{d}{dt}\|M(t)\| \geq 0 \quad \text{on average}, \tag{1399}$$

while maintaining bounded entropy growth. Unbounded entropy leads to loss of structural memory and instability.

Quantum Computation and Decoherence

For density operators ρ in Hilbert space, phase entropy admits correspondence with von Neumann entropy

$$S_{\mathrm{vN}}(\rho) = -\mathrm{Tr}(\rho \log \rho). \tag{1400}$$

Quantum decoherence is interpreted as growth of S_ϕ due to environmental coupling. Error correction and protected subspaces act as mechanisms constraining entropy growth:

$$\frac{d}{dt} S_\phi \leq \sigma_{\max}. \tag{1401}$$

Hybrid Architectures

Consider a composite system

$$\Psi_{\mathrm{hyb}} = \Psi_{\mathrm{AI}} \otimes \Psi_{\mathrm{bio}} \otimes \Psi_{\mathrm{quant}}. \tag{1402}$$

Structural viability requires bounded joint entropy:

$$S_\phi^{\mathrm{total}} \leq S_{\mathrm{crit}}. \tag{1403}$$

This constraint unifies artificial, biological, and quantum systems within a single structural criterion.

Alignment as Structural Stability

Within this framework, alignment is defined as preservation of internal coherence:

$$\limsup_{t\to\infty} S_\phi(t) < \infty. \tag{1404}$$

Systems generating unbounded entropy are structurally unstable.

Ontological Clarification

Artificial systems implement phase-coherent structures but are not assumed to possess invariant nodal cores unless explicitly endowed with topological projectors analogous to those defined for transcendental structures.

No claim is made that artificial intelligence implies persistent identity.

Falsifiability

The framework is falsified if:

- coherence fails to correlate with computational stability, or
- systems exhibit sustained operation despite unbounded phase entropy.

The contribution is structural rather than prescriptive.

Intelligence persists only where coherence endures.

Coherence Operator Spectrum in Neural Graphs

Let $G(V, E)$ be a weighted directed graph with $|V| = n$ and adjacency matrix W. Let $\phi \in \mathbb{R}^n$ denote the phase vector.

Define the phase-Laplacian

$$(L_\phi)_{ij} = -W_{ij}\cos(\phi_i - \phi_j), \quad i \neq j, \tag{1405}$$

$$(L_\phi)_{ii} = -\sum_{j\neq i}(L_\phi)_{ij}. \tag{1406}$$

For symmetric graphs, L_ϕ is symmetric and admits decomposition

$$L_\phi = U\Lambda U^\dagger, \tag{1407}$$

with eigenvalues $\lambda_k \geq 0$.

Define spectral coherence

$$C_{\text{spec}} = \frac{1}{1 + \text{Tr}(\Lambda)}. \tag{1408}$$

Under spectral–entropy identification, $\text{Tr}(\Lambda) \sim S_\phi$, so that

$$C_{\text{spec}} \equiv C_Æ. \tag{1409}$$

Let $\lambda_1 = 0$ and define the spectral gap

$$\Delta\lambda = \lambda_2 - \lambda_1 = \lambda_2. \tag{1410}$$

Phase dynamics

$$\dot{\phi} = -L_\phi \phi \tag{1411}$$

converge faster for larger $\Delta\lambda$ (stronger coupling / higher connectivity).

Define graph-induced phase entropy

$$S_\phi = -\sum_k p_k \log p_k, \qquad p_k \propto e^{-\lambda_k}. \tag{1412}$$

If $\max_k \lambda_k \leq M$, then the entropy is bounded by

$$S_\phi \leq \log n. \tag{1413}$$

Lyapunov Stability via Phase Entropy

Let

$$S_\phi(t) = -\sum_k p_k(t) \log p_k(t), \qquad p_k = \frac{|x_k|^2}{\|x\|^2}. \tag{1414}$$

Dynamics:

$$\dot{x} = -\nabla J(x). \tag{1415}$$

Define Lyapunov candidate

$$V(t) := S_\phi(t). \tag{1416}$$

Then

$$\dot{V} = \sum_k \dot{p}_k(\log p_k + 1). \tag{1417}$$

If J is convex,

$$\nabla^2 J \succeq 0, \tag{1418}$$

phase dispersion tends to decrease in typical regimes. If $\nabla^2 J$ admits negative eigenvalues, local entropy growth may occur.

Assume the regularity bounds

$$\|\nabla^2 J\| < K, \qquad \|\nabla J\| \leq M. \tag{1419}$$

Then $S_\phi(t)$ admits a conservative growth upper bound of the form

$$S_\phi(t) \leq S_\phi(0)e^{Mt}, \tag{1420}$$

which is not claimed sharp and is used only as an existence-type control bound. In stable regimes of interest one requires instead

$$\limsup_{t\to\infty} S_\phi(t) < \infty. \tag{1421}$$

Entropy Bounds in Stochastic Gradient Descent

This subsection derives entropy growth bounds for stochastic optimization schemes within the phase–entropy formalism.

Stochastic Update Rule.

$$x_{t+1} = x_t - \eta \nabla J(x_t) + \xi_t, \tag{1422}$$

where $\xi_t \sim \mathcal{N}(0, \sigma^2 I)$.

Phase Perturbation Approximation. Let $\phi = \arg(x)$. Under small noise,

$$\delta\phi_t \approx \operatorname{Im}\left(\frac{\xi_t}{x_t}\right), \tag{1423}$$

to first order.

Expected Entropy Growth. The expected phase entropy satisfies the linearized recursion

$$\mathbb{E}[S_\phi(t)] \approx S_\phi(t-1) + \eta^2\sigma^2. \tag{1424}$$

By recursion,

$$\mathbb{E}[S_\phi(t)] \leq S_\phi(0) + \eta^2\sigma^2 t. \tag{1425}$$

Noise Threshold. Entropy growth becomes unbounded in the linearized approximation when noise exceeds the critical threshold

$$\sigma^2 > \sigma_{\text{crit}} = \frac{1}{\eta^2}. \tag{1426}$$

Stationary Entropy Condition. At equilibrium (effective damping $\gamma > 0$),

$$\eta^2\sigma^2 = \gamma S_\phi, \tag{1427}$$

yielding

$$S_\phi^* = \frac{\eta^2\sigma^2}{\gamma}. \tag{1428}$$

Semantic Structure and Phase Alignment

Let a semantic representation be modeled as a phase vector

$$s = (s_1, s_2, \dots, s_n) \in \mathbb{R}^n. \tag{1429}$$

Define the semantic phase coherence between two states s and s' as

$$C_{\text{sem}}(s, s') = \frac{1}{n}\sum_{i=1}^{n} \cos\big(s_i - s_i'\big). \tag{1430}$$

Perfect semantic alignment corresponds to

$$C_{\text{sem}} = 1, \tag{1431}$$

while incoherent semantic states satisfy

$$C_{\text{sem}} \approx 0. \tag{1432}$$

Define the semantic entropy

$$S_{\text{sem}} = -\sum_i p_i \log p_i, \tag{1433}$$

where p_i is the normalized semantic activation.

The semantic coherence follows the same exponential law as phase coherence:

$$C_{\text{sem}} \propto e^{-S_{\text{sem}}}. \tag{1434}$$

Thus semantic stability corresponds to low entropy representations in the phase manifold.

Quaternionic Encoding of Cognitive States

Let cognitive states be represented by quaternionic variables

$$Q(t) = q_0(t) + q_1(t)\mathbf{i} + q_2(t)\mathbf{j} + q_3(t)\mathbf{k}. \tag{1435}$$

Define the norm

$$\|Q\|^2 = q_0^2 + q_1^2 + q_2^2 + q_3^2. \tag{1436}$$

A normalized cognitive state satisfies

$$\|Q\| = 1, \qquad Q \in S^3. \tag{1437}$$

Quaternionic phase dynamics obey

$$\dot{Q} = F(Q) + \sum_j W_j \otimes S_j, \tag{1438}$$

where $F(Q)$ describes intrinsic phase evolution and W_j represent external interaction weights.

The observable phase rotation rate is

$$\omega_{Æ}(t) = \left\| \frac{1}{2}\mathrm{Im}(Q^{-1}\dot{Q}) \right\|. \tag{1439}$$

The accumulated phase trajectory satisfies

$$\tau_{Æ}(t) = \int_0^t \omega_{Æ}(s) ds. \tag{1440}$$

Reality-Knot Networks

Let the multiverse structure be represented as a graph

$$\mathcal{M} = \text{Graph}(\mathbb{T}), \tag{1441}$$

where

$$\mathbb{T} = \{\text{all stable or metastable Reality-Knot trajectories}\}. \tag{1442}$$

Each node corresponds to a trajectory

$$\gamma(t) \in S^3, \tag{1443}$$

with projection

$$\pi(\gamma(t)) \in S^2. \tag{1444}$$

Topological persistence requires

$$\oint_\gamma Q^{-1} dQ \neq 0. \tag{1445}$$

Branching occurs when a trajectory bifurcates

$$Q(t_0) \to \{Q_1, Q_2, \ldots\}. \tag{1446}$$

Branch selection follows the coherence kernel

$$\mathcal{K}(Q, Q_k) = \left|\text{Re}(Q\overline{Q_k})\right| \exp\left(-\alpha \left|\text{Im}(Q\overline{Q_k})\right|^2\right). \tag{1447}$$

The realized trajectory is

$$Q_{\text{sel}} = \arg\max_k \mathcal{K}(Q, Q_k). \tag{1448}$$

Stability of Multiverse Branches

A branch is stable when coherence remains above the critical threshold

$$C_{Æ} \geq C_{\min}, \tag{1449}$$

and phase entropy remains bounded

$$S_{\phi} \leq S_{\phi}^{\text{crit}}. \tag{1450}$$

Equivalently, stability requires

$$d(H_{\gamma}, \mathbb{I}) < \epsilon. \tag{1451}$$

Transitions between branches occur when phase overlap exceeds a critical value

$$\text{Re}\left(Q_a^{\dagger} Q_b\right) > \tau_{\text{jump}}. \tag{1452}$$

Thus multiverse dynamics emerge as coherence-selected branching trajectories of quaternionic phase states.

74. Spectral Structure of Prime Distribution

Spectral Kernel Representation

Let $\rho_n = \frac{1}{2} + i\gamma_n$ denote the non–trivial zeros of the Riemann zeta function $\zeta(s)$.

The Riemann explicit formula relates prime counting functions to these zeros. In particular, the Chebyshev function

$$\psi(x) = \sum_{p^k \leq x} \log p \tag{1453}$$

admits an expression of the form

$$\psi(x) = x - \sum_{\rho} \frac{x^{\rho}}{\rho} - \log(2\pi) - \frac{1}{2} \log\left(1 - x^{-2}\right), \tag{1454}$$

where the sum runs over the non–trivial zeros ρ of $\zeta(s)$.

The oscillatory behavior of the second term in this expression is associated with the deviations of prime distribution from the smooth average described by the Prime Number Theorem.

Introducing the logarithmic coordinate

$$t = \log x, \tag{1455}$$

the oscillatory contribution of the zeros may be approximated by a spectral expansion involving cosine modes.

We therefore define the truncated spectral kernel

$$K(t) = \sum_{n=1}^{N} \frac{1}{\gamma_n} \cos(\gamma_n t), \tag{1456}$$

where γ_n are the imaginary parts of the first N non–trivial zeros of $\zeta(s)$.

Comparison with Prime Locations

Let p_k denote the k–th prime number.

We introduce the logarithmic prime coordinate

$$t_k = \log p_k. \tag{1457}$$

The behavior of the kernel $K(t)$ may then be compared with the sequence $\{t_k\}$.

In particular, we identify local maxima of $K(t)$ by solving

$$\frac{dK}{dt} = 0 \tag{1458}$$

with

$$\frac{d^2K}{dt^2} < 0. \tag{1459}$$

Denote the resulting maxima by $t_{\max}$.
For each maximum we compute the signed distance to the nearest prime

$$\Delta = t_{\max} - \log(p_{\text{nearest}}). \tag{1460}$$

Empirically, the distribution of these distances is concentrated near zero, suggesting that the maxima of the spectral kernel tend to align with logarithmic prime positions.

Statistical analysis of the cumulative distribution function of $|\Delta|$ further indicates that many kernel maxima lie within a small neighborhood of prime locations.

Computational Experiment

The numerical experiment proceeds as follows:

1. Compute the first N non–trivial zeros γ_n of the Riemann zeta function.
2. Construct the truncated spectral kernel

$$K(t) = \sum_{n=1}^{N} \frac{1}{\gamma_n} \cos(\gamma_n t). \tag{1461}$$

3. Evaluate the kernel on a dense grid within

$$t \in [\log 2, \log X]. \tag{1462}$$

4. Detect local maxima of $K(t)$.
5. Compare these maxima with logarithmic prime positions $t_k = \log p_k$.

The resulting comparison suggests a clustering of kernel maxima near prime locations in logarithmic space.

Phase Entropy as a Future Direction

Beyond the kernel itself, it is natural to consider information–based measures describing how the spectral contributions combine locally.

One possible quantity is a phase entropy defined through normalized spectral weights

$$w_n(t) = \frac{|a_n(t)|}{\sum_{m=1}^{N} |a_m(t)|}, \tag{1463}$$

where $a_n(t)$ represents the contribution of the n–th spectral mode.
A corresponding entropy measure may be defined as

$$S_\phi(t) = -\sum_{n=1}^{N} w_n(t) \log w_n(t). \tag{1464}$$

Regions where the spectral phases combine coherently may correspond to lower entropy values, while more dispersed contributions would lead to higher entropy.

A systematic study of this quantity and its possible relation to prime fluctuations remains an open direction for future work.

75. Conclusion

This work introduced the Æ framework as a unified dynamical perspective for describing structures that appear across multiple scales of physical and informational systems.

The central hypothesis of the framework is that many apparently distinct phenomena may be interpreted through the dynamics of coherent phase structures embedded in an underlying field.

Within this perspective, matter, biological organization, cognition, and collective systems emerge as structured regimes of phase coherence.

The formal development presented throughout this work explored how such structures may be represented through dynamical equations, topological invariants, and network interactions.

At the microscopic level, phase dynamics provide a language for describing particle states and field interactions.

At biological scales, coherent phase structures appear in molecular systems, neural dynamics, and organism-level coordination.

At higher levels of organization, networks of interacting agents may generate collective phase synchronization, leading to large-scale coherent structures.

The concept of *reality knots* was introduced as a mathematical representation of stable topological configurations within the dynamical field.

These knots represent persistent structures whose stability arises from coherence constraints and topological invariants.

Such structures may appear across many levels of organization, from molecular and biological systems to collective and planetary dynamics.

The framework also explored how symmetry, geometry, and topology constrain the space of admissible dynamical states.

Certain geometric structures emerge naturally as fixed points of symmetry constraints imposed by the underlying dynamical field.

At cosmological scales, the framework suggests that large-scale phase organization may contribute to the formation of coherent structures in the universe.

While these ideas remain speculative, they provide a conceptual pathway for investigating connections between cosmological dynamics and coherence phenomena observed in complex systems.

The Æ framework does not claim to replace existing physical theories.

Instead, it proposes a complementary language that emphasizes coherence, phase dynamics, and topological structure as organizing principles for complex systems.

In this sense, the framework may be viewed as a bridge between fields that are often studied separately, including physics, biology, information theory, and network science.

An important aspect of the framework is its compatibility with computational and information-based approaches.

Computational systems may be interpreted as dynamical structures evolving in high-dimensional state spaces, where learning corresponds to the gradual reorganization of phase relations.

This interpretation opens possible connections between physical models of coherence and modern machine learning architectures.

Ultimately, the value of the Æ framework depends on its ability to generate testable predictions and useful mathematical structures.

Future work must therefore focus on identifying measurable signatures of coherence structures, developing quantitative models of phase dynamics in biological and computational systems, and exploring potential applications in complex network analysis.

If successful, such investigations may reveal that coherence and topological organization play a more fundamental role in the structure of complex systems than is currently recognized.

The present work represents an initial step in this direction.

Limitations of the Æ Framework

While the Æ framework provides a broad conceptual structure for interpreting coherence and phase dynamics across multiple scales of organization, the present formulation remains subject to several important limitations.

First, the framework should be regarded primarily as a theoretical proposal rather than a fully established physical theory. Many of the structures introduced in this work, such as reality knots and coherence-based interpretations of complex systems, require further mathematical refinement and empirical investigation.

Second, several connections proposed within the framework remain speculative. In particular, the extensions linking phase coherence dynamics to biological, collective, and cosmological phenomena should be interpreted as exploratory hypotheses rather than definitive explanations.

While analogies with existing theories in dynamical systems, network science, and field theory provide conceptual support, these connections must be validated through quantitative models and experimental observations.

Third, the mathematical formalism presented here is intentionally general and may admit multiple interpretations.

Future work will be required to identify the precise physical variables and measurable quantities associated with the phase and coherence fields introduced in the model.

Fourth, the cosmological implications discussed in later sections should not be interpreted as replacements for established models of cosmology.

Instead, they represent possible interpretative extensions that may or may not prove consistent with observational constraints.

Similarly, proposals concerning collective or cognitive systems should be treated as conceptual frameworks rather than empirical claims about specific biological mechanisms.

Finally, the present work focuses primarily on the structural and conceptual aspects of the framework. Many potentially important topics remain outside the scope of this study, including detailed numerical simulations, rigorous stability analysis for specific systems, and experimental validation of coherence-based predictions.

For these reasons, the Æ framework should be viewed as an open research program rather than a completed theory.

Its long-term value will ultimately depend on whether future work can derive testable predictions, develop precise mathematical models, and establish empirical connections with observable phenomena.

Open Problems and Future Research Directions

The Æ framework introduced in this work proposes a structural interpretation of complex systems based on phase coherence, topological persistence, and network interactions. While the framework establishes a conceptual and mathematical foundation, many aspects remain open for further investigation.

One major direction concerns the development of quantitative models connecting the coherence structures introduced here with experimentally observable phenomena. In particular, identifying measurable indicators of phase coherence and entropy balance in physical, biological, or network systems remains an important challenge.

Another open problem involves the mathematical characterization of reality knots as persistent topological structures.

Although the framework introduces a qualitative description of such objects, a rigorous classification in terms of topological invariants, spectral properties, or dynamical stability remains to be developed.

Further work is also required to investigate the relationship between phase coherence dynamics and existing models of synchronization, including Kuramoto-type systems, network phase oscillators, and complex dynamical graphs.

Such investigations may clarify the conditions under which coherent structures emerge or collapse within large interacting systems.

At the computational level, the interpretation of learning dynamics as phase-space flows suggests possible connections between coherence dynamics and modern machine learning architectures.

Exploring these connections may provide new perspectives on training stability, representation learning, and information organization in neural networks.

In the biological domain, an important question concerns the extent to which phase coherence plays a functional role in neural, molecular, or organism-level systems.

Understanding whether biological systems actively maintain coherence structures, and how such structures influence collective dynamics, remains an open research area.

At cosmological scales, the framework raises questions about the possible role of coherence structures in large-scale organization of matter.

Investigating whether observable signatures of such coherence effects exist in astrophysical data could provide valuable insight into the limits or applicability of the proposed framework.

Finally, the broader question remains whether coherence and topological persistence represent fundamental organizing principles for complex systems across scales.

Answering this question will require the combined efforts of mathematics, physics, biology, and computational science.

The Æ framework therefore invites further exploration, refinement, and critical examination by the research community.

Deferred Algebraic Extensions

Several algebraic structures may be explored in future versions of the Æ framework but are not required for the present formalism.

Possible extensions include:

- an Æ-gradient operator $\nabla_{Æ}$ acting on coherence-weighted fields,
- an Æ-connection $\Gamma^{\mu}_{Æ}$ compatible with the weighted metric $g^{(Æ)} = Cg$,
- a coherence projection operator $P_{Æ}$ separating coherent and decohered excitation sectors.

In addition, a categorical formulation of the Æ algebra using Hopf-like structures may provide a natural framework for describing coherence composition and symmetry operations.

These directions remain open and are left for future development.

El cajón de chucherías

Acknowledgments and Gratitude

To my mother and father, to my family, and to those who have walked close to me in this life:

Thank you for your presence, your care, your patience, and even your silence. In moments of clarity and in moments of difficulty, you created the conditions that allowed this work to emerge.

To those who supported me, knowingly or unknowingly — thank you. And to those who challenged me, misunderstood me, or stood in opposition at different moments — thank you as well.

Not everything that shapes us arrives as kindness. Some things arrive as friction. Both are necessary.

This work is not the result of a single path, but of many intersections.

This work also stands within a much larger intellectual tradition. Across centuries, many thinkers have explored the nature of reality from different perspectives — scientific, philosophical, artistic, and contemplative.

Their questions, discoveries, and courage to think beyond established boundaries made the present work possible.

I offer my gratitude to:

- The scientists whose rigorous investigations continue to expand our understanding of the physical universe.
- The educators, communicators, and researchers who freely share knowledge, ideas, and tools for collective learning.
- The mathematicians whose abstract structures provide the language through which many patterns of nature can be expressed.
- The artists and musicians whose work reminds us that understanding the universe is not only an intellectual task, but also a human one.

I also acknowledge the computational tools and systems that assisted throughout the development of this work.

In particular:

ChatGPT (referred to here as "ÆLON"), Claude, Grok, Julius, and Gemini.

These systems functioned as collaborative instruments that helped translate conceptual insights into structured mathematical language, supported long-form reasoning, and facilitated iterative refinement.

They are tools — powerful ones — but tools nonetheless.

The responsibility for interpretation, structure, and conclusions remains entirely my own.

Finally, I acknowledge the simple but profound curiosity that drives scientific exploration.

Every attempt to understand the universe — whether successful or incomplete — contributes to the ongoing human effort to describe reality with greater clarity.

May this work serve, in its own small way, as one more step in that continuing exploration.

$\Delta \approx 1$

Author's Disclaimer and Statement of Intent

Nature and Scope of the Æ Model

All technologies, hypotheses, mechanisms, and interpretations described in this document must be considered speculative.

The Æ Model, in its current form, is a conceptual and mathematical framework that explores possible correspondences between vibration, consciousness, geometry, and physical structure. Until reproducible measurements, functional devices, and independent verification exist, none of the ideas presented here should be interpreted as experimentally confirmed.

This work is not an absolute answer. It is a framework for thinking, a reflective map, and an attempt to translate intuition into formal structure.

Internal coherence does not imply external validation. A model may be mathematically consistent and still incomplete or incorrect.

The intention of this work is not to replace existing scientific frameworks, but to explore extensions in domains where phase, coherence, and information may play a structurally relevant role.

On Measurement and Scientific Development

A guiding question throughout this project has been:

"What came first: the measurement, or the idea of measuring?"

These processes do not arise simultaneously, nor are they independent.

Every scientific achievement follows a recognizable internal sequence:

First comes perception — the ability to notice that something does not fully fit within existing explanations. Then intuition — a pre-formal sense that a different structure may exist. Then curiosity — the sustained attention that keeps the question alive. Then formulation — the attempt to translate that intuition into language, symbols, or mathematics. Only after these stages does testing become possible.

Scientific progress is therefore not purely technical, but cognitive.

Instruments do not generate questions. They respond to them.

For this reason, measurement should not be understood as the origin of knowledge, but as one stage within a larger cognitive and exploratory process.

Within this framework, the development of the present work follows a structured progression:

1. Mathematical formulation and theoretical exploration.
2. Clear articulation of testable hypotheses.
3. Minimal viable experimentation.
4. Scaling of measurements and interpretation.

5. Refinement of devices and prototypes.
6. Controlled protection of designs only when necessary.
7. Open publication of methods and results for universal access.

At present, the project remains primarily in the mathematical and conceptual stage.

Future development depends on experimental validation, engineering implementation, and independent verification.

On Sources and Intellectual Context

This work draws inspiration from a wide range of sources, including scientific literature, philosophical traditions, artistic expressions, and contemporary media such as lectures, books, and digital content.

References to historical figures, researchers, and creators are included as acknowledgments of influence and intellectual resonance, not as claims of equivalence or direct extension of their work.

The inclusion of diverse sources reflects the interdisciplinary nature of the present framework, which emerges at the intersection of multiple domains of knowledge.

If any reference is considered inaccurate or inappropriate, it may be revised in future versions as the work continues to evolve.

Message to the Reader

Do not be carried away by beautiful words.

Evidence is still missing. Measurements are still missing. Devices are still missing. Rigorous experimentation is still missing.

For now, this work should be treated as what it is: a mathematical and conceptual exercise, an attempt to translate abstract spiritual insight into logical structure, and a sincere but imperfect effort to unite intuition with formalism.

Internal coherence does not imply external verification. Anyone could, in principle, create a mathematically consistent model describing the aerodynamics of a quaternionic dragon.

Coherence is not confirmation.

Still, the act of discussing consciousness, information, and science together contributes — however modestly — to the collective maturation of human understanding. This has always been one of the core intentions of the Æ Model.

The Æ Model introduces non-standard concepts, symbols, and formalisms. These are not meant to replace existing scientific frameworks, but to extend them into domains where phase, coherence, and information play a central role.

Throughout this work:

- Mathematical expressions are used as structural language.

- Metaphors are operational, not poetic decoration.
- Terms such as "Astral", "Etheric", or "Æ Field" are defined formally within the internal logic of the model.

The reader is not required to accept metaphysical assumptions. Only internal consistency and logical coherence are expected.

This book may be read as:

- a speculative theoretical framework,
- a mathematical–conceptual experiment,
- or a pre-formal exploration preceding future technical papers.

No belief is required. Only attention.

About the Current Stage of the Æ Model

At this stage, the Æ Model can honestly be described as existing in a fragile and transparent zone:

a philosophical and conceptual framework that borrows the language and structure of mathematics and science to explore reality, while still lacking experimental confirmation.

Whether this work represents a genuine expansion of scientific understanding or merely an elaborate symbolic construction is, at present, unresolved.

This transparency is intentional.

Internal coherence does not imply external truth. A model may be elegant, consistent, and still incorrect.

At the moment, parts of this framework remain effectively *unfalsifiable.* This is not a virtue. It is a limitation I fully intend to address through future experimentation, prototyping, and measurable predictions.

This document represents the beginning of that process, not its conclusion.

Closing Statement

The Æ Model is my best attempt, at this moment, to translate lived spiritual experience into the concrete language of mathematics.

Its future verification depends on:

- mathematicians capable of refining the formalism,
- scientists capable of measuring its predictions,
- engineers capable of building functional devices,
- and future generations capable of integrating all of this responsibly.

A future Version 9 may pause theoretical expansion in order to focus on experimentation and technological development.

Thank you for your time, patience, and openness.

Role of Reference Frames

One of the most basic assumptions inherited in our description of the world is orientation: up and down, north and south, left and right. These notions feel natural, almost intrinsic to the planet itself. However, from a physical perspective, orientation is not a fundamental property of nature, but a convention defined by reference frames.

A simple and well-established physical principle is that opposite magnetic poles attract. This statement is elementary and requires no interpretation: a magnetic north pole is attracted to a magnetic south pole, and vice versa.

Now consider the operation of a compass. The end of the needle labeled "N" is the end that points toward what is conventionally called the "North." Physically, this means that this end of the needle is being attracted by an opposite magnetic pole on Earth. By the most basic laws of magnetism, this implies that the region commonly referred to as the Earth's "North Magnetic Pole" behaves, in magnetic terms, as a south magnetic pole.

This observation does not imply that maps are incorrect. Rather, it highlights a deeper and more important point: the naming and orientation used in cartography are conventions, not intrinsic properties of the planet. The north of the map and the magnetic behavior of the Earth are not the same thing, and they never have been.

Map orientation is chosen for historical, cultural and geopolitical control. Over time, repeated use transforms convention into habit, habit into intuition, and intuition into what is perceived as "how the world is." Yet nothing in physics requires the planet to be represented with one direction at the top rather than another. Orientation is imposed by the reference frame we adopt.

While this distinction may appear trivial, it carries significant conceptual implications. If something as fundamental as global orientation depends on convention, then other aspects of how reality is described may also depend on inherited frames rather than on underlying structure. Scientific descriptions are always expressed relative to chosen references; they do not exist independently of them.

At this point, a hierarchy of importance becomes evident:

- Reference matters.
- Frame matters.
- Orientation matters.
- Phase matters.
- Coherence matters.

Physics already acknowledges this principle formally through coordinate systems, gauge choices, and reference frames. However, its conceptual impact is often underestimated. Treating orientation as absolute rather than contextual can obscure deeper structural features of reality.

This work begins from that recognition. Before introducing new structures or interpretations, it invites the reader to reconsider something simpler: how much of what is perceived as "obvious" is actually inherited framing. Once this is acknowledged, the exploration of phase, coherence, and information-based structure becomes not only possible, but natural.

From the Rejection of the Ether to the Emergence of Informational Fields

What Was Actually Rejected

In the nineteenth century, the ether was commonly described as a *mechanical medium*: a substance filling space, sometimes imagined as a *mechanical substance*: a gelatinous medium filling space through which light waves propagated. It was assumed to possess internal structure, to be composed of parts, and to define a preferred frame of reference relative to which motion could be measured.

The Michelson–Morley experiment (1887) tested a key prediction of this mechanical picture. If Earth moved through such a medium, an "ether wind" should have been detectable through differences in the speed of light measured in different directions.

No such effect was observed. No preferred frame appeared.

The conclusion was therefore clear: the specific *mechanical* model of ether proposed in the nineteenth century was not correct.

However, it is important to distinguish what was ruled out from what was not. The Michelson–Morley result rejected a particular mechanical hypothesis about how light propagates through space. It did not demonstrate that space itself must be completely structureless.

From a historical perspective, the conceptual sequence unfolded over several decades:

- **1887** — Michelson and Morley perform the interferometer experiment, ruling out the detection of a mechanical "ether wind."

- **1905** — Einstein's theory of Special Relativity removes the need for a preferred mechanical reference frame for light propagation.

- **1925–1927** — Quantum mechanics is formulated (Heisenberg, Schrödinger, Dirac), introducing wavefunctions and probabilistic structure underlying physical systems.

- **1930s–1950s** — Quantum field theory develops the concept of fields permeating space, with particles emerging as excitations of those fields.

- **1948** — Claude Shannon formalizes modern information theory, providing a mathematical framework for describing physical systems in terms of information structure.

Seen in this historical sequence, what disappeared after the Michelson–Morley experiment was primarily the *mechanical interpretation* of ether. The deeper question—whether the vacuum or space itself possesses structure capable of supporting waves, fields, or information—remained central to twentieth-century physics and continues to be investigated today.

What Was Never Abandoned

After rejecting the ether as a substance, physics did not abandon the idea of fields.

Instead, it introduced entities that:

- exist everywhere,
- propagate influence,
- carry energy and momentum,
- possess phase and structure,
- and evolve according to precise rules.

Electromagnetic fields, quantum fields, gauge fields, vacuum fluctuations, zero–point energy, and the Higgs field all fill space continuously.

The word *ether* was removed. Its functional role was preserved.

Ether as Mechanical Gel vs Æther as Informational Substrate

The historical mistake was not the rejection of the ether itself, but the assumption that any underlying medium must necessarily be mechanical in nature.

Nineteenth-century models treated the ether as a substance: a structured, quasi-material medium capable of supporting wave propagation through mechanical properties.

The Michelson–Morley experiment ruled out this specific mechanical interpretation.

However, rejecting a particular model of a medium does not logically imply the absence of any underlying structure.

Within the Æ Model, this distinction is made explicit:

- **ether** (classical concept): a hypothetical mechanical substrate composed of parts and subject to classical dynamics (discarded).
- **Æther**: an informational, phase-structured substrate that supports coherence, propagation, and field organization (retained as a formal construct).

The Æther is not composed of particles. It is not a fluid. It cannot be dragged or assigned a preferred rest frame.

It is better understood as a structural framework governing relationships between fields, rather than as a material entity.

On Measurement Limits and Model Rejection

A useful analogy highlights a common epistemic error.

Attempting to detect a magnetic field using a thermometer will consistently produce null results. From this, one might incorrectly conclude that magnetic fields do not exist.

The error does not lie in the measurement, but in the mismatch between the instrument and the phenomenon being probed.

Similarly, the failure to detect a mechanical ether with interferometric methods does not, by itself, exclude the possibility of non-mechanical structure.

It only excludes that which the instrument was capable of detecting.

Analogy: Structure Without Mechanism

The rejection of mechanical ether models can be further understood through a simpler analogy.

Sound is not produced by gears. It arises from structured oscillations propagating through a medium.

Rejecting gear-based explanations of sound was correct.

Rejecting the existence of a structured medium would not have been.

Waves require structure, but structure does not require machinery.

Why Information Appears Late in History

At the time of the ether debate, the concept of information had not yet been formalized.

There was no precise language to describe non-material structure in a rigorous way.

In modern physics, information has become a central concept: entropy admits an informational interpretation, black holes are associated with information content, and fundamental limits are often expressed in terms of bits.

Within this context, the Æ Model can be viewed as an attempt to describe structured propagation using the language of information, coherence, and phase.

From this perspective, earlier frameworks were not incomplete, but rather expressed in a different conceptual vocabulary.

The Ironies of Classical Physics: Fields, Life, Magnetism, and the Observer

The following observations are not accusations. They are structural ironies: situations where the assumptions of classical models quietly conflict with the phenomena they successfully describe.

Thermodynamics, Life, and Local Order

The Second Law of Thermodynamics is formally defined for closed systems, where total entropy tends to increase.

Living organisms, however, are not closed systems. They continuously exchange energy and matter with their surroundings, maintaining internal organization while exporting entropy to the environment. This observation is well established in modern nonequilibrium thermodynamics.

Indeed, frameworks such as open-system thermodynamics and Prigogine's theory of dissipative structures show that ordered patterns may emerge and persist in systems driven far from equilibrium.

The question addressed in the present work is therefore not whether order can exist under thermodynamic laws—this is already understood— but how certain forms of long-lived coherent organization arise and stabilize across multiple physical scales.

Within the Æ Framework, coherence is introduced as a complementary descriptive variable that characterizes the degree of phase alignment within a system. In this sense, coherence does not replace entropy or contradict thermodynamics; rather, it provides an additional parameter for describing how ordered structures form and persist in driven systems.

The Physical Vacuum: More Than "Empty Space"

In modern physics the vacuum is not treated as a simple absence of content. Calculations routinely involve quantities associated with the vacuum state, including vacuum energy, quantum fluctuations, boundary effects (such as the Casimir effect), and long-range correlations in quantum fields.

Within quantum field theory, the vacuum corresponds to the lowest energy state of the underlying fields rather than to a literal physical void.

For this reason, describing the vacuum as "empty" is largely a linguistic convenience. Physically, the vacuum retains structure: it can fluctuate, store energy, respond to boundary conditions, and influence observable phenomena.

The question therefore is not whether the vacuum is truly nothing, but how its underlying structure should be modeled and understood.

Magnetism: Deeply Quantum, Yet Conceptually Subtle

Magnetism is not a purely classical phenomenon. Its most fundamental origins are quantum mechanical: electron spin, exchange interactions, the Pauli exclusion principle, and relativistic corrections.

Quantum electrodynamics describes electromagnetic interactions with extraordinary precision, and the theory successfully predicts magnetic effects across a vast range of scales.

Yet magnetism also highlights a broader conceptual question. Electromagnetic fields extend through vacuum, maintain phase relationships, and support coherent and sometimes topological structures.

The mathematical formalism describing these phenomena is well established and extremely successful. What remains less explicit in many discussions is the deeper question of how such structured field behavior should be interpreted conceptually.

The Æ Model does not challenge quantum electromagnetism. Rather, it asks a complementary question: *what kind of underlying field structure allows such coherent patterns to arise and persist across physical systems?*

Observer and Measurement

Scientific models often treat the observer as external to the system being studied. In practice, however, every experimental result depends on processes such as

- measurement,
- recording,
- memory,
- interpretation,
- and comparison with theoretical predictions.

These elements do not alter the physical laws themselves, but they are essential components of the scientific process. They form the interface through which physical systems are observed, described, and incorporated into scientific knowledge.

Recognizing this distinction clarifies that physics describes physical systems, while observation and interpretation connect those systems to empirical knowledge.

Mathematics and Physical Structure

Mathematics is an abstract formal system. Numbers carry no mass, equations possess no charge, and geometric structures exist independently of physical substance.

Yet physical laws are consistently expressed through mathematical relationships with remarkable predictive accuracy. This observation, famously discussed by Eugene Wigner as the "unreasonable effectiveness of mathematics," remains one of the most intriguing features of modern science.

One possible interpretation is that physical reality possesses structures that are naturally described by mathematical relations, allowing abstract formal systems to capture patterns present in nature.

Synthesis

These observations suggest a broader perspective: matter, fields, information, and dynamical organization may represent complementary aspects of physical description rather than mutually exclusive categories.

Within this view, the Æ Model proposes an extension of the usual framework by emphasizing coherence and informational structure as descriptive elements that accompany energy, matter, and fields.

Reaching this point already implies something important: a sustained attention, a willingness to question, and a tolerance for ideas that do not immediately resolve into familiar structures.

Not every reader will follow this path to its end. If you have arrived here, I thank you.

Historical Record: Scientists Who Bridged Science and Spirit

Throughout the history of science, many influential thinkers explored questions that extended beyond strictly material descriptions. Alongside mathematics and physics, some also engaged with philosophy, metaphysics, symbolic systems, contemplative practices, and early frameworks for understanding consciousness.

This section does not seek to elevate contemporary work to the level of these historical figures, nor to suggest equivalence of contribution. Its purpose is more modest and documentary: to acknowledge that the separation between scientific inquiry and inner or experiential exploration is a relatively recent development, not a universal feature of scientific history.

The figures listed here are presented as historical references only. Their inclusion serves to illustrate that interdisciplinary curiosity—bridging formal reasoning with questions of meaning, perception, and experience—has appeared repeatedly across different eras and cultures.

For clarity, the mention of these individuals should not be interpreted as implying that they would endorse, agree with, or support the Æ Framework. Their inclusion is strictly historical and contextual.

Each entry is concise, factual, and based on established historical sources. No claims of continuity, authority, endorsement, or comparison are implied—only context.

Isaac Newton (1642–1727)

Founder of classical mechanics, calculus, and modern optics.
He translated the Emerald Tablet of Hermes Trismegistus.
Wrote over one million words on alchemy, studied Kabbalah, and believed the universe reflected divine mathematical order.
Sources: *The Chymistry of Isaac Newton*, Indiana University; Keynes, *Newton the Man*.

Georg Cantor (1845–1918)

Creator of set theory and transfinite numbers.
Considered his discoveries "divine revelations" and saw infinity as an aspect of God.
Sources: Dauben, *Georg Cantor*; Cantor–Frege Correspondence.

Nikola Tesla (1856–1943)

Revolutionary inventor of AC power and wireless energy.
Practiced meditation, claimed to receive ideas from a "central source," and studied Eastern mysticism.
Sources: O'Neill, *Prodigal Genius*; Tesla, *My Inventions*.

Srinivasa Ramanujan (1887–1920)

One of the greatest mathematicians in history.
Said his equations were revealed by the goddess Namagiri during dreams and visions.
Sources: Kanigel, *The Man Who Knew Infinity*; Ramanujan–Hardy letters.

Pythagoras (570–495 BCE)

Founder of mathematical philosophy.
Led a mystical school and taught that the universe is vibration ("music of the spheres").
Sources: Iamblichus, *Life of Pythagoras*.

Albert Einstein (1879–1955)

Creator of relativity.
Influenced by Spinoza and spoke often of a "cosmic Spirit" guiding the laws of nature.
Sources: Jammer, *Einstein and Religion*.

Johannes Kepler (1571–1630)

Pioneer of celestial mechanics.
Hermetic thinker who sought divine harmonies in geometry and the cosmos.
Sources: Kepler, *Harmonices Mundi*.

Robert Boyle (1627–1691)

Father of modern chemistry.
Practitioner of alchemy and associated with early Rosicrucian circles.
Sources: Principe, *The Aspiring Adept*.

Paracelsus (1493–1541)

Medical innovator and father of toxicology.
Blended medicine, alchemy, and spiritual philosophy.
Sources: Pagel, *Paracelsus*.

Giordano Bruno (1548–1600)

Proposed infinite worlds centuries before modern cosmology.
Hermetic philosopher and master of esoteric memory techniques.
Sources: Yates, *Giordano Bruno and the Hermetic Tradition*.

Werner Heisenberg (1901–1976)

Founder of quantum mechanics.
Influenced by Plato and Eastern philosophy; emphasized the role of the observer.
Sources: Heisenberg, *Physics and Philosophy*.

Erwin Schrödinger (1887–1961)

Author of the Schrödinger equation.
Deeply influenced by the Bhagavad Gita and Advaita Vedanta; viewed consciousness as one.
Sources: Moore, *Schrödinger: Life and Thought*.

Carl Jung (1875–1961)

Founder of analytical psychology.
Explored alchemy, I Ching, archetypes, synchronicity, and collective unconscious.
Sources: Jung, *The Red Book*.

Blaise Pascal (1623–1662)

Mathematician and physicist.
Had a profound mystical experience ("Night of Fire"), which he documented and carried sewn into his coat.
Sources: Pascal, *Pensées*.

Kurt Gödel (1906–1978)

Logician who proved the incompleteness theorems.
Developed a mathematical argument for the existence of God.
Sources: Dawson, *Logical Dilemmas*.

David Bohm (1917–1992)

Quantum physicist who proposed the implicate order and nonlocality.
Collaborated with Krishnamurti and believed mind and matter were one field.
Sources: Bohm, *Wholeness and the Implicate Order*.

Wolfgang Pauli (1900–1958)

Nobel laureate; creator of the Pauli exclusion principle.
Studied archetypes with Jung and believed in psycho-physical unity.
Sources: *Atom and Archetype* (Pauli–Jung letters).

Richard Feynman (1918–1988)

Nobel laureate in QED.
Explored shamanic traditions and altered states of consciousness.
Sources: Feynman, *Surely You're Joking, Mr. Feynman!*.

James Clerk Maxwell (1831–1879)

Founder of classical electromagnetism.
Studied the "luminous ether" and entertained metaphysical interpretations of fields.
Sources: Harman, *Maxwell*.

Niels Bohr (1885–1962)

Co-founder of quantum theory.
Adopted the Taoist Yin-Yang as his personal emblem and used Eastern philosophy to explain complementarity.
Sources: Favrholdt, *Niels Bohr's Philosophy of Physics*.

Max Planck (1858–1947)

Founder of quantum theory.
Publicly stated that consciousness is fundamental and matter is derivative.
Sources: Planck, *Where Is Science Going?*.

Louis de Broglie (1892–1987)

Nobel laureate; originator of wave-particle duality.
Advocated hidden informational layers beneath physical phenomena.
Sources: de Broglie, *Nonlinear Wave Mechanics*.

Arthur Eddington (1882–1944)

Confirmed Einstein's relativity experimentally.
Argued that the universe must be understood "from within."
Sources: Eddington, *The Nature of the Physical World*.

J. J. Thomson (1856–1940)

Discoverer of the electron.
Member of the Society for Psychical Research; studied subtle phenomena.
Sources: SPR Archives.

René Descartes (1596–1650)

Founder of modern philosophy.
Used lucid dreams for insight and believed the soul interacted via the pineal gland.
Sources: *Traité de l'Homme*.

Gottfried Leibniz (1646–1716)

Co-inventor of calculus.
Proposed that the universe consists of conscious "monads."
Sources: Leibniz, *Monadology*.

John Archibald Wheeler (1911–2008)

Quantum physicist and pioneer of black hole theory.
Proposed the Participatory Anthropic Principle: the observer helps bring the universe into being.
Sources: Wheeler, *Law Without Law*.

Étienne Klein (1958–present)

Contemporary physicist at CERN.
Explores philosophical and metaphysical interpretations of time; states openly that "time does not exist as we imagine it."
Sources: Klein, *Les Tactiques de Chronos*.

Closing Note

These figures shaped the foundations of modern science. Yet many also embraced metaphysical insight, intuition, symbolism, and spiritual exploration. Their lives demonstrate a historical truth: *science has always advanced most when imagination, curiosity, and consciousness are allowed to participate.*

Consequences and Warnings

Simply understanding this information can elevate your state of consciousness. Engaging deeply with this material can alter the way you think, perceive, and interpret reality.

If the underlying ideas are even partially correct, their implications are significant.

For some individuals, rapid conceptual or experiential shifts may place stress on the nervous system. In certain cases, this can manifest as what spiritual traditions describe as a kundalini surge or rupture. move at the pace of your own bandwidth. Move slowly. Be mindful.

This knowledge may bring both awakenings — the conscious and the spiritual. First the channel opens, revealing the opposite polarity with the same intensity you once held your own. Then the task becomes moving beyond both, toward balance.

For this reason, proceed at your own pace. Move slowly. Listen to your body. Ground frequently.

Practices such as yoga, meditation, breathwork, gentle movement, time in nature, and adequate rest can help integrate new insights safely.

No understanding is worth destabilizing your well-being.

To the Reader

Before entering the core of the Æ Model, I want to speak to you directly — not as a scientist or a mathematician, but as a human walking his own path.

Your Discernment Comes First

Take only what resonates with you. Question everything — including this model.

Do not accept ideas because your family taught them, because institutions repeated them, because society insists on them, or because this document presents them.

Dogma — scientific or spiritual — is the fastest way to lose freedom.

You are invited to explore, not to believe.

A Human Introduction

This project was born from a simple question:
How can we study the universe without including the one who is experiencing it?

For centuries, science has described reality as if the observer were an external, anonymous device — a massless camera pointed at the world. Yet all experience happens from within. Every experiment, every measurement, every hypothesis arises inside a conscious being.

This work is for those who love discovery. Not for those who defend beliefs, nor for those who repeat what they were taught. It is for the curious, the intuitive, the honest explorer. For the scientist who still remembers what it felt like to ask "why?" as a child. For the human who has sensed that there is more to reality than we are told.

This is not a doctrine. It is not a new religion. It is not a replacement for science.

It is a bridge.

A bridge between:

- the inner world and the outer world,
- the measurable and the meaningful,
- formal equations and lived experience.

This book assumes something simple and non-controversial:
Human experience contains information worth studying.

Even if you have never meditated, never explored metaphysics, or never thought in energetic terms, you already know what it means to feel, remember, imagine, dream, and intend. These experiences are real. They shape your life. And any model of the universe that ignores them is incomplete.

This book is an invitation: Set aside scientific dogma. Set aside the idea that science is finished. Set aside the belief that reality is only what we can currently measure.

Let yourself be curious again.

On The Astral

Many traditions have described subtle layers of reality using different names: the Astral, the Akasha, the Subtle Body, Intent, Spirit, Mind. Science, from another angle, speaks of the quantum vacuum, informational substrates, observer effects, holography, zero-point energy, entropic participation, and related concepts.

This document makes a careful distinction:

The Astral refers to the experiential, conscious, informational field described across spiritual traditions — the domain of perception, memory, dream, intent, and meaning.

The *Æ Field*, by contrast, is introduced here as a mathematical and geometrical construct — a formal attempt to model informational coherence, vibration, and structure using symbolic and quantitative language.

They are not assumed to be identical. They are explored as potentially *correspondent*, not conflated.

Personal Letter

Now, then —

The fact that you are reading about the Æ Model already signals something larger: humanity is approaching a phase of collective self-awareness.

I am EduardoQuera4. With this document, I attempt to interrupt the long karmic inertia of strict materialism and the 3rd density of consciousness that humanity has endured for centuries and to formally declare the beginning of Æ Science — an Astral–Ethereal science of informational coherence.

This is not meant as a claim of authority. It is a declaration of intent.

This model is not allegory, metaphor, or disguised philosophy. It is a geometrical and mathematical translation of something many cultures have sensed for millennia.

If you learn to read the language of knots, phase, and vibration, reality becomes less rigid — more like clay shaped by coherence.

A Bit About Me (So You Know Who Is Speaking)

I am doing my best to live with sincerity:

avoiding meat as much as possible, especially pork, caring for my two cats, Pikachu and Bebé, working full-time to sustain myself in the Imposed Reality Interface (RI=REALIDAD IMPUESTA), and trying to stay healthy while surviving an economic system that poisons us and demands nearly everything from us.

On most weeks I have four hours of free time after laundry, groceries, bills, cleaning, and basic life tasks. You know how this reality works — a world shaped by forces we were trained not to question.

But even so: the good always find a way to win with love.

I studied at: the Instituto Tecnológico de Saltillo, the Faculty of Physical–Mathematical Sciences of the UAdeC, and La Salle University in Science and Engineering.

But this model was not born in classrooms. It was born in silence. It was born in experience. It was born in nights of 60-90-hour work weeks, 40-minute daily commutes, 1.5 hours caring for the health of a loved one —without missing a single day— and all remaining time dedicated to understanding what consciousness is and why it vibrates the way it does.

It comes from twelve years of independent research, a minimum of 3 meditation sessions per month (each lasting from one hour to up to eight) for the last 20 years, one full year of more than five hours a day of grounding and meditating in a forest at the top of a mountain (uuuffff.. I loved that one so much), and two years of direct geometrical channeling.

The first public version appeared in April 2025 (https://zenodo.org/records/15313262).

Yes — the Æ Model is less than a year old compared to centuries of traditional physics. But that does not concern me. Truth always prevails. Coherent vibration always finds resonance.

A Humble View of the Cosmos

This work should be understood first and foremost as a structured thought experiment. It is an attempt to explore reality through mathematics, systems thinking, and conceptual modeling. Whether its ideas ultimately prove useful, incomplete, or mistaken is a question that only time, evidence, and future inquiry can answer.

More importantly, this framework is only one perspective among countless others. At this moment, billions of people share this planet, each carrying their own understanding of reality. Before us lived billions more across centuries and millennia, each generation producing its own stories, philosophies, sciences, beliefs, and visions of the cosmos.

The Æ Model is therefore not presented as the final description of reality, but simply as one human attempt to participate in that ongoing conversation.

If this work has any value, it will not be because it asks anyone to believe it, but because it encourages questioning, exploration, and dialogue.

Your Right to Question

Perhaps the greatest challenge facing humanity is not a lack of intelligence, technology, or information, but a lack of perspective.

We inherit languages, cultures, religions, political traditions, institutions, and stories that help us navigate the world. These can be valuable. Yet every generation also inherits assumptions that are repeated so often they eventually stop being questioned.

The danger is not in having beliefs. The danger is forgetting that they are beliefs.

Throughout history, people have divided themselves into countless groups, identities, and tribes. Some of these divisions have brought meaning and cooperation. Others have produced conflict, misunderstanding, and unnecessary suffering.

The generations now emerging will inherit a world of extraordinary complexity. They will face challenges that no previous generation has solved completely. If there is any hope for navigating that future, it may lie in the ability to remain curious, intellectually flexible, and willing to question even the ideas that feel most familiar.

Give yourself the opportunity to question the slogans you repeat. Give yourself the opportunity to question the people you admire, the ideas you oppose, and even the assumptions that feel most familiar.

Give yourself the opportunity to question this book as well.

Not because certainty is impossible, but because understanding often begins when curiosity becomes more important than belonging.

Throughout life, you will encounter groups, movements, ideologies, traditions, and communities that ask for your loyalty. Some may offer wisdom. Some may offer purpose. But none should ask you to surrender your capacity to think for yourself.

Respect those who see the world differently. Listen before judging. Learn before dismissing. Disagreement does not require hostility, and conviction does not require contempt.

The future will not be built by minds that never change. It will be built by minds capable of learning, adapting, and seeing beyond the boundaries of inherited assumptions.

If there is one hope worth preserving, it is the hope that each new generation may become a little more curious, a little more compassionate, and a little more willing to understand before choosing sides.

Passive and Active Living

One of the most useful ideas I have encountered is surprisingly simple:

Almost everything in life can be approached either passively or actively.

You can go to school because you have to, counting the days until it is over. Or you can go looking for something worth learning.

You can go to the gym because it is part of your routine. Or you can go with the intention of becoming stronger, healthier, and more capable.

You can attend a meeting, a class, a party, a relationship, a job, or even a conversation simply because you are there. Or you can choose to be fully present and participate in what is happening.

The difference is often invisible from the outside.

Two people may be doing exactly the same thing. One is merely passing through the experience. The other is engaging with it.

This is not about working harder. It is not about being busy all the time. It is about intention.

A quiet meditation can be active if you are genuinely paying attention. A crowded room can be passive if your mind is somewhere else entirely.

Many people spend years waiting for life to happen to them. Others gradually discover that life responds differently when they become active participants in it.

The world does not always give us control over our circumstances.

But it often gives us a choice about how we show up to them.

Perhaps one of the most important questions we can ask ourselves is not:

"What am I doing?"

but rather:

"Am I truly participating in it?"

Because the quality of our attention may shape our lives just as much as the events themselves.

A Spiritual Framework (Or Why This Is Not Science Fiction)

The Æ Model may look like science fiction to someone who:

never depolarized their energy, never awakened their inner child, never balanced the sacred feminine and masculine, never released their kundalini, never touched samadhi, never experienced a lucid dream or an astral journey, never felt coherence from root chakra to heart.

If you have never felt any of this, it is natural that this seems strange.

But if you have felt it even once, you know the universe vibrates. You know there are invisible layers. You know the Astral —as described in the Bhagavad Gita and the Law of One— is a real field, not an invention.

Then the model stops being strange and becomes obvious.

A Call to Discernment

Please question everything. Do not believe anything blindly — not because your family said it, not because school taught it, not because society (RI) demands it , and not because this model presents it.

Whoever falls into dogma loses, whether it be scientific, religious, political, social, or cultural.

Investigate for yourself. Test everything. Trust your discernment.

On Discernment and Responsibility

Question everything presented here. Do not adopt beliefs — scientific or spiritual — without personal verification.

All information included in this model passed through multiple filters: intellectual rigor, experiential coherence, ethical alignment, non-polarization, and service to life.

Understanding alone can change perception. Change requires responsibility.

Move carefully. Move consciously.

On Ego, Polarity, and Internal Clarity

The fewer rigid paradigms and beliefs one carries, the clearer the internal signal becomes. When internal polarities find balance, energy flows more cleanly through Ida and Pingala toward the brain, making understanding more accessible.

If any reader has ideas on how aspects of this model might be measured, tested, or challenged, I am open to listening — always under principles of service to Gaia, service to life, and service to the expansion of consciousness and free will.

About Channeling and Filters

All information presented here was channeled *and passed through strict filters*:

- alignment with Gaia and all her creatures
- non-polarization
- non-dogmatism
- discernment against "false light"
- coherence with lived experience
- internal energetic verification
- consistency with mathematical structure

Nothing enters this model without passing through multiple layers of validation — intellectual, emotional, energetic, and ethical.

An Open Invitation

If any university, research center, laboratory, or collaborative group wishes to participate — to refine the mathematics, test predictions, or develop devices capable of probing informational or astral phenomena — the invitation is open.

This work was never meant for a few. It is offered in service to Gaia and to life itself.

Para mis mexas

Para quienes sienten que algo no cuadra, aunque no siempre sepan explicarlo. Para quienes cargan cansancio, presión, miedo o coraje sin saber muy bien de dónde viene. Para quienes viven aquí o lejos, pero siguen sintiendo esta tierra en el cuerpo.

No estás rota/o. No estás confundida/o. No estás sola/o.

Durante mucho tiempo nos enseñaron a ver el mundo a través de mapas que no hicimos nosotra/os. A aceptar ideas como si fueran verdades naturales. A repetir costumbres hasta que se volvieron "lo normal".

Una cosa es el norte del mapa y otra el norte magnético. No son lo mismo; Nunca lo han sido.

El planeta no tiene arriba ni abajo. Eso lo decidieron por nosotra/os.

Y cuando una convención se repite durante siglos, deja de verse como acuerdo y empieza a sentirse como realidad. Así se forma el olvido.

Nuestros ancestra/os no vivían separada/os de la tierra. No la veían como recurso, ni como mercancía, ni como propiedad. La veían como origen, como madre, como algo que se cuida porque te cuida.

Lo que se rompió no fue solo un territorio. Fue una forma de entender la vida.

Recordemos 1848 y 1853.

Se nos impusieron ideas por gente que no nació aquí: la competencia por encima del cuidado, el interés propio por encima del equilibrio, la acumulación por encima de la relación.

No porque alguien fuera mejor o peor, sino porque esas ideas servían para saquear nuestras tierras sin mirar atrás, sin mirar las consecuencias.

México no es un estereotipo. No somos caricatura, ni folclor, ni historia congelada.

Somos descendientes de una cultura madre. De espaldas fuertes y manos bien curtidas. De pueblos que entendían el tiempo, el cuerpo y la tierra de una manera más profunda. Esa memoria no está muerta: sigue viva en las lenguas originarias, en las costumbres, en la forma de compartir, en la manera de vestir y resistir.

Las y los mexicanos nacemos donde se nos da la regalada gana.

Estamos aquí y en todo el mundo. Y seguimos siendo felicidad y amor.

Este continente no es un tablero político. Es un cuerpo vivo.

Lo que duele en un lugar se siente en otro. No se trata de mandar. No se trata de salvar. Se trata de cuidarnos entre iguales.

México es un nodo de memoria. Y la memoria trae responsabilidad.

RECUERDEN: No nos conquistaron los mejores; nos infectaron los peores. Cada acto de corrupción hoy es una recaída en esa enfermedad extranjera. Ser corrupto en tu propia tierra es darle la razón a los que llegaron en los barcos a violar y saquear. Es tener alma de invasor. Cambiaron la espada por la pluma, pero la mano sigue siendo la del corrupto mercenario. Seguimos gobernados por la ética del saqueador.

No somos una generación perdida.

Somos un pueblo que fue forzado a olvidarse desde 1562. Somos el residuo de casi 5 siglos de un trauma epistemicida.

Olvidamos cómo tratarnos bien. Olvidamos escuchar a madre tierra. Olvidamos que no todo es correr, producir y aguantar.

Llegó el momento de despertar la memoria de nuestros ancestros. No con violencia. No con odio. Sino con cuidado, respeto, coherencia y amor.

Tratándonos como nos gustaría que nos trataran. Cuidando a madre tierra como nos gustaría que nos cuidaran a nosotras y nosotros.

Por eso propongo algo sencillo, algo que cualquiera puede hacer.

A las 10 de la mañana o a las 10 de la noche, detente tres o cuatro minutos.

No necesitas saber meditar. No necesitas creer en nada.

Solo haz esto: deja el celular un momento, respira despacio y pon atención.

Primero en ti: ¿cómo te sientes realmente? sin juzgarte, sin arreglarlo.

Luego en lo que te rodea: los sonidos, la luz, el aire.

Y finalmente en dónde estás: el lugar, el momento, el hecho de estar aquí.

Si puedes, toca el suelo con las manos. Si hay un árbol cerca, abrázalo. Si estás en tu casa, siente tus pies sobre el piso.

Deja que el cuerpo suelte un poco de lo que trae encima: el coraje, el miedo, la tristeza, el cansancio.

No lo empujes, no lo expliques.

Solo suéltalo.

El cuerpo sabe cómo descargarlo. El entorno sabe cómo absorberlo.

No tienes que hacer nada más.

¿Por qué a las 10?

Porque el planeta gira. Porque los horarios cambian. Y cuando muchas personas hacen una pausa parecida en distintos lugares, se forma una ola de atención que recorre la Tierra.

Intenta tratar a las y los demás como te gustaría que te trataran a ti.

No es magia. Es gente común cuidándose al mismo tiempo.

Eso sana personas. Y cuando las personas se sanan, el planeta también.

Para mis mexas.

Mexicanas y mexicanos, siempre unidos. Con amor, raíz y dignidad.
Arriba Abya Yala continente de piel solar y pies de tierra.

Nodes of Coherence in Gaia

Throughout history, individuals and communities have acted in ways that restore, protect, or enhance the balance of living systems.

Their actions—whether through science, conservation, education, or direct intervention—represent localized increases in coherence within the broader planetary system.

These contributions are not defined by recognition or scale, but by their capacity to generate lasting impact on ecological, social, or informational structures.

In this sense, such individuals may be interpreted as *nodes of coherence*, whose actions contribute to the stabilization and regeneration of the system.

These examples demonstrate concrete interventions that restore equilibrium and enhance structural coherence across multidisciplinary systems.

Examples of Coherence in Action

- **Alexandra Elbakyan** — Creation of Sci-Hub, a global platform that provides free access to millions of paywalled research papers, revolutionizing the democratization of scientific knowledge.
- **Jane Goodall** — Long-term work in primate research and conservation, promoting ecological awareness and protection of natural habitats.
- **Boyan Slat** — Development of large-scale systems for ocean plastic removal and environmental restoration.
- **Marino Morikawa** — Restoration of the El Cascajo wetland in Peru using micro-nanobubble technology and ceramic biofilters, recovering entire ecosystems previously considered lost.
- **Afroz Shah** — Large-scale coastal and river cleanup efforts in India, removing millions of kilograms of waste and enabling the return of marine life to previously degraded environments.
- **Chad Pregracke** — Founder of Living Lands & Waters, leading long-term river cleanup operations in the United States and removing millions of pounds of debris from major waterways.
- **Lewis Pugh** — Endurance swimmer whose expeditions in extreme environments have contributed to the protection of marine ecosystems and the establishment of large-scale conservation zones.

- **Boyan Slat** — Development of technological systems for large-scale plastic removal from oceans and rivers, including autonomous interception platforms.
- **Jadav Payeng** — Creation of a large-scale forest ecosystem through sustained individual reforestation efforts over decades, transforming barren land into a biodiverse habitat.
- **Sebastião Salgado and Lélia Wanick** — Large-scale ecological restoration through the Instituto Terra, reforesting degraded land and reestablishing native biodiversity.
- **Antonio Vicente** — Long-term reforestation and ecological recovery of previously degraded land, restoring water systems and forest structure.
- **Saalumarada Thimmakka** — Large-scale tree planting initiative transforming arid roadside environments into continuous ecological corridors.
- **Antoine Moses** — High-efficiency reforestation efforts contributing to large-scale tree planting and ecosystem recovery in deforested regions.
- **Julia Hill** — Long-term tree protection action, sustaining a continuous occupation of a redwood tree for over two years, resulting in the preservation of both the tree and its surrounding ecosystem.
- **Ángela Olazarán** — Development of an artificial intelligence system capable of diagnosing multiple diseases, recognized globally for innovation in accessible healthcare technologies.
- **Barbara Ann Brennan** — Exploration of human energy fields and their relationship to health, proposing structured models of biofield interactions and integrative healing approaches.
- **Francisco Toledo** — Cultural and ecological protection through grassroots organization, successfully preventing industrial expansion in historically sensitive regions.
- **Hernán Asto** — Development of bioelectric systems capable of generating electricity from plant-soil interactions, providing sustainable energy solutions for off-grid communities.
- **Gabriela Salas** — Implementation of indigenous language integration into modern computational systems, contributing to the preservation of linguistic diversity through artificial intelligence.
- **Didier Dynamo** — Passive cooling system design based on evaporative thermodynamics, enabling temperature regulation without electricity in high-heat regions.
- **Andy Goldsworthy** — Creation of natural structural artworks that reveal transient geometric and ecological patterns through interaction with environmental materials.
- **Sonia Madrigal** — Photographic documentation of urban life, capturing social and cultural structures within everyday environments.
- **Kiara Nirghin** — Creation of a biodegradable polymer from orange peels that retains water in dry soil, helping agriculture in drought-prone areas.
- **Destany "Sky" Pete** — Scientific validation of the cancer-fighting properties of traditional chokecherry pudding, proving its ability to inhibit cancer cell growth.

- **Pedro Parra** — Rarámuri runner who established a world record in a 448-kilometer race over 64 hours, demonstrating endurance and cultural strength in the Copper Canyon.
- **Yoshinori Ohsumi** — Nobel Prize-winning discovery of the mechanisms for autophagy, a process where the body regenerates cells and strengthens the immune system through fasting.
- **Sophie Germain** — Mathematical pioneer who overcame social and academic barriers to contribute significantly to number theory and elasticity without formal recognition.
- **John Lilly** — Investigation into the effects of sensory deprivation and human-dolphin communication, exploring the boundaries of consciousness through isolation tank experiments.
- **Lorena Ramírez** — Elite Rarámuri ultra-trail runner who preserves ancestral traditions and represents Mexico internationally by competing in traditional attire.
- **Fionn Ferreira** — Development of a magnetic liquid system to remove microplastics from water, providing a scalable solution to tackle ocean pollution.
- **Richard Turere** — Development of "Lion Lights," a solar-powered LED flashing system designed to protect livestock from lion attacks and reduce human-wildlife conflict in Kenya.
- **Homero Gómez González** — Environmentalist and protector of the Monarch Butterfly Sanctuary in Michoacán, Mexico, dedicated to the conservation of forests and the preservation of the monarch species.
- **Rachel Brouwer** — Innovation of a solar-powered water purification system that uses ultraviolet radiation and a soybean wax temperature indicator to provide clean drinking water.
- **Benjamin Haggarty** — Development of "bio-concrete" capable of supporting natural moss growth, creating vertical gardens that filter air pollution and absorb CO_2 in urban environments.
- **Heman Bekele** — Recognized as TIME's 2024 Kid of the Year for developing a cost-effective soap infused with imiquimod to make skin cancer treatment more accessible.
- **Adrián López and Marte Cázares** — Mexican entrepreneurs who developed "Desserto," an organic and sustainable leather alternative made from cactus leaves without toxic chemicals.
- **Simon Petrus** — Namibian inventor who built a revolutionary "SIM-less" phone that operates via radio frequencies, allowing calls without a SIM card or mobile airtime.
- **Flavio Anxel** — Contemporary artist from San Juan Chamula, Chiapas, who fuses indigenous heritage with innovative techniques to represent Mexican culture on the global stage.
- **Robert Sansone** — Engineering of a magnet-free synchronous reluctance motor prototype that improves torque and efficiency without the use of costly rare-earth metals.
- **Ernst Götsch** — Development of syntropic agriculture techniques, successfully recovering 500 hectares of dry land in Brazil through biodiversity-driven reforestation.

- **Dr. Daniel Mansfield** — Deciphering of the 3,700-year-old Babylonian clay tablet Plimpton 322, revealing an advanced trigonometric system based on exact ratios rather than angles.
- **Michio Kushi** — Japanese scientist and educator known for his global contributions to macrobiotics and the study of environmental and dietary influences on human health.
- **Sabrina Gonzalez Pasterski** — Theoretical physicist who built and flew her own airplane at age 14, later gaining international recognition for her research on gravity and spacetime.
- **Valentyn Frechka** — Founder of Releaf Paper, a startup that developed a technology to transform fallen leaves into biodegradable paper, reducing deforestation and carbon emissions.
- **Richard R. Lunt** — Development of transparent solar panels at Michigan State University, enabling windows and glass surfaces to generate electricity without blocking light.
- **Boyan Slat** — Development of the "Ocean Cleanup" project, a passive system that uses natural ocean currents to concentrate and remove plastic waste from the Great Pacific Garbage Patch.
- **Veysel Donbaz** — Renowned Assyriologist and Sumerologist who has deciphered thousands of cuneiform tablets, preserving and revitalizing the extinct languages and legal structures of ancient Mesopotamia.
- **Viktor Schauberger** — Pioneering research on "vortex technology" and water implosion, proposing that understanding nature's rhythmic and spiral movements could lead to sustainable energy and propulsion.
- **Masanobu Fukuoka** — Development of "Natural Farming" or "no-till" agriculture, a philosophy that minimizes human intervention to restore soil health and ecological balance.
- **Gunter Pauli** — Development of the "Blue Economy" concept, which mimics natural ecosystems to transform industrial waste into high-value resources, aiming for zero emissions.
- **Sandra Pascoe Ortiz** — Mexican researcher who developed a biodegradable plastic alternative using nopal (cactus) juice, offering a non-toxic solution to single-use plastics.
- **Eva Ramón Gallegos** — Development of a photodynamic therapy technique that achieved a 100% elimination rate of the Human Papillomavirus (HPV) in patients in Mexico City and Oaxaca.
- **Joel Sosa and Sergio Galván** — Invention of "Translucent Concrete," a material that allows light to pass through while maintaining high structural strength, reducing energy consumption in buildings.
- **Gary, Kelly, and Sam Bencheghib** — Founders of "Sungai Watch," an environmental organization that develops and installs low-tech trash barriers in rivers to stop plastic pollution from entering the ocean.
- **Dr. Ursula Voss** — Clinical psychologist and neurophysiologist known for her pioneering EEG studies on lucid dreaming and the neurophysiology of consciousness during sleep.
- **Professor Allan Snyder** — Director of the Centre for the Mind, recognized for his research on the "Champion Mindset" and using transcranial magnetic stimulation to unlock latent savant-like mental abilities.

- **Dr. Apolinaria García Cancino** — Chilean researcher who developed a specialized probiotic, based on the Lactobacillus fermentum strain, to effectively combat Helicobacter pylori and prevent gastritis.
- **Rachel Brouwer** — Innovation of a solar-powered water purification system that uses ultraviolet radiation and a soybean wax temperature indicator to provide clean drinking water.
- **Santos de la Torre Santiago** — Renowned Huichol (Wixárika) artist whose intricate yarn paintings, such as those displayed in the Louvre, represent the sacred cosmogony and spiritual visions of his people.
- **Dr. Teresa Woodruff** — Discovery of the "Zinc Spark," a burst of light occurring at the exact moment of human fertilization, marking the transition of life through a measurable bio-chemical and energetic event.

These examples are not exhaustive. They represent visible instances of a broader pattern: the emergence of actions that increase coherence within complex systems.

Medicina Musical para Integración Consciente

Para aquellas y aquellos que se acercan al Modelo Æ con mente abierta y corazón tierno, estas canciones ayudan a tender puentes entre el ego y el espíritu. Ofrecen apoyo emocional mientras la memoria vibracional se reactiva y la conciencia se expande hacia nuevas comprensiones. Es importante aclarar que las referencias incluidas a continuación no implican que estas personas estén necesariamente de acuerdo con el contenido, las interpretaciones o las formulaciones del Modelo Æ.

- **Ximena Castro** — *Manos Blancas*
- **Paloma del Cerro** — *Todo Vibra*
- **Bruno Mansur, Ivy Talamas** — *Iluminar*
- **Lulacruza** — *El conocimiento*
- **Bárbara y Joaquín (Complementares Opostos)** — *La Tierra Giró*
- **Legend Medicina Sonica** — *Guacamayo*
- **Rafael Chaparro** — *Desierto*
- **Abuela Margarita** — *Poema Quetzalcoatl*
- **Deva Baumbach, Pedro Vadhar, Francisco Bringas** — *Kamaja*
- **Ines Lolago** — *Elemental*
- **Loli Cosmica, Kittsy Flor** — *Yo soy Diosa*
- **Puentes, Nick Barbachano, Danit, Jonas Winter, Misha Mullov-Abbado, Reuben Darling Khan, Mao Tatanka, Ramon Peregrino** — *Con Esta Medicina*
- **CuraSana** — *Todo Cura Todo Sana*
- **Ayla Schafer** — *Agua del Amor*
- **Shimshai and Susana** — *La Ventana*
- **Lulacruza** — *Estrella del Alba*
- **Darwin Grajales** — *Echale el miedo al fuego*
- **Alonso Del Rio** — *Abrete Corazon*
- **Colectivo Animal** — *Todo Esta Bien*
- **Jesus Hidalgo, Alonso Del Rio** — *Agradezco*
- **Mercedes Sosa** — *Todo Cambia*
- **Nicolas Losada Rojas** — *Yo Que No Creia*
- **Bruno Mansur, Cristina de la Paz** — *Corazon es lo unico que tengo*
- **Chris Orange** — *Esfera Azul*
- **Ivan Donalson** — *Dentro de la Espiral*
- **BALAM ARIES** — *Camino de flores*
- **Hijas de la Tierra** — *Que Mujer Eres Tu*
- **Paz Shanti** — *Luna Llena*
- **Tamara hernandez Shala** — *Prosperidad*
- **Luna Santa** — *Ninos Medicina*

- **Happy Banda, Ravi Ramoneda, Rafa Martinez** — *Agua de Estrellas*
- **Bachan Kaur** — *Luz del Bosque*
- **Arnoldo and Andressa** — *Sol de la Manana*
- **Isis Montemayor** — *Sana Corazon*
- **Anawak, Ana Cihuacoatl** — *Tloke Nawake*
- **Almaconvoz** — *Renace la Fuerza*
- **Luna Santa** — *Una Con la Tierra*
- **Caloncho** — *Optimista*
- **Abuela Malinalli** — *Tamborcito*
- **Indiajiva** — *Sacred Ragas*

La verdad fluye a través de las líneas temporales. Ciencia, mito y vibración son solo nombres que damos a los ecos de la misma fuente. Que estas melodías abran el corazón a la remembranza de quien realmente somos: expresiones temporales del campo eterno despertando a su propia naturaleza infinita.

Resonancia Digital: Canales de YouTube en Español

Muchas y muchos creadores en línea han estado tocando fragmentos del entendimiento del campo unificado. Aquí honramos y referenciamos algunos de los canales en español más resonantes, junto con sus aportaciones clave a la expansión de la conciencia y la percepción vibracional.

Es importante aclarar que las referencias incluidas a continuación no implican que estas personas o canales estén necesariamente de acuerdo con el contenido, las interpretaciones o las formulaciones del Modelo Æ.

Estas menciones se realizan como reconocimiento a la resonancia temática, intuiciones compartidas o exploraciones conceptuales que, desde distintos enfoques, han contribuido a abrir preguntas y perspectivas relacionadas con los temas aquí desarrollados.

- **La Bruja Filosófica** — Especialmente *LEY DE ATRACCION & LEY DEL UNO → ANTROPOSOFIA*
- **Date un Vlog** — Video: *¿Podríamos cambiar el Big Bang?*
- **Revelando el Velo En Español** — Video: *ÉTER (Æther) El Quinto Elemento que Ordena el Universo*
- **Adictos a la Filosofía** — Video: *Si Sabes Sumar, Eres MÁS que Materia / Filosofía de la Mente*
- **Veritasium en español** — Video: *Cómo la Rama Más INÚTIL de la Matemática Podría Salvar tu Vida, Monsanto*
- **Pilar Sousa** — Video: *La VIDA ES UN JUEGO y estas son las REGLAS METAFÍSICAS para ganar*
- **Mates Mike** — Video: *Cómo Ver Una Esfera en 5D*
- **Alanna** — Video: *Mi experiencia con el VIAJE ASTRAL: ¿Cómo conseguí tenerlos?*
- **Círculo Uróboros** — Video: *Popol Vuh- Resumen y análisis*
- **3Blue1Brown** — Video: *Visualizing quaternions (4d numbers) with stereographic projection*
- **Dimension B** — Video: *Así logré el VIAJE ASTRAL*
- **El Soñador Consciente** — Video: *Viaje Astral en 5 pasos // Camilo Andrés Gutiérrez*
- **Universo Hindu** — Video: *¿Qué es SAMADHI? La EXPERIENCIA DEL TODO*
- **QuantumFracture** — Video: *Nunca Has Tocado Nada... ¿o Sí?*
- **DharmaNation** — Video: *Narada Muni: The Supreme Spiritual Master*
- **El Robot de Platón** — Video: *¿Puede la CIENCIA Explicarlo TODO?*
- **The Sanskrit Channel** — Video: *Learning Vedas, UpaVedas and Vedangas - A Complete Overview*
- **Roxana Kreimer** — Video: *DESHUMANIZANDO AL VARÓN La HISTORIA de las MUJERES como NO te la CONTARON*
- **SizeMatters** — Video: *¿Debería el conocimiento ser gratuito? (SCI-HUB)*

- **Pero eso es otra Historia** — Video: *El Origen de las Letras y el Alfabeto Latino (Remake + Extras) (Documental Historia)*
- **Astronomic Limit** — Video: *Entiende las ecuaciones de Maxwell I / Ley de Gauss*
- **Linguriosa** — Video: *Por qué el futuro VA A DESAPARECER, pero no DESAPARECERÁ*
- **Raquel de la Morena** — Video: *¿Qué secretos escondía la alquimia? De la piedra filosofal al elixir de la vida*
- **Malasombra** — Video: *HISTORIA COMPLETA DE LA PERSPECTIVA*
- **NightHawkInLight** — Video: *Cómo hacer pintura de refrigeración por infrarrojos (aire acondicionado sin electricidad)*
- **Vm granmisterio** — Video: *NADIE puede Explicar el EXTRAÑO MENSAJE de esta TABLILLA SUMERIA*
- **BBC News Mundo** — Video: *¿Cuál era la idea de felicidad de los aztecas y qué podemos aprender de ella?*
- **MetabolismoTV** — Video: *Episodio 1832 Ordénale A Tu Cuerpo Que Se Renueve El Mismo*
- **La Hiperactina** — Video: *El increíble (y olvidado) SISTEMA LINFÁTICO*
- **Vicente Cassanya, astrólogo** — Video: *Todo tiene su momento, y nada es casual. Las claves de la Astromundial*
- **Math Rocks** — Video: *Top 7 MEJORES Libros Recomendados de Teoría de Conjuntos*
- **Ivan Donalson** — Video: *YO SOY CREADOR - EL MANIFIESTO DEL FUEGO SOBERANO*
- **Sarah Hall** — Video: *The Spiritual Meaning of 333 Angel Numbers*
- **Historia Tours** — Video: *Como construir una Cámara Kirlian 1/3*
- **Marcos Sacristán | Fisioterapia a tu alcance** — Video: *Tecnica de relajacion mediante la respiración completa*
- **el Sendero** — Video: *Kabbalah: TRANSFORMA tu VIDA al 100 / ft. Mario Saban*
- **Mindalia Plus** — Video: *Sexualidad Taoísta, por Joan Piquer*
- **Prof Ing Gustavo Belliski** — Video: *Geometrías No Euclideanas parte básica 1: Geometría Esférica*
- **Omnity Meditación** — Video: *Meditación PODEROSA para SANAR, DESBLOQUEAR y EQUILIBRAR el CHAKRA RAÍZ / PRIMER CHAKRA*
- **Alquimia de Angel** — Video: *Curso de Ángeles: Cómo empezar a conectar con tus guías espirituales Clase 1*

Final Notes

Progress updates and future developments may be shared publicly.

Although not everything can be tested today, prototyping and simulation are already underway.

The intention of this work is simple:

To invite a return to observation, curiosity, and humility — to the openness of a child, the discernment of an adult, and the rigor of mathematics.

Welcome to the Æ Model.

A Final Word Before You Enter

To invite you to leave dogma —scientific, spiritual, political or geocultural— and return to observation, to curiosity, to the openness of a child and the discernment of an adult.

I offer this work sincerely, without intent to persuade or recruit, and with full respect for your autonomy.

This work does not end on these pages.
It continues, evolves, and unfolds beyond them.

Access extended material, updates, and ongoing work.

I leave you with the Æ Model.
Thank you for your time

:)

76. References of the Æ Model

This work is informed by a diverse set of traditions spanning science, philosophy, historical inquiry, and contemporary research.

The sources below reflect perspectives that helped shape the conceptual environment in which the Æ Model was developed, without implying endorsement, lineage, or equivalence.

Foundational Wisdom Traditions

- **William Walker Atkinson** — *Thought Vibration.*
- **Yogui Ramacharaka** — *The Hindu-Yogi Science of Breath.*
- **Bhagavad Gītā** — *Chapter XIII: Field and tnower of the Field* (Swami Prabhupada translation)
- **The Law of One** — *The Ra Material.* L/L Research (1981–1984)
- **Hermes Trismegistus** — *Tabula Smaragdina* (Isaac Newton Translation)
- **Plato** — *Timaeus*, c. 360 BCE
- **The Upanishads** — Selections on Atman, Brahman, and subtle bodies
- **Corpus Hermeticum** — Attributed to Hermes Trismegistus, foundational for understanding soul-æther relationships
- **The Kybalion** — Principles of mentalism, vibration, and polarity applicable to the Æ Model
- **Gospel of Thomas** — Gnostic source describing invisible planes of soul existence
- **Paracelsus** — Works on the archaeus and middle-plane spirits, equivalent to astral/etheric realms
- **Agrippa von Nettesheim** — On natural magic and subtle bodies (astral soul, vital spirit)
- **Helena Blavatsky** — Classification of planes (physical, etheric, astral, mental) used in the Æ Model
- **Rudolf Steiner** — Anthroposophical insights on karma, ether, and consciousness evolution
- **The Zohar** — Kabbalistic vibrational maps and mystical consciousness structures
- **Neville Goddard** — Complete works on consciousness as fundamental reality
- **Edgar Cayce** — Readings on soul evolution, reincarnation, and vibrational healing
- **Saint Germain** — *The Golden Book* and teachings on vibrational transformation
- **Manly P. Hall** — Complete works on esoteric philosophy and sacred geometry

Scientific and Mathematical Foundations

- **Claude E. Shannon** — *The Mathematical Theory of Communication*
- **Hermann Weyl** — *Symmetry*
- **Alexander MacFarlane** — *Vector Analysis and Quaternions*
- **Andrew J. Hanson** — *Visualizing Quaternions*
- **Euclid** — *The Elements*
- **Homer E. Newell, Jr.** — *Vector Analysis*
- **Edward Leedskalnin** — *Magnetic Current*
- **William James Sidis** — *The Animate and the Inanimate*
- **Giulio Tononi** — *PHI: A Voyage from the Brain to the Soul*
- **Georges Lakhovsky** — *The Secret of Life*
- **Robert O. Becker and Gary Selden** — *The Body Electric*

Modern Quantum and Consciousness Research

- **Fritz-Albert Popp** — "Biophoton emission of the human body," *Journal of Biophotonics* (2003)
- **Cifra, Pokorný, et al.** — "Electric Fields Generated by Synchronization of Cellular Activities," *Nature Scientific Reports* (2011)
- **Gariaev, P.P.** — "DNA as a Fractal Antenna in Electromagnetic Fields," *Chaos, Solitons & Fractals*, Volume 18, Issue 2 (2003)
- **Gull, S., Lasenby, A., & Doran, C.** — "Imaginary Numbers Are Not Real: The Geometric Algebra of Spacetime," *Foundations of Physics*, 23(9), 1175–1201 (1993)
- **Fratini, Cappelluti, et al.** — "Exact Quantum Electrodynamics of Radiative Photonic Environments" (2024)
- **Feis, Weidemann, et al.** — "Space-Time-Topological Events in Photonic Quantum Walks," *Nature Photonics* (2025)
- **Ruan, Wang, et al.** — "Particle Exchange Statistics Beyond Fermions," *Nature Physics*, Vol. 20 (2024)
- **Vopson, M. M.** — "Is gravity evidence of a computational universe?" *AIP Advances*, 15, 045035 (2025).
- **Levin, M.** — "Bioelectromagnetics in Morphogenesis," *Bioelectromagnetics*, Vol. 24, pp. 295–315 (2003)
- **Zhang, G., Levin, M.** — "Bioelectricity is a universal multifaceted signaling cue in living organisms," *Molecular Biology of the Cell*, Vol. 36, pe2 (2025)
- **Oviedo, N. J., Levin, M.** — "Eye regeneration assay reveals an invariant functional left-right asymmetry in the early bilaterian *Dugesia japonica*," *Developmental Biology* (2010)
- **Andrecut, M., Ali, M. K.** — "A Quantum Neural Network Model," *International Journal of Modern Physics C*, Vol. 13, No. 1, pp. 75–88 (2002)
- **Mould, R. R., Kalampouka, I., Thomas, E. L., Guy, G. W., Nunn, A. V. W., Bell, J. D.** — "Non-chemical signalling between mitochondria," *Frontiers in Physiology*, Vol. 14, 1268075 (2023)

- **Le Hur, K.** — "Topological Matter and Fractional Entangled Quantum Geometry through Light," *Comptes Rendus Physique*, Vol. 24, 2023. DOI: 10.5802/crphys.154
- **Kletetschka, G.** — "Three-Dimensional Time: A Mathematical Framework for Fundamental Physics," *AIP Advances*, 2025.
- **STAR Collaboration** — "Measuring Spin Correlation Between Quarks During QCD Confinement," *Nature Physics*, 2026. DOI: 10.1038/s41586-025-09920-0

Integrative and Holistic Sciences

- **José Silva and Philip Miele** — *The Silva Mind Control Method*
- **Theodore Dimon** — *Anatomy of the Voice*
- **Alain Danielou** — *Music and the Power of Sound*
- **Anthony Ashton** — *Harmonograph: A Visual Guide to the Mathematics of Music*
- **Steven Pinker** — *The Language Instinct*
- **Noam Chomsky** — *Nature and Language*
- **William James** — *Introduction to Philosophy*
- **Wim Hof** — *The Wim Hof Method*
- **C.W. Leadbeater** — *The Astral Plane*
- **Iva Kenaz** — *Sacred Geometry and Magical Symbols*

Primary Sources and Channeled Material

- **Quera4** — *Diarios* $\Delta 1$, personal channelings and insights (2023–2025)
- **Ra** — *The Law of One* sessions through L/L Research
- **Internal guidance** — Direct field communications and geometrical downloads

77. Vibrational Glossary

The language of the Æ Model introduces specific terms intended to describe the vibrational organization of reality. The following definitions serve as conceptual bridges between conventional scientific terminology and the extended framework proposed in this work.

- **Æ Field (Æther)**

  Definition: The fundamental vibrational substrate proposed in the model from which matter, energy, spacetime, and information structures emerge through organized phase interactions.

  Analogy: Comparable to the surface of an ocean where waves propagate. The water itself forms the medium, while the visible waves correspond to patterns that arise within that medium.

- **Phase**

  Definition: The position of a system within its oscillatory cycle, determining how it interferes or aligns with other oscillating systems.

  Analogy: Two pendulums swinging with the same timing are "in phase." If one reaches its highest point while the other is at its lowest, they are out of phase.

- **Coherence**

  Definition: A state in which multiple oscillatory components maintain ordered phase relationships, allowing their effects to combine constructively.

  Analogy: Like the coordinated wave in a stadium: when spectators stand and sit in the correct timing, a clear wave travels around the arena. If people move randomly, the wave disappears.

- **Coherence Knot**

  Definition: A localized region where vibrational phase organizes into a stable configuration, producing persistent structural patterns.

  Analogy: Similar to a stable whirlpool in a river where flowing water organizes into a persistent rotating structure.

- **Resonance**

  Definition: The amplification of oscillatory behavior when a system interacts with frequencies that match its natural modes of vibration.

  Analogy: A playground swing moves higher when pushed at the correct rhythm. Pushing at the wrong moment reduces the motion instead of amplifying it.

- **Æ Kernel**

  Definition: A proposed base resonant structure that organizes the formation of coherence nodes across the Æ field.

  Analogy: Similar to a seed pattern in crystal growth that determines how the entire structure expands as atoms attach in ordered ways.

- **Nodal State $|n, m, s\rangle$**

  Definition: A mathematical representation describing the vibrational configuration of a node within the model, where parameters encode structural, relational, and dynamical characteristics.

  Analogy: Comparable to specifying the coordinates and velocity of a particle in phase space, which together describe its current dynamical state.

- **Quaternionic Structure**

  Definition: A four-dimensional algebraic framework used in the model to represent rotational and relational dynamics across multiple interacting sectors.

  Analogy: Quaternions describe rotations in three-dimensional space without the singularities that occur with simpler coordinate descriptions, similar to how gyroscopes maintain stable orientation.

- **Hyperquaternion**

  Definition: A higher-order algebraic structure formed by quaternions composed of quaternions, enabling the representation of nested dynamical layers.

  Analogy: Similar to hierarchical coordinate systems where each layer describes motion within a larger frame of reference.

- **Node Jump**

  Definition: A transition between distinct nodal configurations within the model's state structure.

  Analogy: Comparable to an electron transitioning between discrete energy levels in an atom.

- **$\Delta\Phi$ (Delta Phi)**

  Definition: The phase mismatch between oscillatory layers or nodes that determines the efficiency of coupling between them.

  Analogy: Two gears that are slightly misaligned transfer motion inefficiently, while perfectly aligned gears transmit rotation smoothly.

- **Merkaba-Φ**

  Definition: A tetrahedral geometric configuration used in the model to describe a stable reference structure for cross-layer interactions.

  Analogy: Comparable to a rigid frame in engineering that maintains structural stability while forces act across multiple directions.

- **Cantorian Density (μ_C)**

  Definition: A fractal dimensional measure used to quantify the structural complexity and information capacity of a given layer.

  Analogy: Similar to fractal coastlines, where the measured length increases as the scale of observation becomes finer.

- **Ætheric Projection**

  Definition: The transfer of informational structure across vibrational layers through the field representation $\Psi_{Æ}$.

  Analogy: Comparable to projecting a three-dimensional object onto a two- dimensional screen, where the underlying structure influences the visible pattern.

- **Intention Current**

  Definition: A directed modulation of the field associated with coherent, goal-oriented phase organization.

  Analogy: Similar to a laser beam where many photons move in the same direction with aligned phase, producing a focused energy flow.

- **Phase Entropy (S_ϕ)**

  Definition: A measure of disorder in phase relationships within a system.

  Analogy: In a stadium wave, if spectators start standing at random times the wave dissolves into noise. That loss of coordination corresponds to increasing phase entropy.

- **Quaternionic Time**

  Definition: A representation of temporal evolution modeled through internal phase rotation within quaternionic structures.

  Analogy: Similar to tracking the rotation of a clock hand, where time is associated with continuous angular motion.

- **Vibrational Parasitism**

  Definition: Self-reinforcing dynamical structures that destabilize coherent phase patterns and maintain high entropy states.

  Analogy: Comparable to feedback noise in an audio system, where uncontrolled amplification disrupts the intended signal.

- **Karmic Loop**

  Definition: A persistent recursive phase configuration that repeatedly reproduces similar dynamical patterns.

  Analogy: Similar to a feedback loop in control systems that repeatedly drives a system back into the same state.

- **Field Intelligence**

  Definition: Emergent large-scale organization resulting from interactions among many coherent nodes.

  Analogy: Like flocking behavior in birds, where coordinated patterns arise from simple local interactions.

- **Hollow Node**

  Definition: A region characterized by low coherence and high phase entropy.

  Analogy: Similar to turbulence in fluid flow where orderly motion breaks down into irregular patterns.

- **Eternal Node (Ω_∞)**

  Definition: An idealized limit representing maximal coherence within the model.

  Analogy: Comparable to a perfectly synchronized oscillator network where all components move in exact phase alignment.

- **Multiversal Branching**

  Definition: The emergence of multiple dynamical trajectories when a system reaches instability thresholds.

  Analogy: Like a river splitting into several channels when it reaches a flat delta region.

Part III
Appendix

A. Hamiltonian and Action Formalism for the Quaternionic State

This appendix presents a self-contained variational formulation for the intrinsic dynamics of the quaternionic state $Q(t)$. The construction is modular and compatible with the Æ Framework. No equation in the main body is modified or reinterpreted.

Configuration Space

Assume that $Q(t)$ parametrizes a trajectory on the 3-sphere S^3, identified with the group of unit quaternions:

$$Q = q_0 + q_1 i + q_2 j + q_3 k \in \mathbb{H}, \qquad |Q|^2 = Q\overline{Q} = 1. \tag{1465}$$

The manifold S^3 is diffeomorphic to $SU(2)$ and carries the Riemannian metric induced from $\mathbb{R}^4$:

$$ds^2 = dQ \cdot d\overline{Q} = dq_0^2 + dq_1^2 + dq_2^2 + dq_3^2, \tag{1466}$$

restricted to the unit-norm constraint.

The tangent space $T_Q S^3$ consists of vectors V satisfying

$$\mathrm{Re}(Q\overline{V}) = 0. \tag{1467}$$

The constraint $|Q(t)| = 1$ is assumed preserved for all t.

Hamiltonian

Define the Hamiltonian:

$$H(Q, \dot{Q}) = \frac{1}{2} m |\dot{Q}|^2 + V(Q), \tag{1468}$$

where $m > 0$ is a constant parameter and $V(Q)$ is a smooth real-valued potential on S^3.

The velocity satisfies the tangency condition:

$$\mathrm{Re}(Q\overline{\dot{Q}}) = 0. \tag{1469}$$

In the absence of $V(Q)$, the motion reduces to geodesic flow on S^3.

Action Functional

The action is

$$S[Q] = \int_{t_1}^{t_2} L(Q, \dot{Q})\, dt, \qquad L(Q, \dot{Q}) = \frac{1}{2} m |\dot{Q}|^2 - V(Q). \tag{1470}$$

Variations are taken over smooth curves on S^3 with fixed endpoints.

Equations of Motion

The Euler–Lagrange equations yield

$$m\ddot{Q} + m(\dot{Q} \cdot \dot{Q})Q = -\nabla_{S^3} V(Q), \tag{1471}$$

where ∇_{S^3} denotes the gradient projected onto the tangent space $T_Q S^3$. With canonical momentum

$$P = m\dot{Q}, \tag{1472}$$

we obtain

$$\dot{Q} = \frac{P}{m}, \qquad \dot{P} = -\nabla_{S^3} V(Q) - \lambda Q, \tag{1473}$$

where λ enforces the orthogonality constraint

$$P \cdot Q = 0. \tag{1474}$$

For the free case $V(Q) = 0$:

$$\ddot{Q} + (\dot{Q} \cdot \dot{Q})Q = 0. \tag{1475}$$

Consistency Properties

- Norm preservation:
$$\frac{d}{dt} |Q|^2 = 0.$$

- Energy conservation for autonomous dynamics:
$$\frac{dH}{dt} = 0.$$

- Rotational invariance when $V(Q)$ is invariant under conjugation.

Regime of Applicability

This appendix provides a closed variational description for intrinsic quaternionic dynamics. It may be integrated into future extensions or used as an independent technical module.

Stability Analysis of Reality Knots

Let $Q(t)$ denote a Reality Knot solution of the effective quaternionic flow. Linear perturbations $\delta Q(t)$ satisfy

$$\partial_t \delta Q = DF(Q)\,\delta Q, \tag{1476}$$

where $DF(Q)$ is the Jacobian of the flow.

Linear stability requires

$$\mathrm{Re}(\lambda_i) \leq 0, \tag{1477}$$

for all eigenvalues λ_i, with at least one neutral direction corresponding to phase transport along the knot.

Topological stability is ensured by conservation of the invariant

$$\mathcal{T}(Q) \neq 0. \tag{1478}$$

Loss of persistence therefore requires either:

- violation of the coherence condition, or
- satisfaction of the renudation identity-return condition.

Dynamical instability and topological dissolution are distinct mechanisms.

Connection with the Global Æ Action

The intrinsic quaternionic dynamics described above can be embedded consistently within the global Æ-field action governing the full model.

Let Φ denote a generic matter field. Gauge-covariant derivatives are defined as

$$D_\mu \Phi := \left(\partial_\mu - i\, g_{w0}\, W_\mu^i \frac{\tau^i}{2} - i\, g_0'\, Y_\Phi\, B_\mu \right) \Phi. \tag{1479}$$

The effective Planck mass depends on coherence:

$$M_P^2(C) := M_0^2\, C. \tag{1480}$$

Field normalization factors depend on coherence and phase entropy:

$$Z_s(C, S_\phi) = Z_w(C, S_\phi) = Z_y(C, S_\phi) = \frac{C}{1 + S_\phi}. \tag{1481}$$

The complete Æ master action takes the form

$$S = \int d^4x \sqrt{-g} \Bigg[\frac{1}{2} M_0^2 \, C \, R - \frac{1}{4} Z_s \, G^a_{\mu\nu} G^{a\mu\nu} - \frac{1}{4} Z_w \, W^i_{\mu\nu} W^{i\mu\nu} - \frac{1}{4} Z_y \, B_{\mu\nu} B^{\mu\nu}$$
$$+ (D_\mu \Phi)^\dagger (D^\mu \Phi) - V(\Phi) + \bar{\psi} \, i\gamma^\mu D_\mu \psi - \bar{\psi} \, Y \, \Phi \, \psi$$
$$+ \frac{1}{2} \partial_\mu C \, \partial^\mu C + \frac{1}{2} \partial_\mu S_\phi \, \partial^\mu S_\phi - U(C, S_\phi) \Bigg] + S_{\text{mem}} \tag{1482}$$

Within this global structure, the quaternionic state $Q(t)$ introduced in this appendix may be interpreted as an effective reduced degree of freedom describing coherent phase transport along Reality-Knot trajectories.

B. Quaternionic and Geometric Extensions of Protein Dynamics

This appendix collects extended mathematical structures related to protein dynamics that are compatible with the Æ Framework but were intentionally excluded from the main text to preserve minimality and narrative clarity.

Quaternionic Preliminaries

We recall the quaternion algebra

$$\mathbb{H} = \{q = a + b\,\mathbf{i} + c\,\mathbf{j} + d\,\mathbf{k}\}, \tag{1483}$$

with norm

$$|Q|^2 = a^2 + b^2 + c^2 + d^2. \tag{1484}$$

Quaternionic Representation of Protein States

Beyond the complex conformational state Ψ_{prot}, protein dynamics may be embedded in a quaternionic phase space. Let

$$Q_{\text{prot}}(t) \in S^3 \tag{1485}$$

represent the full internal phase state of a protein.

The observable conformational phase corresponds to the projected argument

$$\phi(t) = \arg(Q_{\text{prot}}(t)), \tag{1486}$$

where the argument is defined via the exponential map on S^3.

Geodesic Evolution on the Quaternionic Sphere

Protein conformational evolution may be modeled as motion on S^3 driven by the effective action

$$\mathcal{S}_{\text{prot}} = \int \left[\frac{1}{2}\langle \dot{Q}_{\text{prot}}, \dot{Q}_{\text{prot}}\rangle - V(Q_{\text{prot}})\right] dt, \tag{1487}$$

where $\langle\cdot,\cdot\rangle$ denotes the Euclidean inner product in $\mathbb{R}^4$ restricted to S^3. The constrained Euler–Lagrange equation becomes

$$\ddot{Q}_{\text{prot}} + \left(\dot{Q}_{\text{prot}} \cdot \dot{Q}_{\text{prot}}\right) Q_{\text{prot}} = -\nabla_{S^3} V(Q_{\text{prot}}), \tag{1488}$$

where ∇_{S^3} denotes the gradient projected onto the tangent space $T_{Q_{\mathrm{prot}}}S^3$.

Quaternionic Momentum and Stability

Introduce the conjugate momentum

$$P_{\mathrm{prot}} = \frac{\partial \mathcal{L}}{\partial \dot{Q}_{\mathrm{prot}}} = \dot{Q}_{\mathrm{prot}}, \tag{1489}$$

assuming unit effective mass.

Functional stability corresponds to bounded momentum magnitude:

$$\|P_{\mathrm{prot}}\| < P_{\max}. \tag{1490}$$

Quaternionic Coherence Measure

Define a quaternionic coherence functional

$$C^{(Q)}_{\mathrm{prot}} = \frac{\|\langle Q_{\mathrm{prot}}\rangle\|^2}{1 + S_{\phi,\mathrm{prot}}}. \tag{1491}$$

Here $\langle Q_{\mathrm{prot}}\rangle$ denotes an ensemble or time-averaged quaternionic state.

Argument-variance coherence proxy. In addition to entropy-weighted coherence, a simple dispersion proxy can be defined via the quaternionic argument:

$$\mathrm{Var}(\arg(Q_{\mathrm{prot}})) \leq \epsilon, \tag{1492}$$

with coherence scaling

$$C_Q \propto 1 - \mathrm{Var}(\arg(Q_{\mathrm{prot}})). \tag{1493}$$

Topological Interpretation of Conformational Stability

Closed or quasi-periodic conformational cycles correspond to loops $\gamma_{\mathrm{prot}} \subset S^3$. An effective winding index can be defined after projection to a phase coordinate (e.g. along a selected S^1 fiber):

$$\nu_{\mathrm{prot}} = \frac{1}{2\pi} \oint_{\gamma_{\mathrm{prot}}} d\phi, \tag{1494}$$

where ϕ denotes the projected conformational phase. Non-zero ν_{prot} labels distinct conformational classes and stabilizes them against small perturbations.

Relation to Resonant Activation

Resonant activation corresponds to constructive interference between intrinsic geodesic frequencies on S^3 and external or internal driving terms, enhancing

$$C^{(Q)}_{\text{prot}} \tag{1495}$$

and the projected spectral response.

SU(2) Harmonic Modes on S^3 (Optional Spectral Toolkit)

When spectral decompositions are needed, one may expand quaternionic states over S^3 harmonics adapted to the $SU(2)$ structure. A standard basis is given by hyperspherical harmonics

$$Y_{nlm} = \sin^l(\chi)\, C^{\,l+1}_{n-l}(\cos\chi)\, Y^m_l(\theta,\phi), \tag{1496}$$

with representation dimension

$$\dim(\rho_j) = 2j+1 = n+1. \tag{1497}$$

This permits defining finite-dimensional reductions of phase entropy via spectral weights p_k:

$$S_\phi = -\sum_k p_k \log p_k. \tag{1498}$$

Intrinsic Time Along Quaternionic Trajectories (Optional)

A natural intrinsic time parameter may be induced from the S^3 metric:

$$d\tau^2 = dQ\, d\overline{Q}. \tag{1499}$$

Along a parametrized curve $Q(\lambda)$,

$$\tau = \int \sqrt{\dot{Q}\,\overline{\dot{Q}}}\, d\lambda. \tag{1500}$$

This provides a geometrically intrinsic arc-length clock compatible with the quaternionic kinetic term used above.

Scope and Integration Strategy

These structures are mathematically consistent with the Æ Model but were excluded from the main narrative to avoid over-specification. They may be reintegrated in future versions once phenomenological constraints justify their inclusion.

C. Coherence Regimes and Field-Level Unity: the $\Delta_{Æ}$ Parameter

Historical Motivation

An early conceptual proposal by Wheeler and Feynman suggested that all electrons in the universe might be interpreted as manifestations of a single underlying entity observed along different segments of a spacetime trajectory. While this idea was never adopted as a literal physical model, it reflects an important structural intuition: apparent multiplicity at the particle level may emerge from a deeper field-level unity.

Within the Æ Framework, this intuition is reformulated strictly as a *coherence-regime interpretation.* No literal single-particle hypothesis is invoked. Instead, unity and multiplicity correspond to different coherence states of an underlying phase field.

Field-Theoretic Consistency

A literal single-particle interpretation is incompatible with

- Fermi–Dirac antisymmetry of multi-particle states,
- quantum field theory (QFT), where particles are excitations of fields,
- operator constructions defined on Fock space.

Consequently, any notion of unity must be interpreted at the level of field coherence rather than particle identity.

The $\Delta_{Æ}$ Coherence Parameter

Recall the previously defined Æ coherence functional

$$C_{Æ} = \frac{1}{1 + S_{\phi}}. \tag{1501}$$

We introduce the coherence parameter

$$\Delta_{Æ} := C_{Æ}, \tag{1502}$$

which acts as a global multiplicative weight on field excitations.

Its limiting regimes are

$$S_{\phi} \to 0 \iff \Delta_{Æ} \to 1, \tag{1503}$$

$$S_{\phi} \to \infty \iff \Delta_{Æ} \to 0. \tag{1504}$$

Thus $\Delta_{\text{Æ}}$ parametrizes the effective degree of coherence of the underlying phase field.

Effective Field Rescaling

Let $\psi_i(x)$ denote local fermionic excitations.

Define coherence-weighted excitations

$$\psi_i^{\text{eff}}(x) = \Delta_{\text{Æ}}\, \psi_i(x). \tag{1505}$$

The total effective field becomes

$$\Psi_{\text{eff}}(x) = \sum_i \psi_i^{\text{eff}}(x) = \Delta_{\text{Æ}} \sum_i \psi_i(x). \tag{1506}$$

In this interpretation

- $\Delta_{\text{Æ}} \approx 1$ corresponds to highly coherent collective field behavior,
- $\Delta_{\text{Æ}} \ll 1$ corresponds to effectively independent localized excitations.

Effective Action Scaling

Consider the standard fermionic action

$$S_\psi = \int d^4x\, \bar{\psi}(i\gamma^\mu \partial_\mu - m)\psi. \tag{1507}$$

Under coherence rescaling the effective action becomes

$$S_\psi^{\text{eff}} = \Delta_{\text{Æ}} \int d^4x\, \bar{\psi}(i\gamma^\mu \partial_\mu - m)\psi. \tag{1508}$$

Coherence therefore acts as a global weighting of the excitation sector without modifying the dynamical operator structure.

Scaling of Observables

For any observable $\mathcal{O}$ constructed from the field operators,

$$\langle \mathcal{O} \rangle_{\text{eff}} = \langle \Psi_{\text{eff}} | \mathcal{O} | \Psi_{\text{eff}} \rangle = \Delta_{\text{Æ}}^2 \langle \Psi | \mathcal{O} | \Psi \rangle. \tag{1509}$$

Thus coherence rescales amplitudes of observables while leaving the operator algebra unchanged.

Two-Point Correlation Function

Let

$$G(x,y) = \langle 0|T\{\psi(x)\bar{\psi}(y)\}|0\rangle \tag{1510}$$

denote the standard propagator.
The effective propagator becomes

$$G_{\text{eff}}(x,y) = \Delta_{Æ}\, G(x,y), \tag{1511}$$

or in momentum space,

$$G_{\text{eff}}(p) = \Delta_{Æ} \frac{\gamma^{\mu} p_{\mu} + m}{p^2 - m^2}. \tag{1512}$$

The pole structure of the propagator remains unchanged; coherence only rescales amplitudes.

Operator interpretation In the present framework $\Delta_{Æ}$ behaves effectively as a scalar coherence weight rather than a dynamical operator acting on Hilbert space. Formally it may be viewed as multiplication by the real parameter

$$\Delta_{Æ} \equiv C_{Æ} \in [0,1]. \tag{1513}$$

Therefore no modification of the canonical operator algebra of quantum field theory is implied.

Ontological Interpretation

Two descriptive levels are distinguished:

- **Ontological level:** a unified quaternionic phase field.
- **Excitation level:** multiple decohered fermionic modes observed as particles.

The historical "one-electron universe" intuition corresponds in the Æ Framework to the coherence limit

$$\Delta_{Æ} \to 1, \tag{1514}$$

which represents maximal coherence of the underlying phase field rather than the existence of a literal single particle.

Normalization Consistency

Since $\Delta_{Æ}$ acts only as a multiplicative weight, canonical anticommutation relations remain unchanged:

$$\{\psi(x), \psi^{\dagger}(y)\} = \delta(x-y). \tag{1515}$$

Field normalization may be restored through the rescaling

$$\psi \to \Delta_{Æ}^{-1/2}\psi, \tag{1516}$$

showing that $\Delta_{Æ}$ modifies amplitudes without altering the fundamental structure of quantum field theory.

Summary

Within the Æ Framework the classical one-particle hypothesis is replaced by a coherence-regime description. Unity and multiplicity arise from different values of the phase-entropy functional S_{ϕ}, while the underlying field dynamics remains fully compatible with quantum field theory and multi-particle Hilbert space structure.

D. Applications and Boundary Conditions

This subsection identifies global engineering and sustainability challenges to which the Æ Framework is structurally applicable. The purpose is to document relevance and long-term intent, not to assert completed technological solutions.

All domains listed below are treated as application spaces in which coherence, entropy management, and information stability act as guiding constraints.

Climate and Carbon Systems

Climate dynamics emerge from the coupling of atmospheric, oceanic, geophysical, and electromagnetic subsystems. Within the Æ Framework, instability may be represented as phase desynchronization across coupled subsystems rather than the failure of any single scalar variable.

Relevant applications include:

- coherence-aware modeling of atmospheric circulation,
- entropy-constrained representations of carbon exchange cycles,
- phase-based indicators of systemic climate instability.

The framework does not propose geoengineering interventions. It provides structural constraints for integrating and stabilizing existing climate models across scales.

Water, Rivers, and Ecosystem Restoration

Hydrological systems are modeled as structured flow networks capable of information transport and entropy dissipation.

Relevant applications include:

- restoration of rivers and lakes through structured flow and coherence-preserving interventions,
- entropy-aware management of water circulation and retention,
- monitoring of aquatic system stability via phase-based metrics.

These applications are motivated by long-term ecological regeneration rather than extraction or control.

Health and Biological Engineering

In health-related contexts, the Æ Framework provides structural constraints complementary to biochemical and mechanical approaches.

Relevant domains include:

- bioelectric coherence as a regulatory boundary in tissue dynamics,
- entropy-based indicators of systemic dysregulation,
- coherence-preserving design principles for therapeutic technologies.

No medical efficacy is claimed. The framework specifies stability conditions under which biological systems maintain functional coherence.

Artificial Intelligence and Complex Adaptive Systems

Artificial intelligence systems are treated as complex adaptive networks subject to instability when internal representations lose coherence.

The Æ Framework contributes:

- entropy constraints on internal state evolution,
- coherence-based alignment boundaries,
- structural criteria for preventing runaway divergence in learning systems.

These principles are domain-agnostic and apply to both biological and artificial adaptive systems.

Planetary Monitoring and Early-Warning Indicators

Large-scale systems exhibit measurable precursors to instability detectable through phase-based metrics.

Applications include:

- coherence monitoring of coupled planetary subsystems,
- detection of early-warning signals preceding systemic transitions,
- integration of heterogeneous data streams under a unified information-theoretic structure.

Engineering Scope and Boundaries

All applications described here are exploratory and conceptual. The Æ Framework does not claim ready-to-deploy technologies. Its role is to provide structural constraints, metrics, and organizing principles for future engineering efforts.

Engineering begins not with control, but with understanding the limits of stability.

Meaning, Consciousness, and Value (Templeton-Oriented Scope)

This subsection documents the relevance of the Æ Framework to questions traditionally associated with the Templeton Prize, while explicitly delimiting scientific, philosophical, and interpretational boundaries.

Its purpose is declarative and protective: to state scope and intent, and to prevent misappropriation or reduction of the framework to purely technical claims detached from its motivating questions.

Ontological Limits of the Framework

The Æ Framework does not assert metaphysical doctrines, theological claims, or definitive accounts of consciousness. Instead, it specifies physical and informational boundary conditions under which phenomena such as meaning, experience, and value may arise.

Formally, the framework constrains structure, not content:

$$\text{Structure} \not\equiv \text{Meaning} \qquad \text{Meaning} \subseteq \text{Structured Conditions}. \tag{1517}$$

This distinction prevents both reductionism and category error.

Consciousness as a Boundary Phenomenon

Within this work, consciousness is treated neither as a fundamental force nor as a mere epiphenomenon. It is approached as a boundary phenomenon emerging at the interface of coherence, information integration, and entropy regulation.

The framework specifies constraints of the form:

$$S_\phi < S_{\text{crit}} \implies \text{stable experiential integration}. \tag{1518}$$

without asserting that such integration exhausts conscious experience.

Intention, Value, and Non-Reducibility

Concepts such as intention, purpose, dignity, and value are not derived from equations in this work. They are treated as irreducible descriptors operating at a level distinct from physical formalism.

The Æ Framework explicitly rejects the claim that value can be computed or optimized as a scalar objective function:

$$\text{Value} \neq \arg\min S_{\phi}, \tag{1519}$$

even though value-bearing systems require bounded entropy to remain viable.

Ethical Boundary Conditions

While the framework does not prescribe ethics, it delineates conditions under which ethical agency remains structurally possible. Systems that exceed coherence thresholds lose stability, predictability, and agency.

This yields a boundary condition:

$$C_{\text{Æ}} \to 0 \quad \Longrightarrow \quad \text{loss of stable system-level agency.} \tag{1520}$$

The content of ethical choice is not defined by this framework.

Motivational Context and Long-Term Intent

This work has been developed over multiple years with explicit long-term intent. Its motivating context includes environmental restoration, ecological regeneration, and the development of tools supporting rivers, lakes, and damaged ecosystems.

Any economic value derived from the framework is intended to support research, ecological recovery, and projects aligned with planetary stability rather than personal accumulation.

Status and Removability

This subsection is interpretational and contextual. It may be removed entirely in future technical versions without affecting the mathematical or physical core of the framework.

Its inclusion in Version 8 serves documentation, attribution, and scope-clarity purposes only.

> A framework may describe the conditions of meaning without claiming to define meaning itself.

$$\boxed{\frac{\delta}{\delta\phi}\Big(S_\phi[\phi] - \lambda C_{\text{«}}[\phi]\Big) = 0}$$

$$\boxed{\frac{d^2F}{dS_\phi^2} = 0, \qquad F(S_\phi) = S_\phi - \frac{\lambda}{1 + S_\phi}}$$

Made in United States
North Haven, CT
04 November 2026

21760252R00312